Immunofluorescence in Clinical Immunology

A Primer and Atlas

by
Wulf B. Storch

translated by
R.A. Klein

Springer Basel AG

Author:

Professor Dr. med. Wulf B. Storch
Netztal 32
D-69469 Weinheim
Germany

Translated by:

Dr. R.A. Klein
Institut für Physiologische Chemie
Universität Bonn
Nussallee 11
D-53115 Bonn
Germany

Library of Congress Cataloging-in-Publication Data
Storch, Wulf.
[Immunofluoreszenzfibel. English]
Immunofluorescence in clinical immunology : a primer and atlas / by Wulf B. Storch ; translated by R.A. Klein.
p. ; cm.
Includes bibliographical references and index.
ISBN 978-3-0348-9540-8
1. Fluorescent antibody technique. 2. Fluorescent antibody technique–Atlases. I. Title.
[DNLM: 1. Fluorescent Antibody Technique. 2. Antibodies–analysis. QW 525.5.F6 S884i 2000]
RB46.7 .S7613 2000
616.07'98–dc21

99-059889

Die Deutsche Bibliothek – CIP-Einheitsaufnahme
Storch, Wulf:
Immunofluorescence in clinical immunology : a primer and atlas / by Wulf B. Storch. Transl. by R. A. Klein. – Basel ; Boston ; Berlin : Birkhäuser, 2000
Einheitssacht.: Immunofluoreszenzfibel <engl.>
ISBN 978-3-0348-9540-8 ISBN 978-3-0348-8376-4 (eBook)
DOI 10.1007/978-3-0348-8376-4

The original edition entitled "Immunfluoreszenzfibel – Grundlagen und neue Anwendungen in der klinischen Immunologie" was published in 1997 by Blackwell Wissenschafts-Verlag, Berlin – Wien.

Originally published by Birkhäuser Verlag in 2000
Softcover reprint of the hardcover 1st edition 2000

ISBN 978-3-0348-9540-8

9 8 7 6 5 4 3 2 1

Preface

The release of the English translation of Wulf Storch's primer on immunofluorescence is especially to be welcomed, since this painstaking and clearly organised survey will thus be accessible to the widest possible readership. The author, whom I have known for more than three decades, has worked tirelessly on improving the methods employed in immunofluorescence and has pursued consistently the broad spectrum of possibilities for their use in diagnosis.

The abundance of proven antibodies, many observed for the first time by Wulf Storch, is an expression of the devotion with which the author has applied to his goal of bringing about a complete survey of the current state of knowledge. The fact that the clinical significance of many antibodies has not, or not yet, been described fully is no argument against their existence. It must be left for the future to show whether these possess a defined role in the pathogenesis of disease, or not.

This Primer is an illustration of the relevance of immunofluorescence in the spectrum of modern diagnostic techniques. It is to be hoped that it will enjoy a wide circulation.

Prof. Dr. med. G. Geiler
Leipzig, February 2000

Foreword to the English edition

This, the first English edition, is nearly identical to the second German edition, with the addition of only small changes, e.g. the inclusion of antibodies against enterocytes.

Immunofluorescence still represents a powerful method both for routine work and for research purposes. Advances such as high-intensity microscopy, in particular confocal laser-scanning microscopy and the use of modern fluorescent dyes (fluorochromes) such as Cy 2 and Cy 3, have led to improvements and to even more widespread application.

Although the commercialisation of reagents has, in the meantime, continued apace on a broad front, the basic foundations including simple coupling reactions and purification procedures, have certainly not been discarded. In consequence, the general section is applicable for 'anybody'. In order to avoid metaphorically breaching the dykes in the section concerned with applications, I have limited myself to clinical immunology, in particular diagnostically relevant antibodies and biopsy findings, above all for kidney, skin and liver.

I would like to thank all those people who have helped and supported me by word or deed. I should like to mention in particular the former Dean of the University of Heidelberg, Professor Dr. Heiner Schirmer. Thanks are also due to Dr. R. A. Klein for the translation, to various commercial organisations for their financial support and, last but not least, to the staff of Birkhäuser Publishing who have been involved, especially to my editor Katrin Serries.

Wulf B. Storch

Weinheim
February 2000

Foreword to the 2nd German edition

I was often asked, very soon after the appearance of the 1st edition, to consider producing a 2nd edition. I struggled against doing this for a long time because the Leipzig collection of antisera had been destroyed. The generous support and encouragement of the former Dean of the Medical Faculty, University of Heidelberg, Prof.Dr. Heiner Schirmer, has, more than anything else, caused me to change my mind.

Although antibodies are increasingly determined by methods other than immunofluorescence, a short Primer and Atlas on this subject is still lacking for routine laboratory use. The 2nd edition should fill this gap. The text has been brought up to date and extended. The older literature references have been done away with in general, with the exception of those describing a method for the first time. Interested readers should refer to the 1st edition for these details. The majority of diagrams have been replaced by ones in colour. Commercially available substrates and reagents have almost always been used, especially those from Viramed (Mardx, USA), since I did not have access to cryotome equipment. Two recently discovered novel antibodies, against the Auerbach plexus and against podocytes, have been demonstrated although their clinical relevance is still not clear.

The many new examples of anti-nuclear factors, seen when one uses a cultured cell line as the source of antigen, are evidence that these methods are not outdated. Their versatility, in conjunction with modern high-power microscopy and especially confocal scanning laser-microscopy, as well as the newer fluorochromes such as Cy 3, allow the simultaneous and sensitive detection of a wide range of different substances, as no other method can. In experienced hands no other method is capable of replacing immunofluorescence.

Last but not least I would like to thank all those colleagues who have helped with advice or practical matters, especially the staff of the photographic laboratory and draughtsman's department of the University of Heidelberg for preparing the illustrations ready for publication.

Weinheim, October 1996
Wulf B. Storch

Table of contents

Abbreviations

Ab	antibody
Ab/P	ratio of antibody protein to total protein
Ab/F	ratio of antibody protein to fluorochrome content
ACA	anti-centromere antibodies
ADC	adrenal cortex
AER	antibodies against endoplasmic reticulum
Ag	antigen (antigen substrate)
AMA	anti-mitochondrial antibodies
AMCA	aminomethyl-coumarinyl-acetic acid
ANCA	anti-neutrophil cytoplasmic antibodies
ANF	anti-nuclear factors (antibodies)
APA	anti-podocyte antibodies
ARA	anti-ribosomal antibodies
B	"bursa-derived", thymus-independent
BBA	brush-border antibodies
BCA	bile canaliculi antibodies
BM	basement membrane
BSA	bovine serum albumin
C1-C9	components of the complement cascade
CAH	chronic aggressive hepatitis
CBR	complement-binding reaction
CENP	centromere nuclear protein
CF	complement fixation
CFR	complement-fixing reaction
CTA-3	connective tissue antibodies type 3
DAB	Deutsches Arzneibuch (German Pharmacopoeia)
DASS	defined antigenic substrate spheres
DLE	discoid Lupus erythematosus
DMSO	dimethyl sulphoxide
DNA	desoxy-ribonucleic acid
DNP	desoxy-ribonucleoprotein
ds-(RNA, DNA)	double-stranded RNA or DNA
E	extinction
EDTA	ethylene-diamine-tetraacetic acid
F/Ab	ratio of fluorochrome to antibody protein
F	fluorochrome
F/P	ratio of fluorochrome to total protein
Fab, $F(ab')_2$	fragment antigen binding from immunoglobulin
Fc	fragment crystallisable from immunoglobulin
FDA	fluorescein diacetate
FITC	fluorescein iso-thiocyanate

g_{av}	average accelerational constant when centrifuging
GBM	glomerular basement membrane
HBcAg	hepatitis B core antigen
HBeAg	hepatitis B e-antigen
HBsAg	hepatitis B surface antigen
HLA	histocompatibility leukocyte antigens
IF	immunofluorescence
Ig(G, A, M, D, E)	immunoglobulins G, A, M, D and E
i.m.	intra-muscular
i.v.	intra-venous
kDa	kilodaltons
LE	Lupus erythematosus
LKM-Ab	liver-kidney microsomal antibodies
LMA	liver cell membrane antibodies
M.m.	muscularis mucosae
OD	optical density
P1, P2, P3	first, second and third segments of the proximal tubules in rat kidney
PBC	primary biliary cirrhosis
PBS	phosphate-buffered saline
PCA	parietal cell antibodies
PEP	plateau end-point
PT	plateau titre
RA	rheumatoid arthritis
RNA	ribonucleic acid
RNP	ribonucleoprotein
RPM	revolutions per minute
s	second
s.c.	subcutaneous
SLE	systemic (disseminated) Lupus erythematosus
SMA	smooth muscle antibodies
ss-(RNA, DNA)	single-stranded RNA or DNA
T	"thymus-derived", thymus-dependent
TRITC	tetramethyl-rhodamine-isothiocyanate
v/v	composition by volume
w/v	composition in parts weight per unit volume
w/w	composition by weight
WHO	World Health Organisation

1 Introduction to the basic principles and techniques of immunofluorescence

1.1 Fluorescence, primary and secondary fluorescence

Fluorescence is defined, based upon the physical definitions of the properties of matter, as the ability to emit light without noticeable delay when irradiated. The emitted light has, according to Stoke's law, a longer wavelength than the absorbed light, i.e., the photons are less energetic (Fig. 1).

Fluorescence emission from untreated materials is known as primary, natural, self- or auto-fluorescence. When fluorescence is generated artificially using additional fluorescent substances, one speaks of secondary fluorescence. Haitinger [1959] has named these materials 'Fluorochromes' in order to distinguish them from the other dyestuffs used in visible-light microscopy. As a consequence of the exceptionally low concentrations that are necessary (as low as 1 in 10^6), fluorochromes cause almost no damage to living preparations, so that fluorescence microscopy allows living tissue to be investigated. Fluorochromes should, where possible, be specific for those structures or molecules which are of special interest.

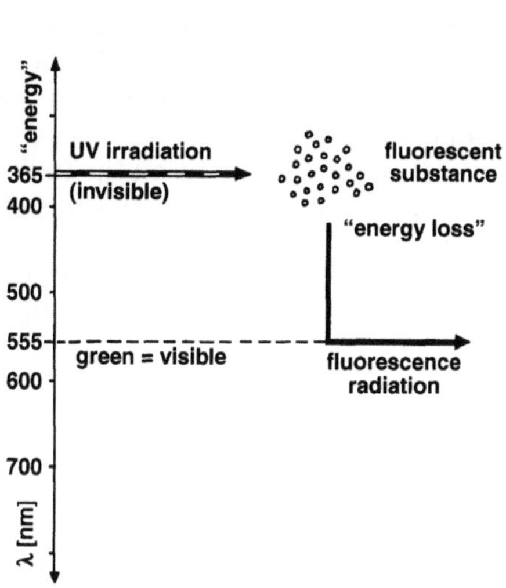

Figure 1. Scheme showing the origin of fluorescence radiation (based on Holz [1975], with modifications).

1.2 Principles of immunofluorescence

Coons et al [1941, 1942, 1950] succeeded in combining the *high sensitivity* of fluorescence microscopy with the *strict specificity* associated with immunological procedures, using antibodies labelled with chemically stable fluorochrome groups. The terms 'immunofluorescence' or the 'fluorescent antibody technique' have been introduced to describe this concept. It is not only relevant *whether* an antigen-antibody complex is seen but, more importantly, *where* it is seen (Holz [1975]).

As a simplification, it is possible to distinguish three different methods (Fig. 2):

a) **the direct method** (Coons & Kaplan [1950]; Gitlin et al. [1953])

In order to demonstrate the presence of an antigen (Ag), the histological preparation is treated with fluorescently labelled (fluorochrome-labelled) antibody, the so-called conjugate. After a predetermined

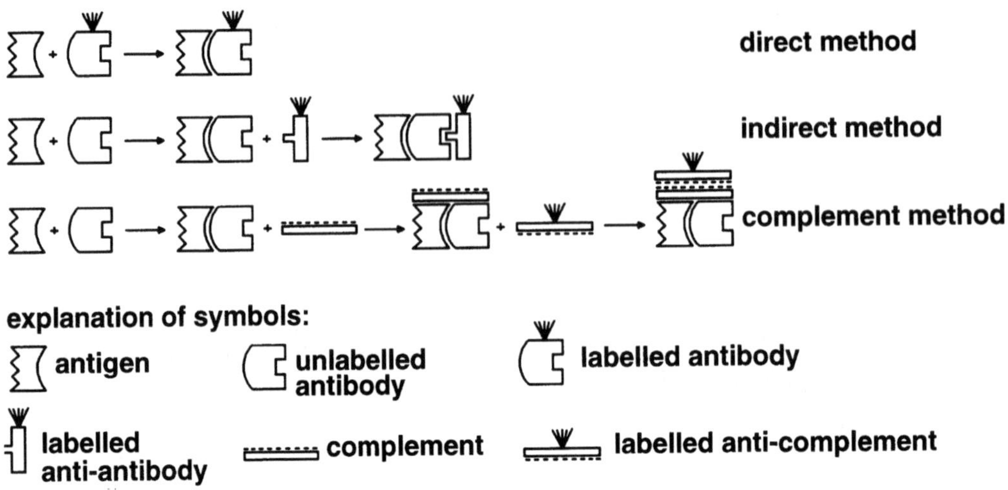

Figure 2. Scheme showing the methods used for immunofluorescence.

reaction time the unreacted or loosely bound antibody is washed off. Ideally, when the reaction is specific, only those parts of the preparation that have bound labelled antibody **and** that contain the required antigen, are visible under the fluorescence microscope. This procedure may also be used 'in reverse' to demonstrate an antibody using a fluorescently labelled antigen.

b) **the indirect method** = two, or more, layer technique (Weller & Coons [1954]; Beutner, Holborow & Johnson [1965])

In the first step an unlabelled antibody (Ab) is allowed to react with antigen (Ag), followed by the addition in the second step of a labelled antibody (Ab*), which reacts with the unlabelled antibody (Ab), a so-called 'anti-antibody'.

Indirect methods are used predominantly to demonstrate humoral antibodies directed against known proteins of cells and tissues. In order to demonstrate antibodies against smooth muscle, for example, a tissue slice which contains the antigen of interest, i.e., the smooth muscle, is exposed to a patient's serum. During the incubation period the serum antibodies against components of smooth muscle bind to those tissue elements which contain, hopefully, only smooth muscle. Excess and loosely bound antibody is removed by washing. At this point the antibody bound to the preparation is not yet visible. It is visualised in a second stage in which fluorescently labelled antibody directed against the first, non-labelled antibody is added. Because human antibodies are immunoglobulins (Ig), anti-human-immunoglobulin serum (AHIgS) must be used. The indirect method has two advantages over the direct method: (i) one AHIgS is sufficient for demonstrating a wide range of different antibodies, and (ii) the method is more sensitive. It does, however, require more stringent controls.

There are many variations on the indirect method (for an overview see Storch [1986]). One in particular should be emphasised, since it is also important for membrane immunofluorescence studies, namely the 'sandwich' method (Fig. 3).

Cells must be fixed prior to demonstrating a specific antibody, in order to prevent the antibody being bleached out during the experiment. After the addition of antigen (Ag), incubation with fluorochrome-labelled antibody (Ab*), raised against the antigen, is carried out. After washing to remove loosely or non-specifically bound antibody, cells which have

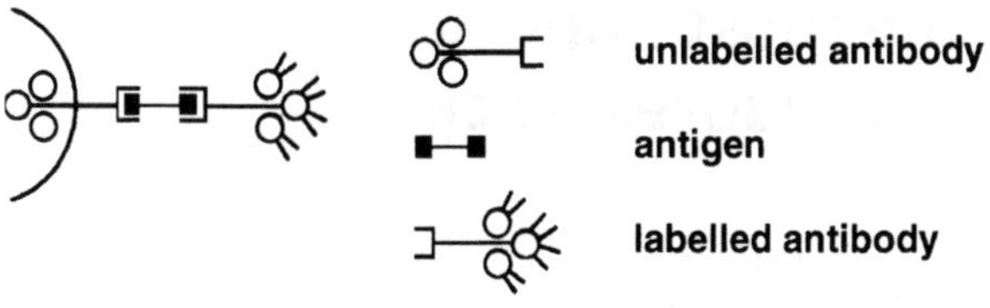

Figure 3. Scheme for the so-called "sandwich" method for detecting a specific antibody using the corresponding antigen and a fluorochrome-labelled antibody – for an explanation, refer to the body of the text (modified based on Roitt [1977]).

bound antigen specifically and are thus able to produce it, can be identified. The name 'sandwich' was originally chosen because the antigen (Ag) lies between two separate antibodies, i.e., the complex is likened to a filled double-layer.

c) **anti-complement methods** (Goldwasser & Shepard [1959]; Klein & Burkholder [1959]; Matuhasi & Usui [1966]; Müller [1960])

These authors showed the particular utility of a technique which relies on the binding of components of the complement system during antigen-antibody (Ag-Ab) complex formation. Direct and indirect procedures are possible. This technique is recommended for the demonstration of Ag-Ab complexes occurring *in vivo*.

Total complement can be demonstrated using the method of Klein & Burkholder [1959], in which a preparation is exposed sequentially to complement (Guinea-pig serum) and, after washing, to a fluorescently labelled anti-GP-globulin antiserum, thus showing that GP-complement is bound by the Ag-Ab complex. If individual components of the complement cascade are of interest, then the best solution is to use the direct method in conjunction with a mono-specific conjugate, for example FITC-labelled anti-human-C3-serum.

In order to demonstrate circulating antibodies in man which bind complement, it is necessary to incubate the substrate antigen sequentially with inactivated human serum, complement and fluorochrome-labelled anti-complement antibody. If no labelled anti-GP-globulin antiserum is available, it is possible to use a non-labelled anti-complement rabbit antibody followed by a labelled anti-rabbit-IgG antiserum.

1.3 Nomenclature used in immunofluorescence studies

Unfortunately there is to date no agreed or widely accepted nomenclature. Mayersbach [1966] even draws a strict distinction, for example, between *staining* and *reaction*. In order to support a uniform nomenclature, that of Beutner et al. [1973] is reproduced below. Based on this nomenclature one speaks of:

1. **Autofluorescence**: synonymous with self or primary fluorescence. The appearance of fluorescence in tissues or materials that have not been treated with labelled conjugates.
2. **Non-specific staining** (NSS): caused as a result of non-immunological staining by fluorochrome complexed to protein or by the presence of free (unbound) fluorochrome.
3. **Specific staining** (SS): this is the consequence of an immunological reaction of fluorescently labelled proteins whereby one can distinguish
   - desired specific staining (DSS), from
   - undesired specific staining (USS).

The different types will be dealt with in detail in section 4.4.

1.4 Short description of the most important procedures for direct immunofluorescence

The most important preparative procedures and techniques for the demonstration of direct immunofluorescence can be summarised as follows:

Preparation of pure antigens

Animal immunisation

Serum production, tests for specificity and Ab-titre

Isolation of immunoglobulins with Ab-activity

Labelling with fluorochrome

Purification of the labelled antibody:
(I) gel filtration over Sephadex to remove free fluorochrome;
(ii) chromatography over DEAE-Cellulose to obtain optimally labelled antibodies;
(iii) where necessary, absorption to remove cross-reacting and heterologous antibodies.

Concentration (reduction in volume)

Specificity testing and determination of Ab-titre, assessment of suitability in the special system for determining the optimal dilution of the labelled antiserum for use, determination of the fluorochrome/protein (F/P) ratio, etc.

Histological preparation or cell suspension

Air drying

Fixation

Washing

Incubation

Washing

Air drying

Embedding

Examination by fluorescence microscopy

2 Production of antibodies and labelling with fluorochromes

2.1 Structure and function of antibodies (immunoglobulins, Ig)

A brief, simplified description of the structure and function of antibodies, the immunoglobulins (Igs), is now given to promote a better understanding of, in particular, the controls required for specificity (Fig. 4).

Immunoglobulins consist of at least four polypeptide chains. Based on differences in molecular weight one talks of light (L) and heavy (H) chains. The distinction between IgG, IgA, IgM, IgD and IgE is made based on the heavy chains. Light chains can exist as κ (kappa) or as λ (lambda) chains. Immunoglobulins can be cleaved using proteolytic enzymes. When an immunoglobulin molecule is treated with papain (E.C. 3.4.4.10), three fragments are produced, two univalent F_{ab} fragments (F_{ab} = fragment antigen binding) with which the antigen can form soluble complexes, and an F_c fragment (F_c = fragment crystallisable). The latter cannot bind antigen but does bind complement and can interact with cells expressing appropriate receptors; it is responsible for the ability to bind to tissue. Cleavage with pepsin (E.C. 3.4.4.1) gives rise to the $F_{(ab')2}$ fragment, which is bivalent. Antibodies have a double function. The F_{ab} portion is responsible for binding to the appropriate antigen (Ag); biological activity of antibody molecules is determined by the F_c portion.

Complex formation between antigen (Ag) and antibody (Ab) is reversible and depends on the interaction energy or

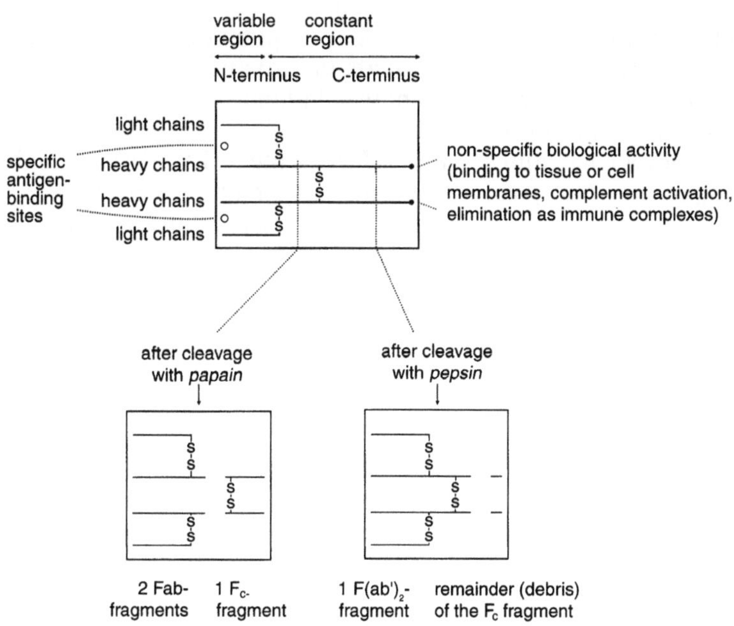

Figure 4. Antibody model consisting of four polypeptide chains, two heavy (H) and two light (L), which are linked together by disulphide (-S-S-) bridges. Cleavage with proteolytic enzymes produces the fragments $F(ab')_2$, Fab, F_c and F_d (part of the heavy chains in the Fab fragment), as well as fragments of the F_c cleavage product from the H-chains.

strength (note the difference between *avidity* and *affinity*), on the valency and on the specificity. One refers to a higher or lower avidity when one means the strength of binding between the antibody and the antigen taken as a whole, i.e., *in toto*. Affinity is taken to refer to the strength of binding for individual regions of the antigen, such as an individual or single antigenic determinant or epitope, or for a so-called hapten which is usually a small molecule incapable on its own of stimulating antibody production but which is able to bind antibody.

Antibodies can cross-react with antigens possessing similar or identical antigenic determinants or epitopes. Binding is, however, generally weaker (Fig. 5). An antibody is also more or less specific for an antigen and can, for example, be mono-, oligo- or polyspecific. This is particularly important for carbohydrate epitopes. The valency or number of binding sites which can react with antigen is also different with IgG, IgD and IgE, having two binding sites, whereas IgA has between two and six and IgM with either five or ten binding sites. When antigen and antibody are present in a ratio such that maximal precipitation of the complex occurs, then we refer to the "region of equivalence". If an antigen possesses at least three binding sites then it is able to generate a three-dimensional, space-filling lattice when it forms an Ag-Ab complex. Depending on the ratio of antigen to antibody (Ag/Ab), the Ag-Ab complexes (immune complexes) may differ in size, stability, solubility and biological activity (Fig. 6).

2.2 Production of antisera

The antiserum quality has a decisive importance for the success or failure of investigations involving immunofluorescence (Knapp & Ludwig [1976]). Even though adequate quantities of sufficiently pure antisera may be available commercially, special antisera will still have to be produced in one's own laboratory. Space does not permit a more detailed discussion of this here. Practical details and advice are available in the literature (Ambrosius [1970]; Huet [1994]; Celis et al. [1994]).

A simple procedure for obtaining anti-human IgG antibody, based on the method of Hijmans et al. [1969], is as follows:

Apparatus and reagents:

1. pure IgG
2. complete and incomplete Freund's adjuvant (Difco Laboratories, Detroit)
3. Sephadex G–25 (Pharmacia, Uppsala)
4. Phosphate-buffered saline (PBS – see 2.4.2)
5. Beakers, mixer, crushed ice
6. 5 ml syringes
7. Rabbits
8. 10% w/v potassium aluminium sulphate (alum) and 1 N sodium bicarbonate solutions.

Procedure: 2 mg IgG and 50 mg Sephadex G–25 are dispersed in 2 ml PBS. 2 ml of Freund's complete adjuvant is then added

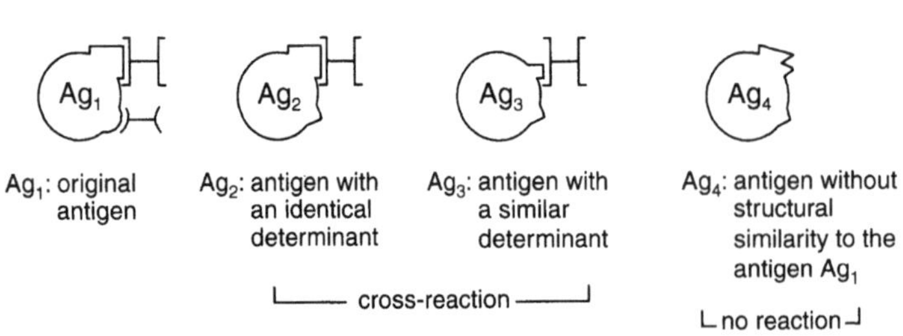

Figure 5. Scheme for explaining antibody specificity and cross-reaction. The avidity of the serum for antigen 1 is greater than for antigens 2 and 3, and is absent altogether for antigen 4. It is in this way that serum demonstrates its specificity. Antibodies are shown as]–[and)–(: modified after Roitt [1977].

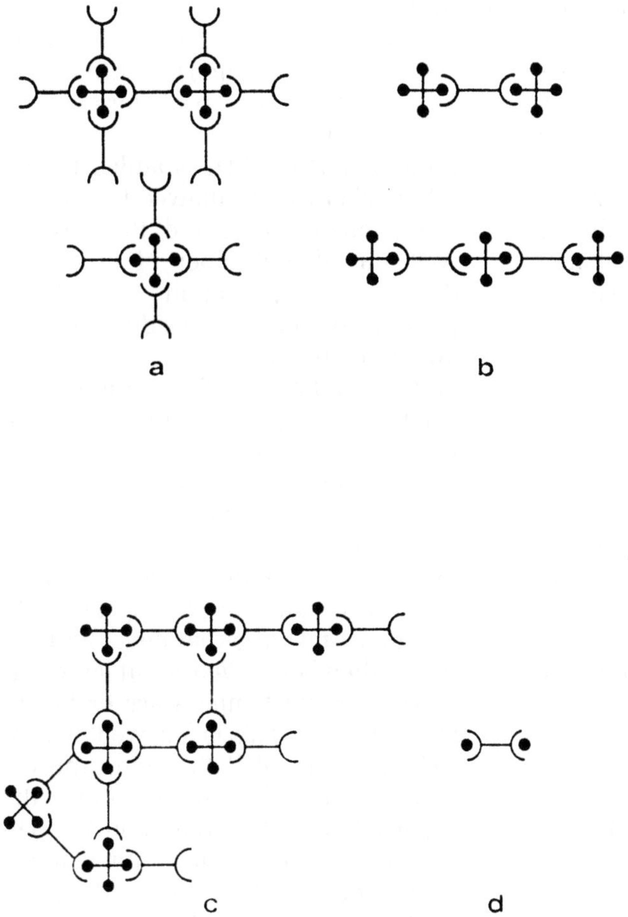

Figure 6. Schematic representation of complexes between a hypothetical tetravalent antigen •‡• and a bivalent antibody)—(at different concentration ratios. In reality it is very unlikely that the antigen valencies would all lie in the same plane, as shown in the diagram.

a) complex formation in the presence of a large antibody excess; the antigen valencies are saturated and the Ab/Ag molar ratio approximates to the antigenic valency.
b) complex formation in the presence of an excess of antigen; in the presence of a very large excess, in which the two valencies of each antibody molecule are rapidly saturated, the complex Ag_2Ab predominates.
c) a large three-dimensional lattice, such as that obtained in a typical immuno-precipitate.
d) monovalent antigen can indeed bind antibody but is incapable of forming cross-linkages (based on Roitt [1977]).

drop by drop, keeping the mixture on ice and well stirred. If no suitable mixer/stirrer is available, then the suspension can be transferred to and fro between two 5 ml syringes connected by tubing, until a stiff emulsion is obtained. A drop of this emulsion, when placed on the surface of luke warm water (body temperature), should not disperse or 'run'. Each rabbit should be inoculated with approximately 0.2 ml of the suspension subcutaneously (s.c.) between the toes, with ~2 ml s.c. under the shaved skin of the back, and the remainder given intramuscularly (i.m.) into the upper thigh. This procedure should be repeated after ten days using half the quantities. If antibodies against L-chains are also required, antigen can be given intravenously (i.v.) after a further 20 days. To do this, dissolve 10 mg IgG in 2.5 ml PBS, mix with 15 ml 1 N $NaHCO_3$ solution and then add 2.5 ml of a 10% w/v solution of potassium aluminium sulphate (alum) dropwise over ten minutes. Allow the suspension to stand overnight at +4°C in a refrigerator. After centrifugation for 30 min at 15,000 g_{av}, the precipitate should be resuspended in 4 ml PBS. The suspension should be given in the following quantities twice weekly over a three week period: (i) Week 1: 0.05 ml and 0.10 ml; (ii) Week 2: 0.10 ml and 0.20 ml; (iii) Week 3: 0.20 ml and 0.40 ml. The first blood samples should be taken and tested by Ouchterlony diffusion (see 3.2.4) no later than 1–2 weeks after the last injection. Serum should be stored at –20°C.

Tip: As a rule of thumb, antibodies produced during the early phase of an immune response (the primary response) have a higher *specificity* but lower *avidity* than those produced during the later phases (the secondary response), when this relationship is reversed. Furthermore, it is often more effective to allow a longer period to elapse (5–6 weeks) after the first injection before giving the second injection.

2.3 Isolation of immunoglobulins from serum

Different methods can be used depending on the purpose for which the immunoglobulins are required. One of the most useful procedures is briefly as follows.

γ-Globulin Purification by Ammonium Sulphate Precipitation (based on the method of Hebert et al. [1972])

Apparatus and reagents:

1. $(NH_4)_2SO_4$ – ammonium sulphate, AR, low in heavy metals for enzyme work
2. 0.145 M (0.85% w/v) NaCl, pH 8 (if necessary adjusted with 0.1 N NaOH)
3. Shaker or magnetic stirrer
4. Ice-bath and refrigerator
5. Refrigerated centrifuge
6. Dialysis equipment
7. Saturated $BaCl_2$ solution
8. 0.1 N HCl
9. Measuring cylinders

Procedure:

1. The preparation of the ammonium sulphate stock solution (pH 5.8) should be carried out 3–4 days before it is required for use.
   55 g pure $(NH_4)_2SO_4$ is added to 100 ml distilled water, allowed to stand at room temperature, and the mixture well stirred a number of times per day. The supernatant should be poured off from the residual crystals and used directly.
2. Working dilutions of the ammonium sulphate solution are prepared just prior to use according to Table 1.
3. The precipitation is carried out depending on the type of animal used (Tab. 2). 10 ml of 70% saturated $(NH_4)_2SO_4$ solution are added dropwise, with constant stirring or shaking, to 10 ml rabbit serum. The mixture is then kept *either* for at least four hours at room temperature *or* for 24 hours at +4°C. This is followed by centrifugation at +4°C (30 min; approx. 1,400 g_{av}). After decantation, the supernatant is discarded and the pellet made up to 10 ml with distilled water, with stirring, followed by the addition of a further 10 ml 70% saturated $(NH_4)_2SO_4$ solution. The precipitate can be taken up in a smaller volume after the third precipitation, if a more concentrated preparation is desired. The globulin solution must then be dialysed against frequently exchanged 0.145 M sodium chloride (pH 8.0) until the dialysate is free of sulphate.

Table 1. Ammonium sulphate working solutions – based on Hebert et al. [1972].

Preparation			Use
Volumes saturated stock solution	Distilled water	% v/v $(NH_4)_2SO_4$ solution	% saturated ammonium sulphate in final solution
10	0	100	50
9	1	90	45
8	2	80	40
7	3	70	35
6	4	60	30

Table 2. Recommended salt concentration and number of precipitations with ammonium sulphate (Jones et al. [1978]).

Animal Species	% $(NH_4)_2SO_4$ × no. precipitations	Total protein mg/ml	Relative γ-globulin	amount albumin	γ-globulin recovery
Rabbit	35 × 3	15	66	1	91
Sheep	35 × 3	24	68	1	84
Chimpanzee	35 × 3	18	76	0	97
Hamster	35 × 3	10	55	0	68
Pig	35 × 3	24	44	1	72
Chicken	35 × 3	15	58	1	73
Horse (1)	30 × 2	14	74	1	63
Horse (2)	30 × 3	8	83	0	45
Mouse	35.40 × 2	7	58	0	75
Guinea pig	35.40 × 2	12	51	0	74
Monkey	30.40 × 2	9	74	1	74
Cow	30.35 × 2	18	90	0	71
Goat	30.45	18	81	1	83

Test for sulphate: acidify a sample with dilute hydrochloric acid and add 1–2 drops of $BaCl_2$ solution. Note whether there is hazy turbidity present – if there is, then the test is positive for sulphate.
Tip: if PBS is used for dialysis, rather than distilled water or a solution of sodium chloride, another sulphate test must be used since PBS and saturated $BaCl_2$ will produce a precipitate of barium phosphate.

2.4 Labelling of antisera with fluorochromes

2.4.1 Fluorescent dyes (fluorochromes)

2.4.1.1 Requirements for a fluorochrome

Ideal fluorochromes should possess the properties shown in Table 3. There are no instances of currently available fluorochromes fulfilling all these requirements. One must take care to use, at very least, materials of high purity. For this purpose, analysis by thin-layer chromatography (TLC) is useful.

2.4.1.2 Fluorescein-*iso*-thiocyanate (FITC)

Source of crystalline Isomer I:	(i) Zentral Institut für Mikrobiologie und experimentelle Therapie Jena (Germany). (ii) Baltimore Biological Laboratory, United States of America. (iii) Sigma Chemical Co., Milwaukee, United States of America – also European subsidiaries.
Molecular weight:	389.4, $C_{21}H_{15}O_5NS$ (Fig. 7a).
Absorption maximum:	490 nm
Emission maximum:	520 nm
Colour of fluorescence:	bright yellow-green (apple green).
Storage conditions:	at +4°C in the presence of desiccant and in a closed container out of contact with air.
Reaction with proteins:	Dye-N=C=S + HN-Protein → Dye-NH-(C=S)-NH-Protein

Table 3. Required properties of fluorochromes – modified based on Chadwick et al. [1958] and Pearse [1968].

1. Straightforward, rapid and stable binding with antibody protein.
2. No modification of the immunological properties of the antibodies, even on prolonged storage – constant affinity and avidity.
3. Intense fluorescence even after labelling, without non-specific staining.
4. Ability to distinguish clearly between auto-fluorescence and other fluorochromes (double labelling).
5. Minimal photo-bleaching.
6. Easily produced, can be stored, good solubility in water, high purity, of testable quality and cost-effective.

2.4.1.3 Tetramethyl-rhodamine-*iso*-thiocyanate (TRITC)

Source:	(i) Baltimore Biological Laboratory, United States of America. (ii) Sigma Chemical Co., Milwaukee, United States of America.
Molecular weight:	376.4, $C_{25}H_{21}O_3NS$ (Fig. 7b).
Absorption maxima:	510 nm and 545 nm (amorphous TRITC) ~555 nm (crystalline TRITC)
Emission maximum:	620 nm
Colour of fluorescence:	red
Storage conditions:	at +4°C, in the presence of desiccant and in a closed container out of contact with air.

2.4.1.4 Further fluorochromes

A survey of available fluorochromes is available in Storch [1986].

Dichlorotriazinyl-fluorescein (DTAF) and, more recently, Cy 2 (Biological Detection Systems Inc., USA) are available as alternatives to FITC, being more photostable. Many substances are suitable as alternatives to TRITC. Lissamine-rhodamine-sulphonyl chloride (LRSC) and Texas-Red-sulphonyl chloride are frequently used. The trend is to use the newer fluorochromes. Not only has Indocarbocyanin (Cy 3, Fig. 8) proved its worth above all for confocal microscopy, but so has Indodicarbocyanin (Cy 5) as well as others, especially Aminomethyl-coumarin-acetic acid (AMCA) (Ernst et al. [1989]; Mujumdar et al. [1989]; South-

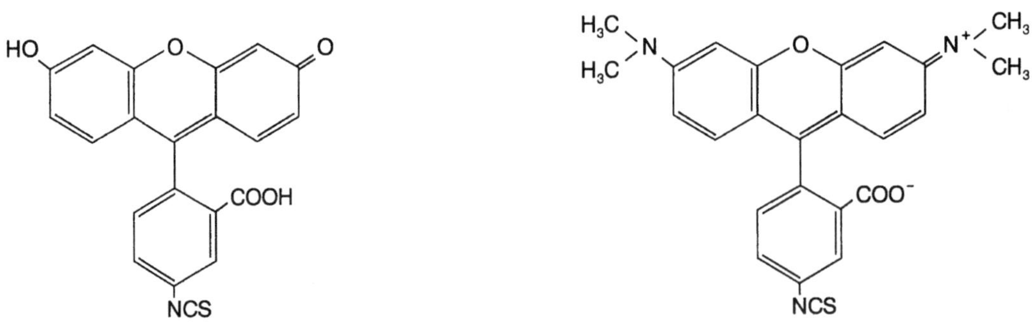

a FITC b TRITC

Figure 7. Structural formulae for (a) FITC, fluorescein *iso*-thiocyanate, and (b) TRITC, tetramethyl-rhodamine-*iso*-thiocyanate.

wick et al. [1990]; Khalfan et al. [1986]; see Fig. 8).

Whether enzyme-labelled fluorescent (ELF) substrates will be successful remains to be seen (Larson et al. [1995]). The absorption and emission maxima for Cy 3 are 550 nm and 570 nm, for AMCA ~360 nm and 450 nm, and for the ELF-phosphatase substrate 365 nm and 515 nm. The current advantage of the red fluorescence of Cy 3 is that the fluorescence yield is high, some 100-fold better than for FITC. The advantage of the blue fluorescence associated with AMCA is the possibility of carrying out triple-labelling experiments. The high photo-stability of the yellow-green ELF-phosphatase substrate, which is some 550-fold more stable than fluorescein itself, is a distinct advantage. When a confocal microscope is available, a combination of Cy 2, Cy 3 and Cy 5 is particularly convenient for visualising triple-labelling.

2.4.2 Coupling with FITC

A number of different recipes exist (for an overview see Wagner [1967]). It is generally necessary to ensure that the pH during conjugation is kept between 9.0 and 9.5 and that the protein concentration lies between 1% and 4% w/v, i.e., 10–40 mg/ml. The amount of FITC is determined by the required fluorochrome/protein ratio, taking account of the purity of the dyestuff and the fact that the relationship with the degree of conjugation is not linear. Empirically determined values are collected together in Table 4. The time for the reaction depends on temperature. At 25°C 30–60 minutes are sufficient ("fast labelling"), whereas at +4°C 24 hours are necessary ("slow labelling").

Table 4. The relationship between FITC quantities and the desired F/P ratio (±20%) – Beutner et al. [1968].

mg FITC per g protein	expected average molar F/P ratio
4	1.0
6	1.5
8	2.0
11	2.5
14	3.0
18	3.5
21	4.0
25	4.5
29	5.0

Apparatus and reagents:

1. Phosphate-buffered saline (PBS) containing 145 mM NaCl, 0.01 M NaH_2PO_4/Na_2HPO_4, pH 7.1–7.2. This requires 8.5 g NaCl, 1.07 g Na_2HPO_4 (anhydrous) and 0.39 g $NaH_2PO_4 \cdot 2 H_2O$, made up to 1 litre with distilled water.

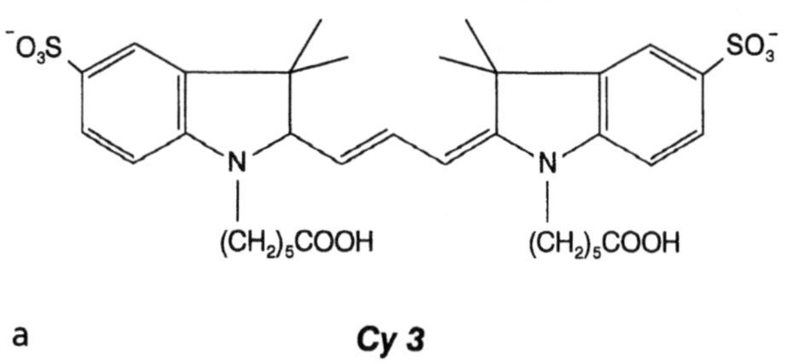

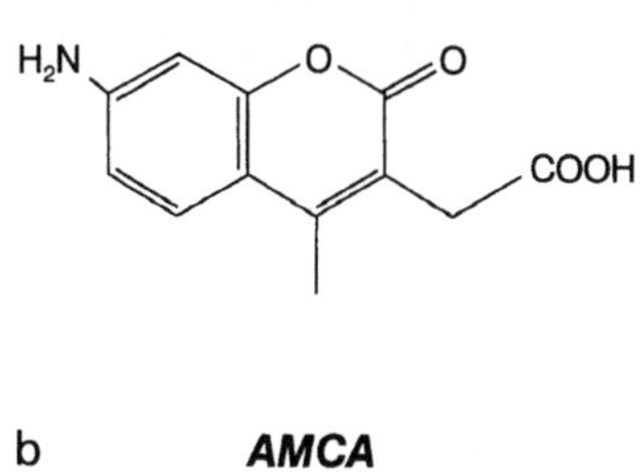

Figure 8. Structural formulae for (a) Cy 3, indocarbocyanin, and (b) AMCA, aminomethyl-coumarinyl-acetic acid.

2. FITC (Isomer I).
3. 0.1 M Na_2HPO_4 solution (MW $Na_2HPO_4.2H_2O$ = 178.05)
4. 0.2 M Na_2HPO_4 solution (MW $Na_2HPO_4.2H_2O$ = 178.05)
5. 0.1 M Na_3PO_4 solution (MW $Na_3PO_4.12H_2O$ =380.1)
6. 0.145 M NaCl solution
7. Water bath (25°C), ice-bath, magnetic stirrer, pH meter, flasks with ground-glass stoppers.

Procedure: for example, suppose that we have 4 ml of a 2% w/v protein solution. The four millilitres therefore contain 80 mg protein. We require an F/P ratio of between 2.5 and 3:1. Using Table 4, this corresponds to 12.5 µg FITC/mg protein and therefore 12.5 × 80 µg = 1 mg of FITC will be required.

The protein solution as well as reagents 3–6 are maintained at 25°C in a water bath. 1 mg FITC is dissolved, with stirring, in 2 ml 0.1 M Na_2HPO_4 over 5–10 minutes, and then placed in the water bath. 1 ml of 0.2 M Na_2HPO_4 is added to the 4 ml of protein solution in a flask with a ground-glass stopper, followed by the FITC solution. The pH should be measured immediately and adjusted to pH 9.5 with a few drops of 0.1 M Na_3PO_4. The total volume is made up to 8 ml, the flask stoppered and then vigorously shaken. It is then placed, without further shaking, in the water bath. Some form of covering is advisable to prevent access by light. After 30 min the reaction mixture should be cooled in an ice-bath with shaking. Subsequently, gel chromatography over Sephadex G–25 is used to remove unbound dye (see 2.5.1).

Tips:
1. If ammonium sulphate precipitation has been used to prepare the globulins, these must be freed of ammonium sulphate before labelling is carried out (see 2.3.1).
2. Amorphous (non-crystalline) FITC preparations are not recommended, since these contain a mixture of isomers I and II and frequently contain impurities, whereby reduced fluorescence intensity and stronger non-specific staining may result.
3. Optimal results are achieved using 4–12.5 mg FITC per gram protein. Labelling with more than 20 mg FITC/g protein is not sensible, especially as the relationship with conjugation is not linear (Tab. 4 and The & Feltkamp [1970]) and conjugates with a high F/P ratio tend to be unusable because of non-specific staining.
4. If lyophilised globulin is to be labelled, account must be taken of the weight excluding that of salts from the elution buffer.
5. If a precipitate is formed during the conjugation reaction, this should be removed by centrifugation (20 min at approx. 2,200 g_{av}).

2.4.3 Coupling with TRITC

Here also the details differ considerably, being determined generally by differences in purity of the main components (Amante et al. [1972]; McKinney & Spillane [1975]). Nairn [1976] has used the R-isomer, recommending the use of 20 µg/mg globulin. Immediately prior to use, TRITC is dissolved in dimethylformamide (DMF) at 16 µg/ml and diluted 20-fold with 0.1 M Na_2HPO_4. Conjugation takes place in an analogous way to FITC (see 2.4.2). Wick et al. [1976] used freshly prepared 2% w/v $NaHCO_3$ (pH 8.2), achieving a labelling of 1 mg TRITC/100 mg protein. 0.6 mg TRITC is needed for 2 ml of a 3% w/v protein solution, i.e., a total of 60 mg protein, having first been dissolved in 2 ml $NaHCO_3$ solution over two hours (the same volume as the protein solution). The TRITC is then added to the protein solution over 30 minutes, with constant stirring at +4°C. Stirring is continued for a further 30 minutes and the solution allowed to stand overnight at +4°C. Goding [1976] has reported good results using Bergqvist & Nilsson's [1974] suggestion of a DMSO solution.

Coupling of IgG with TRITC (based on Goding [1976]):

Apparatus and reagents:
1. Carbonate buffer, pH 9.3 (17.3 g NaH-CO_3 , 8.6 g Na_2CO_3 to a final volume of 1 litre with distilled water).
2. Dialysis equipment, refrigerator.
3. Beakers.
4. Magnetic stirrer.
5. Crystalline TRITC.
6. Dimethylsulphoxide (DMSO).

Procedure:
1. The IgG fraction, previously dialysed overnight against carbonate buffer and containing ~6 mg IgG/ml in a total volume of 10 ml, was placed in a beaker and stirred magnetically at room temperature,
2. Approximately 3 mg crystalline TRITC were dissolved in 3 ml DMSO (1 mg fluorochrome/ml). This was then added dropwise to the IgG solution with continuous stirring (approximately 10 µg fluorochrome/mg IgG, giving a ratio for $OD_{550}/OD_{280} \approx 0.5$, equivalent to a molar F/P ratio of ~2) – see 3.2.1.2.
3. The solution should be allowed to stand for two hours in the dark at room temperature, before gel chromatography over Biogel P–6 (see 2.5.2).

2.5 Purification of the conjugate

Purification consists of the following tasks:
1. Separation and removal of excess "free" dye as well as impurities and degradation products.
2. Selection of optimally labelled Ab with the most appropriate dye/protein ratio – removal of over- and under-labelled Ab.
3. The removal of heterologous and cross-reacting Ab, organic solvent and buffer salts, all of which can upset the immunological reaction.

Gel filtration over either Sephadex G–25 (FITC) or G–50 (TRITC) is the method of choice these days for removing free dye, buffer salts and solvent.

2.5.1 Gel-filtration over Sephadex G–25 and G–50

Principle: separation using cross-linked dextrans or polyacrylamides (Sephadex, Biogel) which, on soaking in water, function as molecular sieves because they form insoluble gels with a pre-determined pore size. Small molecules, such as free FITC, penetrate the pores and are retarded whereas larger ones, such as protein-bound FITC, do not. The Biogel polyacrylamide series should be used for separations in which it is important to avoid carbohydrate contamination or binding problems.

Apparatus and reagents:
1. A water-jacketed chromatography column. A simple glass tube with a sintered glass frit (G2 or G3, Schott), or ½–1 cm of glass wool held in place by glass beads, at the bottom, is sufficient (Fig. 9).
2. PBS pH 7.1 (see 2.4.2).
3. Sephadex G–25 or G–50 medium (Pharmacia, Uppsala).

Procedure:
1. The column should be prepared 1–2 days before use.
   6–10 g Sephadex, depending on the column used (length/diameter 10:1 to 30:1), will be required for the purification of 10 ml labelled serum. The dry powder is dispersed by stirring in sufficient PBS so that, after settling, the supernatant volume is approximately five times that of the gel layer. After soaking for ~5 hours, 'fines' are re-

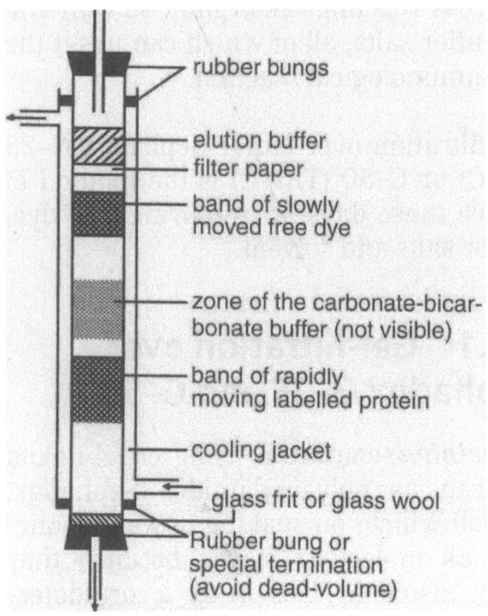

Figure 9. A simply produced arrangement for gel-filtration using Sephadex, for purifying conjugates (modified, based on Steffen [1968]).

moved by washing and decantation a number of times. Degassing is carried out, when required, using a water pump.

The chromatography column, which should be mounted vertically, is washed through with PBS and left containing approximately 1/3 of its volume of liquid. The swollen Sephadex is suspended in an equal volume of buffer and, in one movement, poured into the column. The column stopcock should only be opened after the gel has settled by sedimentation. It is advisable to cover the top of the gel bed with either filter paper or an applicator having a nylon net. A serum sample should be used to test the column for correct size and packing.

2. Gel filtration.

Most of the buffer above the gel should be sucked off. The remainder is allowed to run into the gel and then the conjugate is layered, drop by drop, onto the top of the gel. After the sample has percolated in, a small amount of buffer is added and this in turn allowed to penetrate the gel. When this has been done, more buffer is added and the buffer reservoir connected. Free dye remains at the top of the column, whereas the rapidly migrating labelled protein can be collected as the first fraction in a high degree of purity. A fraction collector can be dispensed with as the separation can be clearly seen with the naked eye. If necessary the relatively dilute eluate can be concentrated (see 2.6).

Tip: the column can be re-used if the top layer of gel containing free dye is removed and/or the remaining dye eluted with PBS. When the Sephadex column is not in use, 0.01% w/v sodium azide should be added to the buffer as a preservative.

2.5.2 Gel Filtration on Biogel P–6 (BioRad Laboratories, Richmond, CA)

Goding [1976] recommends the use of a 0.01 M phosphate buffer for elution (1:100 dilution of a stock 1 M K_2HPO_4 solution) with a flow rate of ~1.5 ml/min.

2.5.3 Fractionation on DEAE-Sephadex A–50 (after Goding [1976])

Principle: fractions with closely similar F/P ratios can be separated by ion-exchange chromatography. Globulins with a low level of labelling are eluted first at low salt concentrations. Higher salt concentrations are required to elute more highly labelled, i.e., more acidic or charged, globulins. The method has the disadvantage of relatively high losses associated with the need to use large volumes.

Apparatus and reagents:

1. Chromatography column or 50 ml plastic syringe with a centrally placed outlet and built-in nylon net.
2. DEAE-Sephadex A–50 (Pharmacia, Uppsala).
3. 0.5 M potassium phosphate buffer, pH 8.0 (1:2 dilution of a stock solution made by adding 1 M K_2HPO_4 and 1 M KH_2PO_4.
4. 0.01 M potassium phosphate buffer, pH 8.0 (1:100 dilution of the above stock solution (3)).
5. Büchner funnels.
6. 0.002% w/v chlorohexidine.
7. Gradient mixer (Pharmacia, Uppsala).
8. UV-monitor.
9. Fraction collector.

Procedure:

1. Preparation of the DEAE-Sephadex gel: 15 g of the commercially available DEAE-Sephadex (chloride form) is exchanged against 0.5 M phosphate by soaking for 24 hours. (The potassium salts should be used as the sodium salts have a tendency to crystallise). ~300 ml of the previously soaked gel are thoroughly washed, supported on a Büchner funnel, using 2 litres 0.01 M phosphate buffer (pH 8.0) and then allowed to equilibrate in this buffer for a further 24 hours in the presence of 0.002% w/v chlorohexidine. It is then ready for use and can be kept at room temperature.
2. Procedure for fractionation: The column, or plastic syringe, is filled with pre-swollen DEAE-Sephadex. After applying the conjugate sample, separation is achieved using a 0.01–1.0 M salt gradient, produced using a gradient mixer. The gradient reservoirs should contain 100 ml 0.01 M phosphate (pH 8.0) (A) and the same buffer but with the addition of 5.8 g NaCl (B). 5 ml fractions are collected. Optical density is measured at 280 nm (protein) as well as at 495 nm (FITC) or 515 nm and 550 nm (TRITC). Based on these values, appropriate fractions can be pooled and, if necessary, concentrated. Dialysis against PBS is advisable.

2.5.4 Adsorption and immunological purification

Whether purification is necessary or not must be assessed beforehand in a special system. There is no universally recommended procedure. The method of Coons & Kaplan [1950], involving the use of an acetone-powder from liver for removing "over-labelled" antibodies (Abs with a high fluorochrome content), can be dispensed with in favour of more modern techniques using DEAE-cellulose or DEAE-Sephadex. Adsorption may, however, be necessary for removing unwanted cross-reacting antibodies. An absolute requirement is, however, that the specific Ab is neither removed, wholly or partially, nor are interfering substances, antigens or complexes introduced during the procedure. Ideally immunosorbents should be insoluble, capable of completely removing interfering Ab as far as the conjugate is concerned, thus resulting in an absolutely pure antibody solution. Simple procedures have been described by Huet et al. [1994], Nairn [1976], Ambrosius [1986], and Beesley [1993].

Gröschel-Stewart [1982] has recommended using a 'reverse' procedure, in which the unwanted Ab is bound and the sought after Ab allowed to pass through or elute. The appropriate conditions must be assessed on an individual basis. Here we merely describe the preparation of a liver homogenate and powder.

Preparation of a liver homogenate and powder

Apparatus and reagents:

1. Syringes, beakers, scissors, forceps. Potter-Elvejhem homogeniser. Ultraturrax macerator, refrigerated centrifuge.
2. PBS with the addition of 0.25 M sucrose (MW 342.3 = 85.575 g/l) and 1 mM EDTA (MW 372.24 = 0.372 g/l).
3. Container with dry ice.
4. Freeze drier, if required.

Procedure:

The liver is removed after bleeding-out the animal, washed free of blood by perfusion through the inferior vena cava with cold PBS, cut into small pieces with scissors and washed thoroughly a number of times with PBS. An appropriate amount of PBS is added depending on the volume of the liver used, followed by maceration using the Ultraturrax for three minutes at 2,000 rpm and finally manual homogenisation with the Potter-Elvejhem homogeniser. The sample is then rapidly frozen at –70°C, thawed, centrifuged at 4,000 g_{av} for ten minutes, resuspended in PBS and centrifuged again. The pellet is resuspended in PBS and centrifuged at ~10,000 g_{av} for fifteen minutes. After resuspension in PBS, the protein content is determined and the material dispensed into small volumes (0.05–1 ml), which are stored at –20°C. The powder is prepared by freeze-drying the homogenate.

Adsorption by liver homogenate or powder

Immediately before use 1 ml of conjugate is added either to a thawed-out sample of homogenate (0.5–1 ml), or to a portion of powder previously moistened with PBS and then incubated at 37°C for 30–60 minutes, with occasional shaking. This is followed by centrifugation at ~10,000 g_{av} for 15 minutes in a refrigerated centrifuge. The supernatant can be used immediately.

Adsorption to other antigens can be carried out in an analogous fashion although a longer incubation period, e.g. 12 hours at +4°C, and repeating the procedure one or more times, may be necessary. Rabbit or bovine liver is particularly recommended for absorbing anti-human IgG antibodies.

2.6 Concentration of dilute protein solutions

There are different procedures currently in use, for example:

1. Lyophilisation.
2. Pressure-filtration or negative pressure dialysis using semi-permeable membranes (e.g., from Sartorius, Göttingen).
3. Adsorption onto protein-A-Sepharose followed by elution with 0.58% w/v acetic acid in 0.15 M NaCl and dialysis against PBS.
4. Dialysis against water-soluble polymers such as polyethylene glycol (Carbowax series).

Dialysis against Carbowax is straightforward and generally applicable.

Apparatus and reagents:

1. Polyethylene glycol (Carbowax 20000 – Serva, Heidelberg) 20–25% w/v in PBS.
2. Dialysis equipment.
3. Refrigerator.

Procedure:

Dialysis tubing is filled with the protein solution to be concentrated and the ends firmly secured with nylon thread. The tubing can then be placed either directly, or stretched on a glass frame, in the previously prepared polymer solution (at least ten volumes). The time required for dialysis depends on the quantities used and the degree of concentration desired, and can be anything from a few hours to a day or two.

3 Characterisation and quality control of antisera and conjugates

3.1 Identification data for conjugates

The customary data for individual identification are laid out in Table 5. The use of conjugates exhibiting the same specificity and having similar properties, is a prime requirement for reproducible results and valid comparisons between laboratories. Appropriate standards or at least reference sera must be available in order to make any standardisation possible. There must also be a consensus as to the methods of analysis to be used.

Table 5. Reference data for conjugates.

1. Specificity of the serum – species, H- and L-chains.
2. Origin of the serum – animal species.
3. Batch and manufacturer.
4. Total protein content.
5. Antibody protein content.
6. Amount and identity of contaminating proteins.
7. Total fluorescence content.
8. Content of protein-bound fluorescence.
9. Content of non-protein-bound fluorescence.
10. Fluorochrome profile = distribution of the dye in the conjugate (protein-bound versus non-protein-bound, equivalent to (8/(8+9)).
11. F/P ratio = molar ratio of fluorochrome to total protein.
12. Ab/P ratio = ratio of antibody protein to total protein.
13. Ab/F ratio = ratio of antibody protein to fluorochrome content.
14. Type and quantities of preservatives or other additives present, for example, the addition of albumin as a stabiliser.
15. Results of any specificity controls carried out in model systems, noting non-specific reactions and cross-reactions.
16. Suitable working dilution for use measured in a standardised assay system.

3.2 Methods for characterising antisera

The methods customarily used at present are collected together in Table 6.

3.2.1 Testing for free dye

Johnson et al. [1978] give details of a very simple and practical method. Sephadex G–25 or G–50 is suspended in phosphate-buffered saline (PBS) and allowed to sediment for about 15 minutes. After removal of the supernatant some of the concentrated suspension is spread as a thin layer on a microscope slide. The slide is then placed on the slope between two Petri dishes, at an inclination of 10°–20°. Strips of filter paper are used to 'connect' the layer of gel on the slide with the two Petri dishes. The higher of the two is filled with PBS. After some minutes have passed, a few microlitres of the conjugate solution are applied to the gel layer approximately 1 cm from its upper end. 15–20 minutes later the labelled protein will be found to have moved a few centimetres, whereas the free dye will have remained at the point of application (the

Table 6. Methods for characterising and at the same time assessing the quality of conjugates (ant-IgGs) – based on Storch [1986].

Parameter	Recommended method	References
1 Fluorochrome concentration	Chromatography	Johnson et al. [1978]
2 Protein concentration	Biuret Lowry-Folin Bradford, Spector	Weichselbaum [1946] Lowry et al. [1951] Bradford [1976], Spector [1978]
3 Distribution of protein and fluorochrome	Cellulose acetate electrophoresis	Hebert et al. [1972]
4 Fluorochrome concentration	Comparison with a standard, e.g., for FITC use fluorescein diacetate (FDA)	McKinney et al. [1964]
5 Antibody activity, specificity and concentration	a. double radial diffusion b. immuno-electrophoresis c. reverse simple radial immuno-diffusion or standard gel double diffusion e. "DASS" system f. immuno-blotting g. biological assay systems	Ouchterlony [1958] Scheidegger [1955] Beutner et al. [1968, 1970] van Dalen et al. [1973] Towbin et al. [1979] e.g., Hijmans et al. [1969]; Schuit et al. [1981]
6 Suitability testing in special assay systems	"Chess-board" titration	Beutner et al. [1968]

chromatographic origin). If this is the case then free dye will have to be removed by gel filtration.

3.2.2 Determination of the fluorochrome/protein ratio (F/P quotient)

The measurement of absolute values, especially for the molar concentration of protein-bound dye, is difficult. This difficulty arises from changes in the absorption maxima and extinction coefficients for the dye, amongst other reasons, brought about by coupling to protein. Differences in dye quality, including impurities and breakdown products, cause additional problems. It is, therefore, not surprising that many different methods and correction procedures have been suggested for determining the F/P ratio (for an overview see Nairn [1976], and also The & Feltkamp [1970], Brandtzaeg [1975], and de Luca [1982]). The procedure recommended by Hebert et al. [1972] has been extensively used for FITC conjugates. It is described in the following section.

3.2.2.1 Determination of the F/P ratio for FITC conjugates

Determination of protein concentration
The protein concentration is determined using either the biuret (Gornall, Bardawill and David [1949]), Lowry et al. [1951] or the Bradford [1976, see also Spector 1978] methods, with extinction measured between 500 and 600 nm, using pure bovine serum albumin (BSA) or a standard protein solution for comparison (all concentrations expressed in mg/ml). Depending on the amount of protein available in the sample, as well as on the amount that can be used for the quantitative determination, either the Biuret method (1–10 mg), or the Lowry-Folin procedure (25–500 µg), or the Bradford dye-binding method (1–10 µg) should be

used. The choice between methods may also be determined by the presence of interfering substances and the absorption maxima used for the determination, which should not interfere with or be interfered with by the fluorochrome present. The following wavelengths are suitable for quantitative work: Biuret (540–560 nm); Lowry-Folin (500 nm – relatively high protein concentrations; 750 nm – low protein concentrations); Bradford dye-binding (595 nm). Practical details for these three methods are given in "Data for Biochemical Research" by Dawson et al., Oxford University Press [1987].

Determination of the concentration of protein-bound FITC (after Hebert et al. [1972])

Principle: the determination is carried out spectrophotometrically. Because FITC is unstable in solution, fluorescein diacetate (FDA) is used as the standard.

Apparatus and reagents:

1. Fluorescein diacetate (FDA) – Eastman Organic Chemicals, USA.
2. 0.1 N NaOH solution.
3. 96% ethanol.
4. Water bath.
5. 500 ml volumetric flask, polyethylene bottles.
6. Spectrophotometer.

Procedure:

1. *Preparation of an FDA standard solution (100* µg/ml). 50 mg FDA are added to 25 ml hot alcoholic sodium hydroxide solution. The alcoholic caustic soda solution is prepared by adding 3 g NaOH to 40 ml 96% ethanol, under reflux with stirring on a boiling water bath, followed by decanting the solution from the excess NaOH which has not dissolved. Heat the alcoholic sodium hydroxide stock solution (25 ml) and the FDA (50 mg) on a water bath until all the FDA has dissolved, then cool to room temperature. Transfer the solution quantitatively to a 500 ml volumetric flask by washing with distilled water and make up to the mark. This stock solution containing 10 mg FDA/100 ml, i.e., 100 µg/ml, can be stored for up to two years in polyethylene bottles if kept cool and in the dark.
2. *Construction of the FDA standard curve.* The FDA stock solution should be diluted with 0.1 N NaOH to give final FDA concentrations in the range 1 to 4 µg/ml (for example, diluted 1:100 = 1 µg/ml; 1:50 = 2 µg/ml; 1:25 = 4 µg/ml). The extinction is then measured

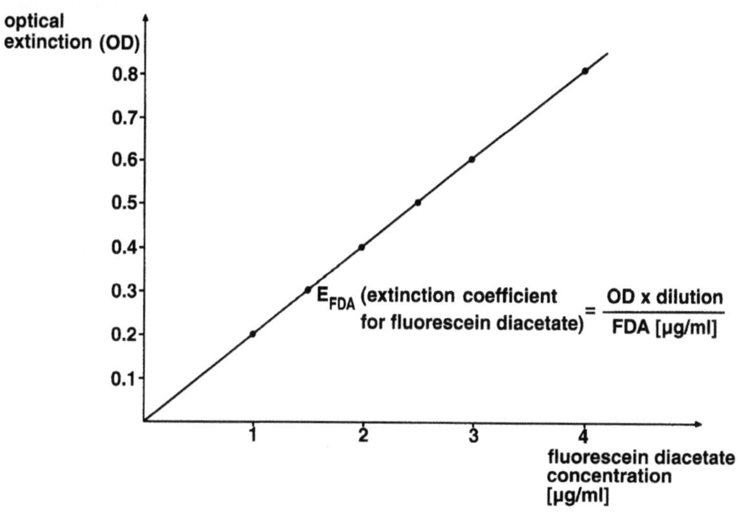

Figure 10. Reference curve for the determination of the FITC concentration using fluorescein diacetate (FDA) as the reference standard.

Table 7. An example of a procedure for determining the FITC concentration using fluorescein diacetate (FDA) as the reference standard.

Extinction E_{490} (1)	FDA concentration µg/ml (2)	Protein-bound FITC concentration µg/ml (3)	Factor (3/2)
0.204	1.000	1.070	1.070
0.303	1.500	1.600	1.067
0.403	2.000	2.130	1.065
0.502	2.500	2.670	1.068
0.612	3.000	3.200	1.067
0.815	4.000	4.280	1.070
		Average =	1.068±0.002

at 490 nm for each standard dilution, using 0.1 N NaOH in the reference cuvette (10 mm glass). FDA concentrations can be converted to FITC concentrations by multiplying by 1.07, as a correction factor for the lower extinction coefficient of FITC, and then the values plotted to give a standard curve (see Tab. 7 and Fig. 10).

3. *Determination of FITC in the conjugate*. Protein solutions must be diluted since only FITC concentrations between 1 µg/ml and 6 µg/ml can be measured accurately. A series of dilutions can be dispensed with, or greatly limited, if a crude estimate of the expected FITC concentration is available beforehand (Tab. 8). For example, with a protein content of 30 mg/ml (3% w/v) and based on a mean value for the F/P ratio, the expected value for the dye content would be $5 \times 10^{-3} \times 30$ mg/ml = 150 µg/ml. A dilution of 1:50 is therefore necessary to bring the concentration into the region where it can be measured accurately, i.e., which is in the middle of the linear range (see also Tab. 8). Dilution should be done with 0.1 N NaOH and the extinction measured at 490 nm. If an undiluted conjugate solution is to be measured, the pH should be adjusted to ~13.
The concentration of protein-bound FITC in µg/ml is obtained by multiplying the appropriate value for the concentration from the FDA standard curve by the serum dilution. In cases like our example (see Fig. 10 and Tab. 7) where no conversion has already been carried out, the value should be multiplied by the correction factor of 1.07.

Calculation and details of the F/P ratio (after McKinney et al. [1964, 1966] and Hebert et al. [1972]):

1. Protein concentrations should be determined by the biuret method and the results expressed in mg/ml.
2. The FITC concentration should be determined using the optical extinction

Table 8. Recommended conjugate dilution for the determination of the FITC concentration (Hebert et al. [1972]).

Protein mg/ml	Recommended dilution of conjugate based on expected F/P ratio					
	5	10	15	20	25	30
1.2	1: 5	1: 5	1: 10	1: 10	1: 20	1: 20
2.5	1:10	1: 10	1: 20	1. 20	1: 50	1: 50
5.0	1:10	1: 20	1: 50	1: 50	1: 50	1:100
10	1:20	1: 50	1:100	1:100	1:100	1:200
15	1:50	1:100	1:100	1:200	1:200	1:500
20	1:50	1:100	1:200	1:200	1:200	1:500
25	1:50	1:100	1:200	1:200	1:500	1:500

at 495 nm in 0.1 N NaOH with reference to a pure FDA standard.
3. The F/P ratio may be expressed as µg FITC per mg protein (‰, a weight ratio or w/w F/P).

Example: FITC content = 90 µg/ml
Protein content = 20 mg/ml

$$w/w\frac{F}{P} = \frac{90\mu g/ml}{20 mg/ml} = \frac{90 \cdot 10^{-3}}{20} = 4.5 \cdot 10^{-3}$$

It is useful to convert this into a molar F/P ratio. For most purposes an average molecular weight for IgG of 160,000 and for FITC of 389 Daltons is satisfactory, giving a factor of 0.411.

Molar F/P =

$$\frac{\frac{FITC}{389}(\mu g/ml)}{\frac{IgG}{160\,000}(mg/ml)} = \frac{FITC \cdot 160\,000}{389 \cdot IgG}$$

$$= w/w\frac{F}{P} \cdot 0.411$$

It is obviously necessary, especially with commercial conjugates to which unlabelled proteins have been added to reduce non-specific staining and to stabilise the preparation, that the F/P ratio is determined beforehand and the type and quantity of the added protein identified.

3.2.2.2 Determination of the molar fluorochrome/protein ratio for TRITC conjugates

Because there is no stable reference standard available for TRITC, it is only possible to use optical density as the basis for quantitative determinations. It should not be forgotten that amorphous TRITC exhibits two absorption maxima at 515 nm and 550 nm, whereas crystalline TRITC has only one maximum (~555 nm).

Amante et al. [1972] give the following expression for calculating the molar F/P ratio of IgG-conjugates with crystalline TRITC:

$$\text{molar}\ \frac{F}{P}\ \frac{\text{cryst.TRITC}}{\text{IgG}}$$

$$= \frac{6.6 \times OD_{555\ nm}}{\text{proteinconcentration (mg/ml)}}$$

The factor 6.6 arises because of the relationship between the extinction coefficient for 1 M TRITC (0.041) and the factor for converting IgG concentration in mg/ml to molarity (0.00625). If the protein concentration is determined photometrically then correction factors should be taken into account to allow for changes in absorption. Brandtzaeg [1975] recommends the following relationship:

$$IgG(mg/ml) = \frac{OD_{280nm} - (a \times OD\ \lambda max)}{1.4}$$

A correction factor, a, of 0.56 for TRITC (Cebra & Goldstein [1965]), 0.95 for rhodamine B isothiocyanate, and 0.32 for rhodamine B 200 sulphonyl chloride, should be used.

3.2.2.3 Determination of the F/P ratio for conjugates with other labelled dyes

Only relationships involving the optical density are cited here, in which the optical density (or extinction) at the absorption maximum of the protein-bound fluorescent dye (see Tab. 28) is divided by that at 280 nm (Brandtzaeg [1975]).

Blakeslee & Baines [1976] use the empirical formula suggested by The & Feltkamp [1970] for determining the molar F/P ratio of DTAF-labelled conjugates:

$$\text{molar}\frac{F}{P} = \frac{(a)\ OD_{495\ nm}}{OD_{280\ nm} - (b)\ OD_{495\ nm}}$$

FITC a = 2.87 DTAF a = 3.55
b = 0.35 b = 0.38

The variables, a and b, are calculated based on a molecular weight of 160,000 for IgG and 492 for DTAF. The optical density (OD_{280}) for 1 mg/ml IgG has been determined to be 1.4, and that for DTAF 127.5.

Wessendorf et al. [1990] have applied the following expression to AMCA-conjugates with IgG:

$$\text{molar}\frac{F}{P} = \frac{(\varepsilon_{280}{}^{IgG}) \cdot A_{350} - (\varepsilon_{350}{}^{IgG}) \cdot A_{280}}{(\varepsilon_{350}{}^{AMCA}) \cdot A_{280} - (\varepsilon_{280}^{AMCA}) \cdot A_{350}}$$

where $\varepsilon_\lambda{}^m$ is the extinction coefficient for substance, m, at wavelength λ.

Because it is difficult to measure extinction coefficients, it is often only those relationships which are based on the measurement of optical density that are cited, for example in the case of AMCA for which the ratio obtained by dividing OD_{350} by OD_{280} (IgG) is used. Cy 3 conjugates are mostly measured at 552 nm.

3.2.3. Determination of the protein distribution using cellulose acetate thin-layer electrophoresis (Hebert et al. [1972]; Storch [1979])

Detection of irrelevant labelled proteins

This serves for the rapid testing of commercial conjugates, because these often contain not only γ-globulins but also other serum fractions (proteins), especially albumin added as a stabiliser.

Principle: protein electrophoresis is carried out, on a micro-scale, on cellulose acetate foil as the support material. FITC and proteins migrate with different velocities in a DC field under alkaline conditions, allowing separation into individual fractions. Fluorescent fractions can be visualised after separation using UV light. Proteins are stained with Amido Black (C.I. 20470; Naphthol Blue Black; Amido Black 10B) and quantitatively estimated using an integrating scanning-densitometer set up for transmission. The two sets of results are compared.

Apparatus and reagents:

1. Electrophoresis chamber with accessories from Sartorius (Göttingen, Germany) – membrane sheets 25 × 130 mm, perforated, etc.
2. Barbiturate-acetate buffer.
3. Dye solution – Amido Black.
4. Destaining solution: 1 part v/v acetic acid and 9 parts v/v methanol.
5. Clearing solution: 70 ml acetic acid and 30 ml ethyl acetate.
6. DC power supply.
7. UV lamp.
8. Scanning densitometer.

Procedure:

- the chamber insert should be so positioned in the separation tank that the side-mounted metal tongues make good contact with the sockets;
- the electrode chambers should each be filled with 250 ml buffer and the large inner reservoir of the preparation stand with 75 ml buffer;
- fill the front cell with 3 ml distilled water to rinse the applicator;
- fix the separation bridge to the preparation stand.

Preparation of membrane sheets and sample application:

- lay the necessary number of sheets on a clean base and mark them with a suitable pen;
- put the sheets into buffer and allow to soak;
- remove excess buffer by laying the sheets between filter paper without application of pressure;
- hang sheets with perforations in the slots of the separation bridge;
- put the application cards ("Dispo" cards) in the holding frame;
- pipette 10 µl of each sample of serum to be analysed into the "Dispo" card cells;
- position the application die firmly and – after testing it on absorbent paper – apply to the sheet.

Electrophoresis is carried out with barbiturate-acetate buffer, at 120 V D.C. with the voltage held constant, for approximately 50 minutes. Staining and destaining both last for five minutes. Examination is under UV light with recording of the fluorescent bands (number, size and position). Mount sheet on a holder with the help of a rubber roller, expose to clearing solution for approximately 3 minutes and then hang vertically to dry at 80°C in a drying oven for 10–15 minutes. After cooling to room temperature photometric analysis should be carried out.

Assessment: by comparing the record of the examination under UV light with the protein profile, both unlabelled protein and non-protein-bound fluorochrome components can be detected. Free dye migrates in front of albumin towards the anode (Fig. 11), i.e., it is more negatively charged. The higher the F/P ratio, the more rapid the migration.

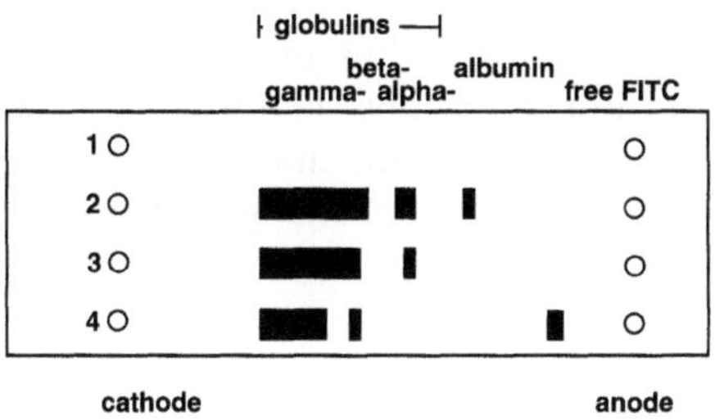

Figure 11. Representation of a typical fluorescence profile obtained on cellulose acetate electrophoresis (based on Hebert et al. [1972]).
1 = no fluorescent bands (unlabelled antiserum).
2 = fluorescent bands: γ-, ß-, α-globulin, and albumin.
3 = fluorescent bands: γ-, ß-, and α-globulins.
4 = fluorescent bands: γ- and ß-globulin-free FITC.

3.2.4 Testing the activity of antisera for activity using double radial immunodiffusion in agar gels (after Ouchterlony [1958])

Principle: antigens and antibodies diffuse through agar and form lines of precipitation in the contact zone. Semi-quantitative estimations are possible if a central well, containing the antiserum to be examined, is surrounded in a circular fashion by a series of wells containing antigen dilutions (see Fig. 12). The precipitation titre yields the estimate of antiserum potency.

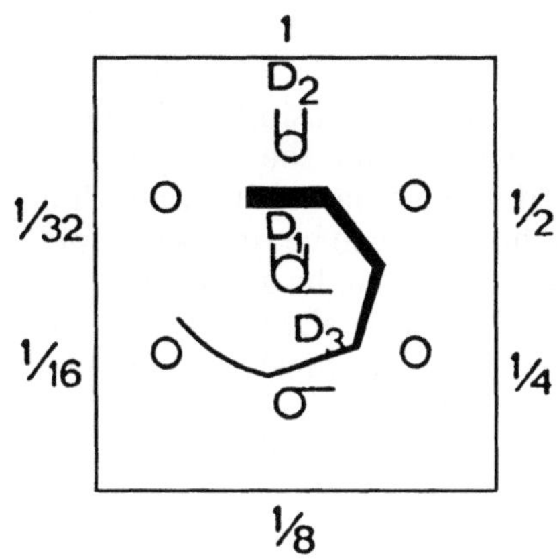

Figure 12. Method for Ouchterlony double-diffusion using slide cover-glasses (5 × 5 cm).
D 1 = diameter of the central hole, 5 mm;
D 2 = diameter of the surrounding holes, 3 mm;
D 3 = distance between the holes, 5 mm; agar 1.5% w/v, titre 1:32 (in the example given, 1:16).

Apparatus and reagents:

1. Phosphate buffer, pH 7.5.
   Stock solution A: $Na_2HPO_4.2H_2O$ = 11.9 g/l or $Na_2HPO_4.12H_2O$ = 23.8 g/l
   Stock solution B: KH_2PO_4 = 9.1 g/l
   For use, mix 85.2 ml solution A and 14.8 ml solution B.
2. Agar (Difco Laboratories, Detroit, USA).
3. 0.5% w/v Amido Black in methanol/ acetic acid 9:1 v/v.
4. Water bath, transparency (slide) glass covers 5 × 5 cm, humidity chamber.

Procedure:

1. 1.5 g Agar and 100 ml buffer are heated on a water bath. 3 ml of this solution is then applied to transparency covers, which have been well defatted and are lying on a flat surface. The glass covers are kept for at least two hours in a humidity chamber at +4°C after the gel has set.
2. The wells are cut, in the pattern shown in Figure 13, using a punch or cannula and sucked out with a water pump. Serial dilutions of antigen and, if necessary, antiserum are prepared using 0.145 M NaCl or PBS (undiluted to a dilution of 1:32).

The antiserum to be examined is added to the central well using a suitable pipette or capillary, and the antigen dilutions placed in the other, radial wells. If necessary the order can be reversed depending on the question to be answered. After diffusion has been allowed to proceed for 24–48 hours, or longer if necessary, non-precipitated proteins are removed by washing with buffer. The gel is then dried at 37°C by laying onto filter paper, stained with 0.5% w/v Amido Black B in methanol/acetic acid (9:1 v/v) for 5–10 minutes, followed by destaining in methanol/acetic acid (9:1 v/v) for five minutes and washing with distilled water.

Assessment: the antiserum titre is the antigen dilution for which a line of precipitation is still detectable.

Tip: many variations are possible (see also Behm [1976]).

3.2.5 Determination of the antibody concentration by reversed radial immunodiffusion (after Beutner et al. [1970])

Principle: in contrast to the usual radial immunodiffusion method of Mancini et al. [1964], it is the antigen (Ag) that is present in the agar, not the antibody. For example, in order to determine IgG antibodies, serial dilutions of an anti-IgG reference standard and the unknown serum sample are placed in the wells. Antibodies diffuse through the agar to different extents depending on their concentration. Precipitation occurs at the optimal concentration ratio between Ag and Ab. The area of precipitation, or the diameter of the corresponding ring of precipitation, is proportional to the Ab content.

Apparatus and reagents: if commercial equipment is not available, then it is possible to prepare plates, e.g., anti-IgG plates, oneself.

1. Pure agar (Difco Laboratories, Detroit, USA).
2. Antigen (IgG).
3. Antibody solution of known concentration (reference serum).
4. PBS, water bath, microlitre syringes, gel-punch, magnifying glass, slides, water pump.

Procedure:

1. Preparation and pouring of agar plates.
   2% w/v pure agar in PBS is heated to boiling and then maintained at 56°C in a water bath. Agar and Ag solutions are mixed in equal parts by volume. Approximately 3 ml, or somewhat more of the now 1% w/v agar solution containing 0.025 mg IgG/ml, are

poured onto a slide or Petri dish, kept flat so that a layer 1.5 mm is produced. 1.5 mm diameter holes, spaced at 17 mm, are punched in the gel after it has set, followed by the addition of at least three serial dilutions of reference serum containing a known amount of Ab and the unknown serum (if the protein concentration is > 2 mg/ml, dilution is recommended), using a Hamilton syringe.

2. Establishing the Ab protein content. After precipitation has taken place (with IgG after ~12 hours, and with IgM or other antibodies after ~24–36 hours), the diameter of each ring of precipitation is measured using a hand lens or microscope with a calibrated graticule. One then plots the antibody concentration for each serial dilution of the standard on the X-axis and the associated precipitation ring diameters on the Y-axis using semi-logarithmic graph paper. These points form a straight line so that the 'unknown' Ab content can be read directly from the reference curve.

3.2.6 Determination of the antibody protein content as "precipitin units" using standard double gel-diffusion (after Beutner et al. [1973])

Principle: two rows, or lines, of wells are punched in a 1% w/v agar gel. Antigen (1 mg/ml) is placed in one set of wells, and serial dilutions of the antiserum to be measured in the other row (see Figure 13). The highest antiserum dilution, e.g., 1 to 32, which gives a detectable zone of precipitation after 24 hours, determines the number of "precipitin units", in this case equal to 32. These units also represent the reciprocal of the otherwise customary precipitation titre (1/32).

3.2.7 Specificity control by coupling antigens to artificial substrates – the so-called DASS method (Defined Antigenic Substrate Spheres) – after van Dalen et al. [1973]

Principle: antigens or antibodies are coupled to an inert carrier (carrier spheres or beads) and then used to bind either Ab or Ag. Bound protein is detected using direct or indirect immunofluorescence (IF). Sepharose 4B or 6B (Pharamacia, Uppsala) are especially recommended as carriers because prolonged storage is possible after coupling with antigen and freeze-drying. Commercially available substrates are particularly worthwhile for a routine clinical immunology laboratory because the activation and deactivation of Sepharose demands experience and different pure antigens are required. The procedure of Haaijman & Bloemmen [1975] and Knapp et al. [1975] is briefly described here, essentially as worked out by Capel [1974].

Apparatus and reagents:

1. Sepharose 4B (Pharmacia, Uppsala).
2. Cyanogen bromide (Fluka, Basel).
3. PBS with 0.5 M NaCl (see 2.4.2).

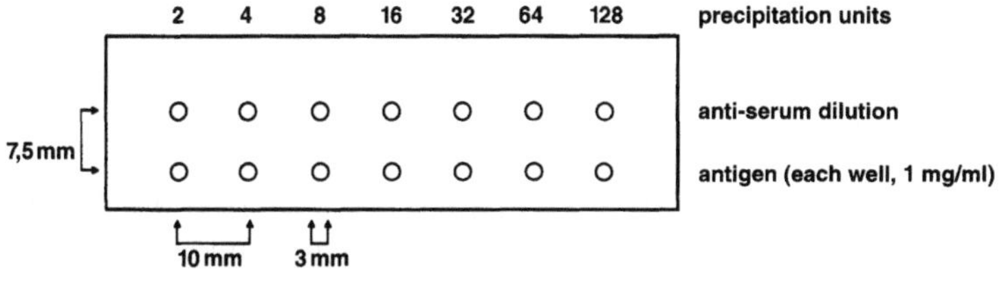

Figure 13. Scheme for the determination of "precipitin units" in 1% agar, with a diffusion time of 24 hours (based on Beutner et al. [1973]).

4. 0.5 M ethanolamine in 2% w/v $NaHCO_3$, pH 9.0.
5. Antigens, e.g., IgG, and controls for testing for non-immunological binding, e.g., albumin.

Procedure:

1. Preparation of Sepharose beads.
   1 ml Sepharose 4B beads are activated with 250 mg cyanogen bromide (fume cupboard) in 10 ml distilled water at pH 10.5–11.0 for eight minutes. 0.1–1.0 mg Ag-protein, dissolved in 1 ml PBS plus 0.5 M NaCl, is incubated overnight at room temperature with the washed and sedimented activated Sepharose beads. These are then washed and deactivated using 0.5 M ethanolamine in 2% w/v $NaHCO_3$, pH 9.0, for four hours. The concentration of the beads in the final solution is estimated from their packed volume per millilitre of suspension.
2. Incubation, washing procedures and production of the preparation.
   The density of the suspension of beads is expressed as a volume percent (% v/v). With the direct method, 50 µl of a 20% v/v suspension of beads are incubated for 2 hours with 50 µl conjugate (diluted if necessary) at room temperature with continuous shaking. The beads are then washed three times with PBS containing 0.65 M NaCl and left to stand overnight in this buffer. After one further wash, one drop of the suspension is mixed with one drop of phosphate-buffered glycerol (glycerol : PBS = 9:1 v/v, pH 7.8) on a microscope slide, covered with a cover-slip and the edges sealed with paraffin. Where possible the bead fluorescence should be determined micro-fluorimetrically and the amount of undesired specific staining and non-specific staining expressed quantitatively. The data should be checked using a biological system, such as monoclonal bone marrow cells.

3.2.8 Specificity control using biological systems

3.2.8.1 Testing on bone marrow cells (after Hijmans et al. [1969])

Principle: plasma cells from patients with multiple myelomatosis produce, in particular, monoclonal immunoglobulins. Ideally, bone marrow cells from healthy patients ("polyvalent") or from those with a plasmacytoma ("monovalent", e.g., IgG, A, D, E, H, L, κ, λ chains) or Waldenström's disease, are sedimented onto a glass slide using a cytocentrifuge, or simply prepared as a smear, and then stored at –70°C, being available for use whenever the need arises.

Preparation of bone marrow cell pellets

(after Johnson et al. [1978])

A few drops of bone marrow aspirate are taken up in ~1 ml EDTA solution (50 mg/ml), mixed and centrifuged (5 min, 900 g_{av}, +4°C). The supernatant is discarded and the lymphocyte-rich layer placed in micro-haematocrit tubes and recentrifuged (5 min, 700 g_{av}, +4°C). The "buffy coat" is then added to 30 ml wash solution in a centrifuge tube. (Wash solution: 20 ml bovine serum albumin (BSA, 200 mg/ml), 60 ml 0.01 M PO_4, pH 7.2, in 0.15 M NaCl, 8 ml EDTA (50 mg/ml), 10,000 units penicillin/ml). Adjust the pH to 6.8 with $NaHCO_3$, mix thoroughly and centrifuge at 175 g_{av} for ten minutes at +4°C. The pellet is resuspended in 0.5 ml of the wash solution and coated onto microscope slides. The slides, after centrifugation and drying, are kept in an air-tight container and stored at –20°C or less. Each slide should have on it at least 10^5 cells, since the plasma cell content of polyclonal bone marrow is ~0.5–1.0% of the cell total. After fixation in 5% v/v glacial acetic acid in ethanol (15 min at –20°C) and washing with PBS for one hour at +4°C, incubation is carried out for immunofluorescence with the test conjugate and with a reference conjugate, which contains another fluorochrome.

In order to detect membrane immunoglobulins, lymphocytes are separated by centrifugation using Ficoll-Isopaque© and used either unfixed or after fixation on 0.04% v/w HCHO in PBS. Unfixed mature mononuclear cells from healthy donors are highly suitable for discovering unwanted F_{ab}-antibodies, since mature monocytes bind to them through their F_c-receptor IgG, so that the F_{ab} fragment remains exposed and able to bind anti-F_{ab} antibodies. In order to exclude activity against H-chains, the anti-L-chain conjugate should be tested against a monoclonal bone marrow cell preparation, with L-chains that are not the same and expressing a different H-chain specificity to the one against which the conjugate is directed.

The specificity of antisera against L-chains can be checked very well using double immunofluorescence, because neither plasma cells nor B-lymphocytes are able to produce both L-chain types (κ and λ) at one and the same time. If double labelling is observed, this does not arise, in most cases, from additional antibodies against another L-chain type but from antibodies against determinants present in the H-chain or F_{ab} fragments.

Of 29 commercially available anti-human-immunoglobulin conjugates tested by Schuit et al., 14 (!!) did not meet the required standard, that is to say they were found to contain adventitious activity against other immunoglobulins (Schuit et al. [1981]). This suspicion should be considered seriously whenever it is found, during determination of the optimal working dilution, that a conjugate reacts with different serum dilutions producing staining which is patchy, or shows a range of intensity.

3.2.8.2 Special tests for membrane immunofluorescence

Different procedures should be used depending on the area of application. The use, for example, of human red blood cells or cultivated human lymphoblast cell lines, on which neither surface immunoglobulins nor F_c or complement receptors can be detected, e.g., Molt 4T, is ideal. If positive membrane immunofluorescence is observed, this indicates non-specifically reacting antibodies. One other cell line (RAJI) possesses F_c and complement receptors but no surface immunoglobulins (Theofilopoulos et al. [1974]). This allows a distinction to be made which is especially important for optical studies of membrane immunofluorescence (see Sedlacek [1976]).

3.3 Suitability testing in special test systems with determination of the optimal working dilution for the conjugate

The most important and at the same time the most reliable methods for checking specificity are described here, especially for the determination of humoral antibodies.

Principle: a cross-titration (also known as a "chess board" titration) is carried out using the preparation with conjugate and a known positive Ab-containing serum (the "positive" control, where possible using an international reference serum), together with further controls including a "negative" control or normal serum control, and a "conjugate" control or saline control using just PBS and the conjugate. All preparations should be identified with a code-number and assessed, where possible, separately by independent investigators ("double blind"). The results should be collated in a table and both the plateau titre and end-point calculated (see below).

Procedure: (after Storch [1974])

1. A so-called "blank" section = an untreated section to determine autofluorescence.
2. Incubation simply with antiserum (conjugate) to establish non-specific staining.
3. Determination of those dilutions which, in comparison with the level of autofluorescence, produce fluorescence that cannot be distinguished from background levels.
4. Incubation of various serial dilutions of normal serum, as well as the determination of the most appropriate dilution taking into account background staining, for establishing the non-specific reaction of normal human serum.
5. Cross-titration with a known positive, Ab-containing serum to establish the so-called *plateau titre* and the *plateau end-point*.

The results for a typical "chess board" titration are shown in Table 9. It is possible to see that:

1. with undiluted or weakly diluted conjugate it is not possible to make a firm decision because of non-specific staining (up to 1:10);
2. the same is true for weakly diluted normal serum, especially in the presence of low dilutions of conjugate. It is only with a dilution of > 1:20 that non-specific staining is no longer visible;
3. at a certain fixed dilution of the reference serum (titre = 1:640) the fluorescence intensity remains constant for a number of conjugate dilutions, the so-called *plateau titre*. In these samples the contrast between specific and non-specific staining is clear;
4. on further dilution of the conjugate (> 1:160) the titre drops off more or less suddenly. Based on the suggestion of Beutner et al. [1967], the range of concentration over which the titre decreases rapidly, either within one or, better, two dilutions, is called the *plateau end-point*.

Beutner et al. [1968] have shown that the *plateau end-point* is independent of the F/P ratio but does depend on the Ab-content of the conjugate. At constant Ab-content, the F/P ratio influences the *plateau titre* (Fig. 14). The choice of the most suitable conjugate dilution for use is, in consequence, limited by:

1. non-specific staining by the conjugate at relatively high F/P ratios;
2. and by non-specific reactions with normal serum at relatively high Ab-content for the conjugate.

The meaning of these parameters is summarised for further reference in Tables 10 and 11.

The dilution of the conjugate used should be approximately 2 to 4 times more than the PEP (plateau end-point). A four-fold concentrate of the working dilution should cause either no non-specific staining in the system used, or only very little (Poetschke [1976]). The optimal Ab-content and F/P ratio must be determined individually for each test system because these depend on many factors, including the way in which the microscope has been set up (transmitted light, reflected light, interference filter – Braveny & Disko [1976]). In order to detect different humoral Abs concurrently, conjugates are used with an Ab-concentration of ~25–50 mg/ml (~$^1/_8$ – $^1/_4$ E/ml) at their working dilution and an average F/P ratio of 1.5 (Storch [1974]).

In order to avoid drawing incorrect conclusions because of changes in the conjugate due to ageing, i.e., a reduction in the Ab-titre with a relative increase in F/Ab amongst other reasons, its suitability should be checked every time a new series of experiments is started using, for example, incubation with PBS, with normal serum ("negative" control) and with "positive" controls, followed by incubation with the working dilution of the conjugate (Table 12).

Table 9. Example of a "chess-board" titration = crosswise titration of conjugate against a positive reference serum, or patient's serum, using both positive and negative conjugate controls to determine the optimal dilution for use. Detection of cytoplasmic antibodies. Antigen, unfixed cryotome sections from kidney; conjugate, anti-human IgG (Ab = 20 units/ml, F/P=1) – based on Storch [1979].

		Conjugate dilution (1:X)												
								↓PEP						
Sample	Dilution	1	2	5	10	20	40	80	160	320	640	1280	2560	5320
Patient's or reference serum	1:2560	+++	±	±	-	-	-	-	-	-	-	-	-	-
	1:1280	+++	++	±	±	±	-	-	-	-	-	-	-	-
PT →	**1:640**	+++	+++	++	++	++	+	+	-	-	-	-	-	-
	1:320	+++	+++	++	++	++	+	+	±	-	-	-	-	-
	1:160	+++	+++	+++	+++	++	++	+	+	-	-	-	-	-
	1:80	+++	+++	+++	+++	++	+	+	+	-	-	-	-	-
	1:40	+++	+++	+++	+++	+++	+++	++	+	+	-	-	-	-
	1:20	+++	+++	+++	+++	+++	+++	+++	++	+	±	-	-	-
Normal serum control	1:20	+++	+++	++	+	±	-	-	-	-	-	-	-	-
	1:10	+++	+++	+++	++	++	+	+	-	-	-	-	-	-
	1:5	+++	+++	+++	+++	+++	+++	++	++	++	++	+	-	-
Conjugate control		+++	++	++	±	-	-	-	-	-	-	-	-	-
Units/ml		20	10	4	2	1	1/2	1/4	1/8	1/16	1/32	1/64	1/128	1/256
Ab-protein µg/ml		4000	2000	800	400	200	100	50	25	12.5	6.25	3.13	1.56	0.78

PT: plateau titre for the patient's serum (1:640).
PEP: plateau end-point for the conjugate (1:80 = ¼ IgG/ml or 50 µg/ml).
+, ++, +++: fluorescence intensity (weakly, moderately, strongly positive).
–: no difference compared with the level of autofluorescence.
±: questionably positive, not clearly differentiated from autofluorescence.

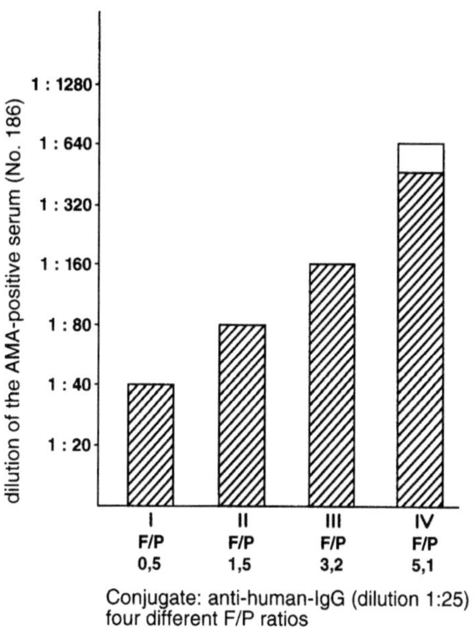

Figure 14. The dependency of the fluorescence titre of a patient's serum on the F/P ratio at constant antibody content. Conjugates III and IV were unusable because these gave non-specific staining and falsely positive results for normal sera at a dilution of 1:20, during testing for their suitability (based on Storch [1974]).

Table 10. Significance of the most important parameters of labelled antiglobulin sera for the staining properties of conjugates (based on Beutner et al. [1973] and Poetschke [1976].

Parameters influencing the staining properties of the conjugate			
F	=	fluorochrome concentration	NSS-Titre = amount of non-specific staining at the working dilution
F/P	=	ratio of fluorochrome to total protein concentration	fluorescence intensity of protein solutions at the same concentration, height of the plateau and titre of the patient's serum PT (plateau titre)
Ab	=	antibody protein concentration (µg/ml)	DSS = desired specific staining working dilution PEP (plateau end-point)
Ab/F	=	ratio of antibody to fluorochrome concentration	ratio of desired specific staining (DSS) to non-specific staining (NSS) = DSS / NSS ease of use

Table 11. Conversion of weight and molar fluorochrome-protein ratios for conjugates and their characterisation (Beutner et al. [1973]).

	Weight ratio	Molar ratio	Parameter
Ab/F	µg/µg	× 1/411	convenience of use
F/P	µg/mg or × 10^{-3}	× 0.411	optical sensitivity
Ab/P	mg/mg or % × 0,1 or Units/1% protein	× 1.0	immunological sensitivity

Table 12. Example of a protocol for the simultaneous detection of antibodies against mitochondria, endoplasmic reticulum, ribosomes, parietal cells, smooth muscle, cell nuclei, bile canaliculi, connective tissue and vascular endothelium. Antigen substrate: unfixed cryotome sections from an organ block containing rat kidney, mouse stomach and rabbit liver.

Section No.	Serum	Dilution	Conjugate anti-human IgG	Results			Interpretation
				Liver	Stomach	Kidney	
1	-	-	HE-stain	-	-	-	all parts visible
2	PBS	-	-	low level autofluorescence			blank
3	PBS	-	1:20	no difference to autofluorescence			conjugate control
4	Normal serum	1:20	1:20	no difference to 2 and 3			serum control
5	Positive control	1:160	1:20	+	+ to ++	+	Positive control
6	Mitochondrial Ab	1:320	1:20	-	-	-	titre 1:160
7	Positive control	1:80	1:20	++	-	+ to ++	Positive control
8	Microsomal Ab	1:160	1:20	-	-	-	titre 1:80
9	Positive control	1:640	1:20	+	(+)	(+)	Positive control
10	Nuclear Ab homogen	1:1280	1:20	-	-	-	titre 1:640
11	Positive control	1:80	1:20	-	+ to ++	+	Positive control
12	Smooth muscle Ab	1:160	1:20	-	(+)	-	titre 1:80
13	New serum	1:20	1:20	-	Muscularis ++	Glomeruli +	Ab against
14	Clinical viral hepatitis	1:40	1:20	-	Muscularis +	Glomeruli +	smooth muscle
15		1:80	1:20	-	Muscularis (+)	-	titre 1:40
16		1:160	1:20	-	Muscularis -	-	

Parameters for the FITC-labelled anti-human IgG conjugate from rabbit (Behring Werke, BRD: batch 572A), determined Dec 1972. FITC content, 53.5 µg/ml; total protein, 38.4 mg/ml; antibody protein, 1.25 mg/ml; F/P, $1.39 \times 10^{-6} \approx 0.6$ (molar); Ab/F, 0.023; Ab/P, 0.033; working dilution, 1:20 (= 62.5 µg Ab/ml); microscope, Zeiss-Nf (transmission).

4 Actual immunofluorescence techniques

An absolute requirement for carrying out successful immunofluorescence, is that the reactants used neither change their properties nor cause cleavage of Ab and Ag during the staining. Recommended conditions for the various areas of application differ widely. For this reason specific instructions are given in those sections of this book concerned with applications. If one can work reproducibly using a particular method, one should, as far as possible, keep the conditions constant.

4.1 Storage of sera

All sera should be kept at –20°C. Repeated freezing and thawing should be avoided because Ab activity is damaged by this. Appropriate quantities of the conjugate, as well as the positive and negative controls, for experimental work, should be kept frozen.

One can proceed as follows in determining humoral antibodies (Ab):

1. A conjugate, e.g., FITC-anti-human-IgG, has been chosen based on suitability testing as particularly appropriate. In order to use as little as possible, and thus to prolong the useful life of the reagent, it should be dispensed in volumes of 100 µl in Eppendorff tubes and frozen. Immediately before use, a tube is taken from the deep freeze and diluted to the working concentration, e.g., 1:40, with PBS. The 4 ml of conjugate so obtained is sufficient for around 40 sections.
2. Positive controls. Serum containing anti-mitochondrial antibodies (titre = 1:800) is diluted 1:10 with PBS, and then frozen in 0.05 ml aliquots. Before use 0.95 ml PBS is added, thus producing an initial dilution of 1:200. Serial dilutions of 1:200, 1:400, 1:800 and 1:1600 are then used as positive controls.
3. Negative controls. A negative control such as normal serum is divided, without dilution, into 0.05 ml aliquots and frozen. The dilution will be 1:20 after addition of 0.95 ml PBS.
4. Serum from patients, where possible not containing products of haemolysis, is frozen in quantities between 0.5–1.0 ml. Each tube or container should be marked with an ID number and entered in a sample register.

Centrifugation may be necessary after thawing to separate impurities or precipitated protein.

4.2 Preparation and production of sections

4.2.1 Sectioning technique

4.2.1.1 Cryotome-sectioning technique (Linderström-Lang & Mogensen [1938]; Coons et al. [1951])

This technique is used predominantly for demonstrating auto-antibodies.

Freezing of antigen substrates
Apparatus and reagents:

1. Fresh tissue – for human tissue, a few hours *post mortem* or taken at operation.
2. Embedding medium, e.g., Tissue Tek OCT Compound (OCT = optimal controlled temperature), Ames, Indiana, USA.
3. Deep-freeze with storage container and transport vessel for dry ice (solid CO_2).
4. Dewar for liquid nitrogen (N_2).
5. Tongs, forceps, scalpel, aluminium foil, plastic bags, safety goggles or glasses.

Procedure for freezing:

- cut the tissue to be frozen into cubes with 3–5 mm edges.
- cover the cubes with Tissue Tek OCT compound, place on aluminium foil and immerse in liquid nitrogen. It is recommended that biopsy material be placed beforehand on either 2 mm thick 10–12% w/v gelatine blocks or on thin cork sheet.
- wrap the cubes in aluminium foil using forceps, and place in small airtight plastic bags. These should be clearly marked, as well as containing a packing slip, showing the date, sample identification, etc., with permanent waterproof ink or pencil.
- material should be transported for use in a Dewar filled with dry ice.

Tip: if liquid nitrogen is not available, one can freeze the samples in *iso*-pentane or *n*-hexane cooled with a dry-ice-acetone mixture to –90°C (Fig. 15). CARE! – these liquids are highly flammable and form explosive mixtures with air *even at very low temperatures*!

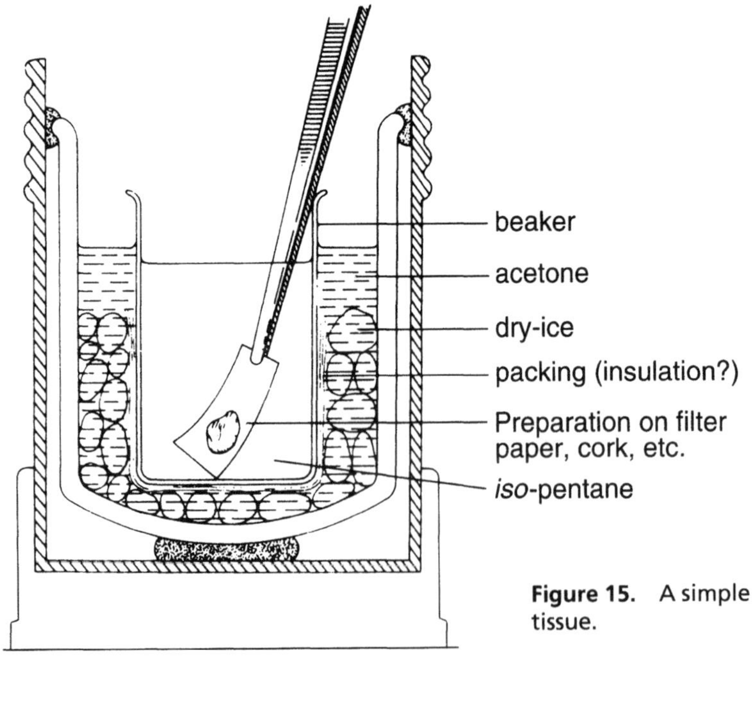

Figure 15. A simple device for freezing tissue.

Tissue should not, under any circumstances, be allowed to thaw out and sudden increases in temperature should also be avoided. Take care to store under airtight conditions. It is also possible to use glass or plastic tubes, so long as these can be sealed, instead of plastic foil bags. The need for accurate and sufficient labelling of the samples is self-evident.

If combination blocks from a number of organs are used, e.g., from liver, stomach, kidney or thyroid, it is recommended that the preparation of the block is done directly on the microtome table. The tissues are bound together with Tissue Tek OCT compound, PBS or gelatine and then, in order to prevent them falling apart, wrapped with a thin strip of stomach wall.

Storage of frozen tissue samples

Storage in liquid nitrogen is ideal (–196°C). Most antigens can be kept for only a few weeks in normal deep freezes (–20°C to –30°C). Low temperature deep freezes are recommended (–70°C to –90°C).

Preparation of sections

Because the quality of the sections decisively influences the success, or otherwise, of the whole method, it pays to have well-trained and experienced staff as well as the best equipment available. In particular, blocks of biopsy material should be trimmed with a scalpel beforehand and reduced to the required size. This makes sectioning easier and protects the microtome knife.

It is necessary to achieve section thicknesses of between 4–6 μm. The optimal cutting temperature lies between –18°C and –30°C, the more fat present, the lower the temperature necessary. Sections are mounted on clean, fat-free slides and, if necessary, these can be stored overnight or even longer at –70°C in a refrigerator.

Many cryotome sections can be stored for months at –20°C essentially without loss of structure or antigen content, so long as they are air-dried beforehand, warmed to 40–50°C for 30 minutes, heat-fixed for 10 minutes at 100°C, wrapped in hot filter or X-ray paper (≤ 100°C) and then sealed or vacuum packed in aluminium-polyethylene foil and protected against water vapour (Rost, personal communication [1976]; Gernand et al. [1978]).

Tip: take care to use clean, fat-free slides. The section can be straightened out or 'stretched' during the thawing process by pressing a finger on the underside of the slide.

Cleaning the slide

One week in chromic acid (CARE – corrosive and carcinogenic!).

½ day washing under running tap water.
Wash x1 with distilled water.
Storage in 96% v/v ethanol.

4.2.1.2 Paraffin wax embedding technique

Embedding in paraffin wax after fixation in the cold with ethanol (after Sainte-Marie [1962] and Heron [1970]).

Apparatus and reagents:

1. Tissue (biopsy material.
2. 99%, 96%, 95% and 70% v/v ethanol, pre-cooled to +4°C.
3. Xylene, pre-cooled to +4°C
4. Paraffin wax.
5. Ice-bath, incubator at 58°C.
7. PBS.

Procedure:

1. Place the biopsy material in 96% v/v ethanol pre-cooled to +4°C and allow to stand overnight at +4°C; then dehydrate with 99% v/v ethanol at +4°C for three hours, followed by clearing (x2) in xylene for three hours, with the second clearing solution being allowed to warm up to room temperature.
2. Embedding in paraffin wax.

3. Cutting sections with a microtome, catching the sections in an ice-bath and mounting them on fat-free slides.
4. Removal of paraffin wax:
   60 sec xylene at +4°C, change x2.
   60 sec 99% v/v ethanol at +4°C, change x2.
   10 sec 96% v/v ethanol at +4°C, change x2.
   10 sec 70% v/v ethanol at +4°C, change x1.
   60 sec PBS, change x3.
5. Customary immunofluorescence procedure (see 4.2.2).

Summary:
Day1: fixation of the biopsy material in 96% ethanol overnight at +4°C.
Day2: dehydration in 99% ethanol, +4°C for three hours, and ×2 xylene at +4°C for three hours. Embedding in paraffin wax at 58°C overnight.
Day3: cut sections, remove paraffin wax, incubate for immunofluorescence.

Disadvantages:
1. The background fluorescence, especially of connective tissue, is stronger as compared with cryotome sections.
2. The detection of C3 is less strong as a result of partial inactivation during fixation and heating during paraffin wax embedding.

Advantages:
1. The tissue block can be stored for months on end.
2. Immunofluorescence and routine histology can be carried out at one and the same time.
3. It is convenient.

Paraffin wax embedding for formalin-fixed tissue and treatment with pronase (after Huang [1975] and Witting [1977]).

Principle: pronase treatment results in a reduction in the non-specific fluorescence, so that antigens may be detected whose reactivity is not influenced by formalin, such as carbohydrate determinants.

Procedure: 4 µm sections from tissue, e.g., kidney or liver, which have been fixed in 4% v/v buffered formaldehyde, typically embedded in paraffin wax and sectioned, are freed of wax with xylene. This is followed by coating with 0.3% v/v collodion (Celloidin) and incubation with 0.1% v/v Pronase® (type VII[1], Sigma Chemical Co. Ltd.) for 15 minutes at room temperature. The reaction is quenched in ice-cold PBS, pH 7.45, after 60 minutes and the immunofluorescence determined as usual by the direct method.

4.2.1.3 Freeze-drying (Marshall [1951]; von Mayersbach [1957]; Pearse [1968])

This method is technically very demanding but, on account of the extremely good morphology that results, permits excellent structural analysis so that application to particular problems is justified and can be useful – as, for example, in the correct demonstration of Hepatitis B antigens (Hess & Arnold [1977]). The tissue is frozen either in liquid N_2 cooled iso-pentane or in Freon©, followed by freeze-drying at –40°C, and rehydration in a humidity chamber and embedding in paraffin wax.

4.2.2 Immunohistological staining

Apparatus and reagents:
1. Slides. The slides used must be fluorescence-free, of equal thickness (1.0–1.2 mm) and well cleaned (see 4.2.1.3). In order to prevent the serum running out, the slides can be sprayed around the edge of the preparation with a fast-drying varnish or plastic.

[1] note the comments in the Sigma catalogue about type VII made prior to 1975. There are now two types XXIV and XXVII with different specificities. Also Pronase is a tradename not a pure enzyme!

The cover glasses should not be thicker than 0.13–0.16 mm.
2. PBS, pH 7.2, 0.01 M – preparation as in 2.4.2.
3. Humidity chamber – a self-made construction, holding 20 slides, has proven to be advantageous (Fig. 16).
4. Staining jars.
5. Pasteur or automatic piston-type pipettes.
6. Sera, complement and conjugate.

4.2.2.1 Carrying out the direct method (see Fig. 16)

- air-dry the sections for 20 minutes; this reduces the danger of them floating off during incubation and washing.
- if necessary, fixation (depending on antigen, for example, 60 sec acetone, methanol, or ethanol etc.).
- wash for 20 minutes. Put the slides in a staining jar filled with PBS, and change the buffer two or three times.
- air-dry. Take the insert with the slides out of the buffer.
- wipe the slides dry around the section, place in the humidity chamber and dry thoroughly with a hot-air blower. Drying out of the sections at a later stage is to be avoided!
- incubation for 30 minutes, as a rule at room temperature. Add conjugate drop by drop to the section with a pipette so that it is completely covered. Finally close the chamber and, if necessary, cover with a cloth to prevent entry of light.

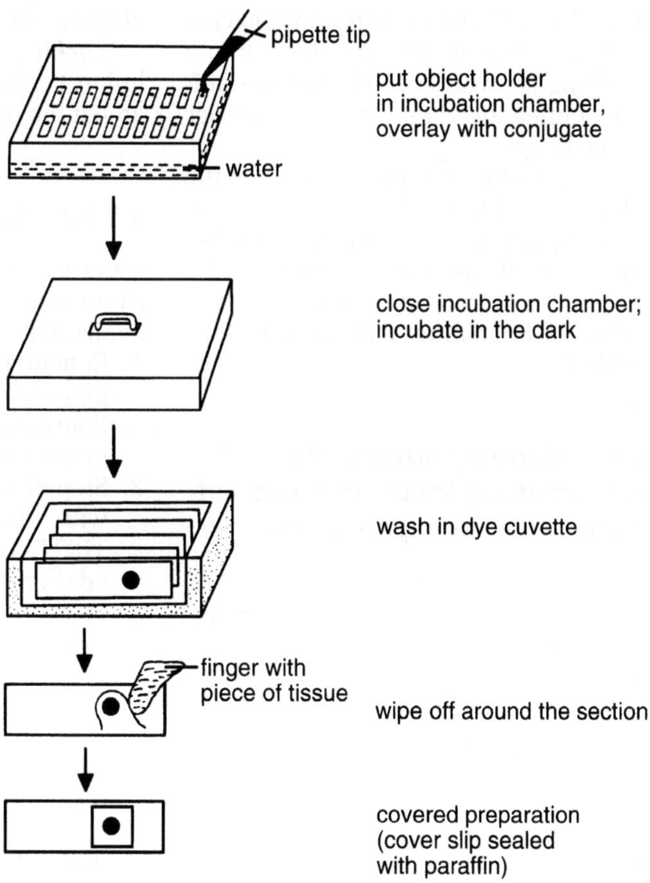

Figure 16. A summary of the procedure for carrying out indirect immunofluorescence.

- wash for 30 minutes (as above).
- embed. Wipe the slide dry around the section. Dry the section with the hot-air blower and add one drop of embedding medium to the section. Put on a cover-slip and, if necessary, surround this with paraffin (Vaseline) or nail-varnish.

4.2.2.2 Carrying out the simple indirect method without fixation at room temperature

- air-dry for 20 minutes.
- 1st incubation (30 minutes). Lay the numbered slides in the humidity chamber and cover the sections according to the protocol with the appropriate serum or diluted serum.
- wash for 30 minutes. Take the slides out of the chamber and allow excess serum to run off. Put the slides into a staining jar with PBS, change the buffer immediately and then a further three times.
- 2nd incubation (30 minutes). Take the slides out of the buffer, dry round the section, put into the humidity chamber and cover the section according to the protocol with conjugate (see Tab. 12).
- wash for 30 minutes (see above).
- embed.

4.2.2.3 Carrying out the "double" indirect method for the detection of complement-binding antibodies

- air-dry for 30 minutes.
- incubate with the serum to be examined for 30 minutes.
- wash for 30 minutes with a barbiturate-buffered wash solution.
- incubate with complement for 30 minutes (fresh human or guinea pig serum).
- wash for 30 minutes.
- incubate for 30 minutes with fluorochrome-labelled anti-complement serum.
- wash for 30 minutes.
- embed.

Tip: "inactivate" all antisera for 30 minutes at 56°C beforehand. All dilutions and washing procedures should be carried out using a barbiturate-buffered wash solution, pH 7.4.

Preparation of the barbiturate-buffered wash solution for determining complement-binding antibodies (after Wick et al. [1976]):

Stock solution – the following are mixed one after another:
83.09 g NaCl
10.19 g sodium 5,5'-diethyl-barbiturate
500 ml distilled water
34.58 ml 1 N HCl
5.0 ml salt solution: (1 M $MgCl_2$ + 0.3 M $CaCl_2$ ≈ 20.3 g $MgCl_2.6H_2O$ + 4.4 g $CaCl_2.2H_2O$ + 100 ml H_2O) made up to 2000 ml with distilled water.

Working solution: dilute the stock solution 1:5 with distilled water. If required, adjust the pH to 7.4 with 0.1 N NaOH.

4.2.2.4 Special methods

Determination of two antigens in the same preparation. Different methods are available:

1. Concurrent incubation with conjugates which are labelled with different fluorochromes, using either the direct or the indirect technique (Fig. 17).
2. Stepwise incubation, for example, first with FITC-anti-A and then with TRITC-anti-B using the indirect method (Fig. 18).
3. In the event of double-labelling with two fluorochromes not being possible, one can manage by examining the preparation after the first incubation, recording the findings, and then bleaching out the fluorochrome completely by irradiation with high intensity UV light. Having done this an additional incubation with the other conjugate is carried out to demonstrate the second antigen or antibody

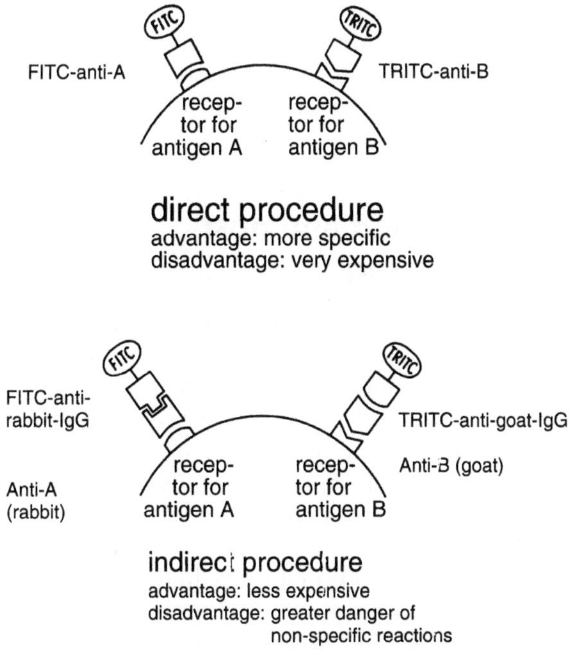

Figure 17. Showing the possibilities for double-labelling in both the direct and indirect immunofluorescence method.

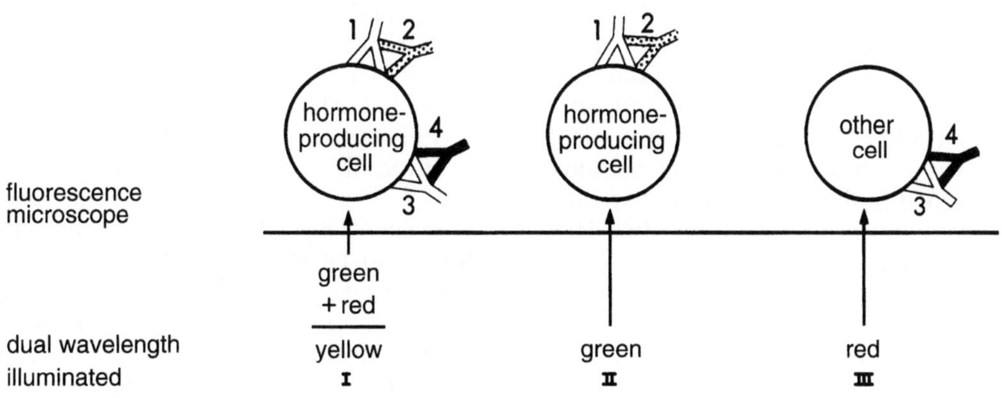

Figure 18. Illustrating the principal of the four-layer double-fluorochrome immunofluorescence method for detecting antibodies against hormone-producing cells. A frozen section is incubated consecutively (i) with anti-hormone immunoglobulin, e.g., with rabbit serum, for 2–6 hours depending on the tissue; (ii) with FITC-labelled anti-rabbit IgG, e.g., with goat serum, for 30–40 minutes; (iii) with patient's serum for 2–8 hours, depending on the tissue; (iv) with TRITC-labelled anti-human immunoglobulin, e.g., from goat, for 30 minutes; the sections are washed in between each stage. Examination by fluorescence microscopy is carried out using the appropriate sets of filters for FITC and TRITC. When the anti-hormone serum and the patient's serum react with the same cells, both red and green fluorescence can be seen. A yellow to orange-red colour appears on double excitation (I). If the anti-hormone antibody and the patient's serum react with different cells, cells show either green (II) or red (III) fluorescence separately from one another (based on Scherbaum et al. [1983]).

using the same preparation. An alternative method is microwaving performed in between the first and second staining cycles (Tornehave et al. [2000]).

4.2.3 Background and post or counter-staining

For increasing contrast and reducing autofluorescence or non-specific staining there are a number of possibilities, for example:

- counterstaining with rhodamine-labelled serum albumin (Smith et al. [1959]), which can be added already diluted (e.g., 1:20) to the conjugate, or with
- Evans Blue (C.I. 23860) or Congo Red (C.I. 22120) (Nichols & McComb [1962]), whereby the preparation is treated with a 0.01%–0.1% w/v solution for 1–5 minutes after incubation, followed by washing with PBS as usual;
- it is also possible simply to add 0.02% w/v Brilliant Cresyl Blue (C.I. 51010) or Eosin (C.I. 45400) to the last wash solution (Potworowski & Nairn [1967]);
- Eriochrome Black (C.I. 14645) is recommended for immunofluorescence with lymphocytes (Fey [1972]), used as a 1:30 or 1:60 w/v solution in distilled water. One drop of this solution is added to the preparation after the last washing procedure and, after 10 sec, washed off with PBS.

In some instances it is useful to demonstrate known structural elements in order to make alignment or orientation easier. Fluorochromes which stain nuclei include ethidium bromide (2,7-diamino–10-ethylphenyl-phenanthridinium bromide), propidium iodide (5-[3-(diethylmethylammonio)propyl]–3,8-diamino–6-phenyl-phenanthridinium di-iodide: Hudson et al. [1969]), or DAPI (4,6-diamidino–2-phenylindol hydrochloride: Russell et al. [1975]) – see Figures 104(a), 107(b) and 108. Ethidium bromide (0.5 µg/ml) has been widely abandoned on account of its carcinogenicity and has been replaced with propidium iodide. The stock solution (2 mg/ml) is kept refrigerated after dissolution in sodium chloride solution. The stock solution is added to the embedding medium to give a concentration of 1 µg/ml. The DAPI solution (stock solution 25 mg DAPI per ml in distilled water; dilute for use, 3 µl of the stock solution in 1 ml PBS) is added five minutes before the end of the last incubation, by adding a drop of the diluted solution directly to the conjugate on the preparation.

4.2.4 Embedding

Although it is possible to examine many preparations without using a cover-slip, we would recommend covering the preparation with a buffered glycerol mixture on account of the marked 'bleaching out' effect. The pH should lie between 7 and 9. In most cases pH 8.6 is ideal (see Storch [1979]). Nairn [1976] recommends the use of lepinal (a barbiturate)-buffered glycerol:

- Lepinal-sodium 10.3 g
- NaCl 6.2 g
- Distilled water to 1 l.
- Adjust pH to 8.6 with HCl
- ***For use:*** 1 part v/v buffer + 1 part v/v glycerol.

It is especially advisable, if the preparation is to be stored or photographed, to add a reagent to reduce 'bleaching out'. Although *p*-phenylene diamine is the most effective (Valnes & Brandtzaeg [1985]), DABCO (1,4-diazobicyclo–2,2',2''-octane) is to be preferred for routine use (Johnson et al. [1982]). The DABCO is dissolved directly in glycerol, kept in a slightly warm water bath, e.g., at 6 g/l.

Tip: the buffer should be kept in a refrigerator and should not be more than three weeks old. In addition, cover-slips should be edged with paraffin (Vaseline) or nail-varnish in order to avoid drying out or movement of the preparation.

4.3 Preparation and production of preparations for membrane immunofluorescence

Because membrane immunofluorescence, used for demonstrating antigenic determinants in living cell membranes, has acquired greater significance and the procedures used are somewhat different, a separate short description is given here.

4.3.1 Membrane immunofluorescence for lymphocytes

4.3.1.1 Isolation of lymphocytes (after Boyum [1976])

Apparatus and reagents:

1. Lymphoprep® solution (Nyegaard & Co, Oslo, Norway).
2. PBS (different from "normal" PBS!!). 8.0 g NaCl, 1.15 g Na_2HPO_4 anhydr., 0.2 g KH_2PO_4. 0.2 g KCl, distilled water to 1 litre, if necessary with the addition of 1% w/v albumin (bovine or human serum albumin, BSA or HSA).
3. Counting chambers, counting pipettes, 3% v/v acetic acid.
4. Pasteur or automatic piston-type pipettes.
5. Refrigerated centrifuge, shaker.

Procedure:

- mix heparin-treated (5 IU heparin/ml) or defibrinated venous blood with the 'special' PBS.
- layer 4 ml of the diluted blood onto 4 ml Lymphoprep® solution in a 10 ml centrifuge tube and centrifuge for 20 minutes at 430 g_{max}.
- take off the lymphocytes (interfacial layer), resuspend in three volumes of PBS and centrifuge for five minutes at 250 g_{max}.
- repeat this wash procedure twice and adjust the cell suspension to 2×10^7 lymphocytes/ml.

4.3.1.2 Incubation

Special dishes (boats) or tubes, as well as a shaker, are required.

Direct method:

- centrifuge 0.25 ml aliquots of the above lymphocyte suspension (2×10^7 cells/ml) for five minutes at 180 g_{max}.
- incubation (30 min). Resuspend the pellet (5×10^6 cells/ml) in 50–100 µl of the working dilution of the conjugate and shake at room temperature.
- wash three times in PBS. Centrifuge for three minutes at 250 g_{max} and resuspend the pellet in three volumes of PBS.
- embed (see 4.3.1.3).

Indirect method:

- centrifuge 0.25 ml aliquots of the lymphocyte suspension (2×10^7 cells/ml) for five minutes at 180 g_{max}.
- incubation (30 min). Resuspend the pellet (5×10^6 cells/ml) in 100 µl unlabelled (or even labelled) serum, for example, goat anti-human-IgM, and shake at room temperature.
- wash three times in PBS (see direct method).
- incubation (30 min). Resuspend the pellet in 50 µl of the conjugate, e.g., FITC-anti-goat-IgG, and shake at room temperature.
- wash three times with PBS (see direct method).
- embed (see 4.3.1.3).

4.3.1.3 Embedding and preparation of permanent preparations

After the last centrifugation the lymphocytes are resuspended in 0.25 ml PBS. A drop is then placed on a microscope slide, a cover-slip applied, and its edges sealed with paraffin (Vaseline). Wick et al. [1976] recommended producing a "long-lasting" preparation in the following way: place one drop of the suspension on a slide, air-dry for ten minutes and then

dry for a further 20 minutes with a fan (cold air!); fix the dried preparation in 96% v/v ethanol for five minutes, wash in PBS, dry with a fan again and cover in buffered glycerol – instead of PBS it is better to use glycine buffer, pH 8.6, see chapter 4.2.4).
Assessment: see chapter 7.3.1.4

4.3.2 Membrane immunofluorescence for liver cells

4.3.2.1 Preparation of liver cells (enzyme free, Hopf et al. [1974])

Apparatus and reagents:

1. Liver tissue from rabbit, rat or man.
2. Eagle's Medium with the addition of 2.5% w/v bovine serum albumin (BSA).
3. Water bath, shaker, nylon gauze filter, oxygen.

Procedure:

- perfuse the liver through the vena cava so that it is completely free of blood.
- cut the liver into small pieces with scissors and wash a number of times in Eagle's Medium so that it is 'blood-free'.
- shake for five minutes at 37°C in Eagle's Medium.
- filter the cell suspension through the nylon gauze, treat the cells with a solution of Trypan Blue (C.I. 23850)[2] and assess cell viability under the phase microscope.
- adjust the suspension to a concentration of 6×10^6 cells/ml.

4.3.2.2 Incubation

Direct method:
e.g., for detecting IgG, F_c, C3 or liver-specific proteins, etc.

- incubate the cell pellet (6×10^6 cells/ml) with 150 µl of conjugate for 30–60 minutes at 37°C.
- wash three times in Eagle's Medium with added BSA.
- cover and mount.

Indirect method:
e.g., for detecting F_c in two stages or for liver cell membrane antibodies.

- incubate the cell pellet (6×10^6 cells/ml) with 150 µl serum for 30 minutes at 37°C.
- wash twice in Eagle's Medium with added BSA.
- incubate the cell pellet with 150 µl of conjugate for 30 minutes at 37°C.
- wash a further three times.
- cover and mount.

Assessment: see chapter 7.3.1.1

4.4 Undesirable (so-called) fluorescence and specificity controls during immunofluorescence investigations

False negatives as well as false positives are, in principle, possible as findings – for an overview see Storch [1986]. These can arise because of the 'unsuitability' of the antigen substrates, the antibodies, the methods used, the microscope or even the investigator. False negatives can only be recognised if pure Ag and Ab, as well as standard or reference substances and defined Ag-substrates, are available. False positives can be caused for many reasons which are not always recognisable. In general one can say that a positive immunological result is characterised by structure-specific fluorescence of typical colour, with faint controls. In individual cases the decision whether a positive or negative result exists can be made more difficult or impossible by the coexistence of unwanted or background fluorescence.

[2] this must be of a suitable grade for tissue culture work.

Such unwanted fluorescence can be caused by (also see 1.3):

1. Autofluorescence[3].
2. Staining of the labelled serum by free dye[4].
3. Non-specific staining of eosinophils, basophils or other structures by relatively acidic labelled or degraded proteins, especially those with high F/P ratios and for conjugates that are relatively insoluble in aqueous solutions[5].
4. Actual non-specific reactions with cross-reacting or polyvalent antibodies (v. Mayersbach [1966]), so-called undesired specific staining[5] (USS, see 1.3).
5. The binding of immunoglobulins to physiological receptors, for example, F_c-receptors.

Nairn [1976] has suggested the following specificity controls:

1. Staining should only occur in preparations which contain antigen. The staining should only affect the antigen.
2. No staining should occur with fluorochrome-labelled *non-immune* serum.
3. On pre-incubation with non-labelled *immune* serum prior to incubation with labelled *immune* serum there should be a reduction in fluorescence intensity but not on pre-incubation with *non-immune* serum (the inhibition or blocking test of Cherry et al. [1960]).
4. If the labelled serum is absorbed beforehand using the appropriate antigen, staining should be inhibited but not when another antigen is used.
5. In order to eliminate binding to F_c-receptors, F_c-free antibody cleavage products should be used, for example, F_{ab} or $F_{(ab')2}$.

v. Mayersbach [1966] and Barthelmes & Sönnischen [1970], amongst others, have pointed out the dubious nature of the absorption test because, as a rule, cross-reacting Ab is removed at the same time as the absorption of conjugates by the appropriate Ag takes place. It is generally simpler to replace the antigen substrate or conjugate. Furthermore it must be stressed that an inhibition test – as described in (3) above – does not often yield a clear explanation. This can arise from exchange between labelled and unlabelled antibodies with unstable Ag-Ab complexes, or from differences in concentration or avidity. It is also possible that antigenic determinants remain "free", i.e., not bound to Ab, after the first incubation. The inhibition test of Moody et al. [1956] generally gives better results. The conjugate is mixed with unlabelled antiserum before incubation, and with normal serum as the control.

Controls of specificity carried out in defined test systems beforehand and the testing of suitability using special test procedures, guarantee the best protection against false positives (see Ch.3). Simple controls based on histochemical criteria are sufficient and these are summarised below.

---

[3] knowledge about autofluorescence with different filter combinations is one of the pre-requisites for immunofluorescence (IF) – for reviews see Haitinger [1959], Hamperl [1934], Hicks & Matthaei [1955] and de Lerma [1958].

[4] it is less common today as a result of gel filtration techniques and causes no problems in general.

[5] this is still a problem – in spite of well characterised coupling and purification methods being currently available. In individual cases the most appropriate conditions must be determined using preliminary experiments covering, for example, the F/P ratio, dilution of the labelled antiserum, adsorption onto tissue homogenates, storage conditions, thickness and fixation of sections, counter-staining of the background, reduction of the isoelectric point of tissue proteins by raising the buffer pH (Grossi & v. Mayersbach [1964], v. Mayersbach & Schubert [1960]) or by blocking free amino groups (Becker & v. Mayersbach [1969]) or, where necessary, raising the salt concentration of the buffer used for washing (Lunzenauer et al. [1966]; Rudolph [1968]).

[6] it is not always possible to avoid this so a wide range of controls for specificity have been recommended (reviewed by Storch [1986]).

4.4.1 Controls during indirect immunofluorescence for the detection of auto-antibodies (see Tabs 9–12)

1. Haematoxylin and eosin (HE) staining for correct orientation and identification of tissue elements.
2. "Blank" sections to establish levels of autofluorescence – incubation with PBS alone.
3. Conjugate controls (incubation with conjugate alone) to establish the levels of non-specific staining especially with high fluorochrome content.
4. Normal serum or 'negative' controls – incubation in the presence of normal serum at low dilutions – to establish non-specific binding by proteins, which do not have a specific Ab function but are able to bind antigen at high concentrations and cannot be removed during washing (generally occurs with a high Ab content for the conjugate).
5. 'Positive' controls – incubation with a reference or standard serum that must be positive at a predetermined titre.

4.4.2 Controls for direct immunofluorescence (for example, at kidney biopsy)

1. Haematoxylin and eosin (HE) or other staining procedure for correct orientation and identification of tissue elements.
2. "Blank" sections to establish levels of autofluorescence – incubation with PBS alone.
3. 'Negative' controls – a preparation which does not contain the sought-after antigens, e.g., immunoglobulins.
4. 'Positive' controls – a preparation which contains the antigens.
5. Incubation with a conjugate directed against other proteins, e.g., anti-human-albumin, which point, for example, to the presence of an exudate.

4.4.3 Controls in the detection of complement-binding antibodies

1. Haematoxylin and eosin (HE) or other staining procedure for correct orientation and identification of tissue elements.
2. Incubation with test serum, inactivated complement (inactivated at 56°C for 30 minutes) and conjugate (labelled anti-complement).
3. Incubation with normal serum, complement and conjugate.
4. Incubation with test serum and conjugate.
5. Incubation with complement and conjugate.
6. Incubation with conjugate alone.

Additional controls must be carried out, or the conjugate changed, if no clarification is achieved using the above controls. The necessity for the type and the number of controls for specificity depends on the particular investigation. One must keep in mind, however, that the specificity and sensitivity of immunofluorescence cannot be better than that of the antisera and conjugates used (Knapp et al. [1975]).

5 Instrumental techniques and the documentation of findings

5.1 Microscope

Any microscope is suitable as long as it is fitted with an appropriate light source and suitable filters. Based on current requirements, however, one should aim for a specialised fluorescence microscope. The optimal adjustment of all the component parts to suit the absorption and emission spectrum of the fluorochrome in question, is a pre-requisite for good results. This is especially so for the transmission properties of the excitation and blocking filters and also, secondarily, for the excitation light source which must emit sufficient radiation in the region of the excitation maximum for the fluorescence being investigated. The quality of the microscopic image is determined mainly by the brightness and contrast of the field. The image contrast depends on the relationship between the fluorescence of structures stained specifically or non-specifically by the conjugate (see 4.4) and by the brightness of the background. Self-fluorescence of optical components, mentioned by Ploem [1973], is no longer an issue today. Moreover, it is self evident that any microscope slides, cover slips, or embedding and immersion media used must also be fluorescence-free.

The sole purpose of the light source is to stimulate fluorescence in the specimen. It should have the highest possible intensity at the absorption maximum of the fluorochrome. Monochromatic laser illumination is ideal because this provides at the same time a constant intensity of illumination. At present for routine use, high-pressure mercury-vapour discharge lamps, for example, HBO 50, are the sources that are principally used. Depending on the particular application one can also use high-pressure xenon or tungsten-halogen sources – for an overview see Ploem [1973] and Nairn [1976]. Xenon lamps are especially necessary if photometric measurements are to be carried out.

The *excitation* or *primary* filter has the function of only allowing light to pass which is able to stimulate fluorescence optimally. At the same time it must also block any unnecessary or interfering light of shorter or longer wavelength. The *blocking* or *secondary filter*, which is situated on the optic axis behind the objective, only allows through fluorescent light of a specified wavelength. Excitation light of shorter wavelength should have been completely blocked. The part of the fluorescence spectrum which is observed can be additionally restricted using special filter combinations – see below and also Figures 19–21.

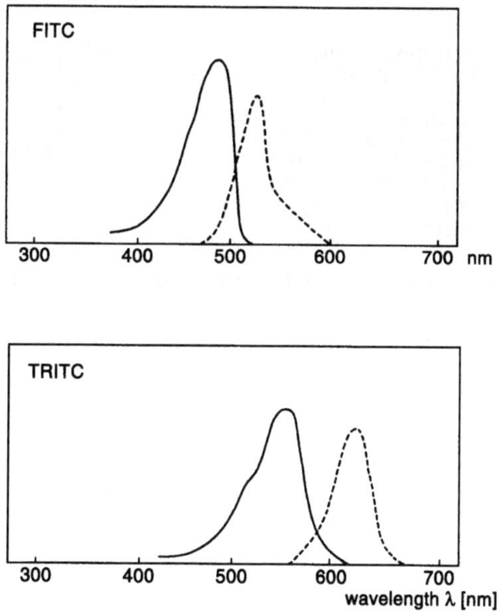

Figure 19. Absorption (—) and emission (- - -) spectra for FITC and TRITC (schematic).

Filter type	Filter response	Designation
Broadband filter A broad band of wavelengths is allowed through, e.g., from 320 nm to 510 nm.	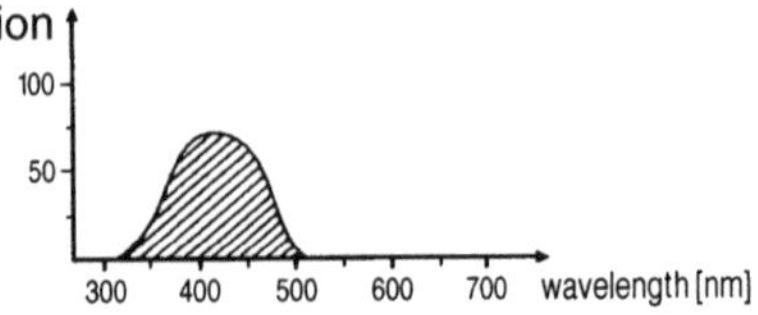	BG 12
Narrowband filter Also known as a *bandpass* filter or *line* filter by analogy to the mercury absorption line; a chosen band of wavelengths is allowed through, e.g., from 540 nm to 560 nm.	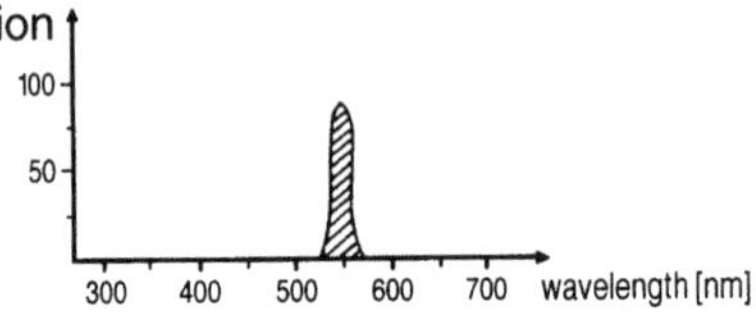	BP 540-560 (RSIF 546)
Shortpass filter Wavelengths shorter that the cut-off limit are allowed to pass, e.g., those less than 490 nm.	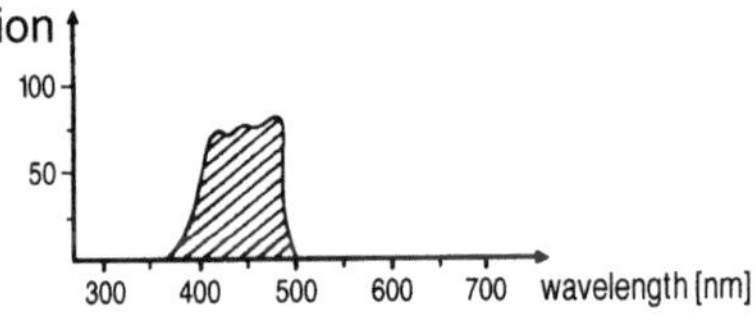	KP 490 = BP 390-490
Longpass filter Wavelengths longer than the cut-off limit are allowed through, e.g., those greater than 455 nm.	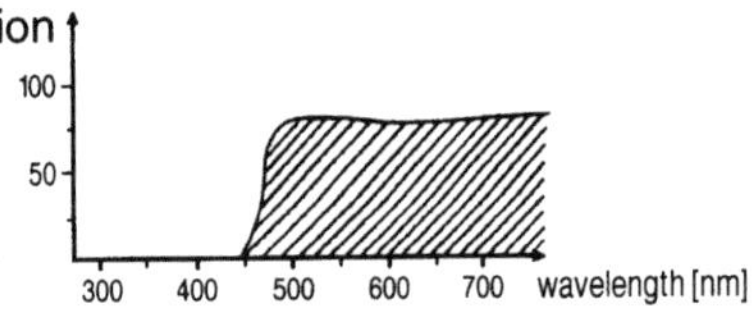	LP 455
Bandpass filtering for 470-490 nm Achieved through a combination of a 490 nm *shortpass* filter (KP 490) and a 470 nm *longpass* filter (LP 470), together with a 520 nm *longpass* filter as a *blocking* filter.	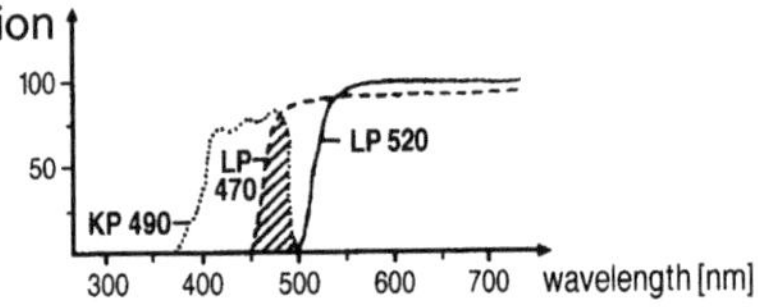	

Figure 20. Explanation of the different filters used for immunofluorescence.

Excitation can be achieved using either transmitted (light- or dark-field), or reflected light (Fig. 22). Transmission excitation is no longer appropriate these days as the method has many disadvantages:

(a) the transmission condenser must be centred; for dark-field excitation the condenser must be of the immersion type which, although capable of producing very high contrast images, is particularly complicated to use.

(b) large object fields, of the size of the image field of a low-power objective, are only weakly illuminated and, as a result, the fluorescence is relatively poorly excited.

(c) the excitation and blocking filters are easily misaligned.

(d) incorrect installation or choice of the filters can lead to the observer receiving the full excitation intensity directly in their eyes.

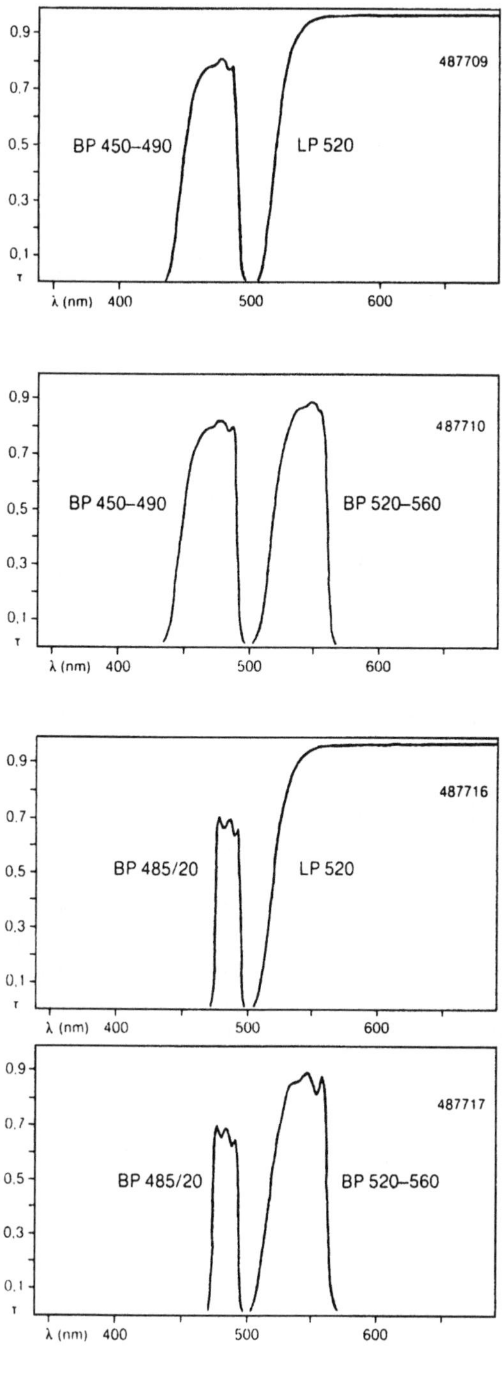

Figure 21. Transmission curves for the excitation and blocking filters used for FITC fluorescence (data sheets obtained from Carl Zeiss, Oberkochen, Germany, No. F 41.350-d). Fluorescence intensity is reduced in practice using narrow-band excitation and blocking filters; the selectivity is, however, increased (a further explanation is given in the text).

Reflectance excitation is currently the method of choice. This method has, compared to transmission excitation, many advantages:

(a) adjustment of the condenser is not necessary since the objective used for observation is also the objective used for illumination of the sample; as soon as the image can be seen sharply in focus, the illumination is already optimal.

(b) the total excitation radiation is concentrated only on the field under observation and the fluorescence is stimulated optimally within the field of view.

(c) the excitation and blocking filters are mounted in a single combined unit together with the dichroic (colour-splitting) mirror and are always positioned in the optic path. Thus the observer's eyes are protected constantly from the effects of intense excitation radiation.

(d) transmission procedures remain available for use with additional microscopic procedures.

The former customary use of broadband excitation using glass absorption filters has in general now been replaced by excitation using interference bandpass-filters. Their main advantage consists of higher optical transmission at the transmission maximum combined with a more selective pass-band with sharp edges (Fig. 20 and 21). The most convenient solution in each area of application can be found by varying and combining different filters and beam-splitting mirrors (Tab. 13 and Fig. 21).

Whether one should only observe the preparation by eye, photograph it, or even make quantitative photometric measurements, depends on the spectrum and intensity of the light source, on the absorption and emission spectrum of the fluorochrome, on the microscopic technique used, for example, the condenser, whether transmitted or reflected light, phase- or interference-contrast microscopy are being used, and on the material

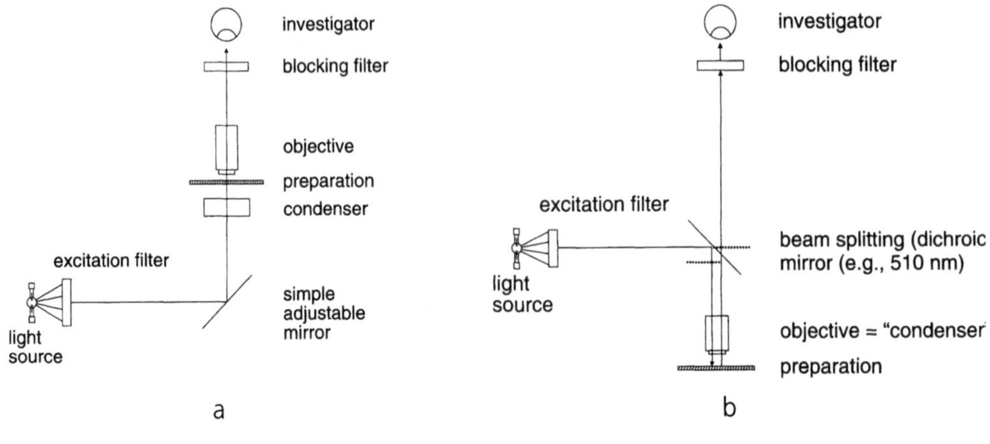

Figure 22. Representation of the optical path in (a) transmission and (b) bright-field microscopy. A dichroic splitting mirror, or beam-splitter, for FITC is shown in the optical path of the reflected light; this mirror reflects light with a wavelength less than 510 nm and allows light with a wavelength greater than 510 nm to pass.

being investigated. Use of narrow-band excitation and a restricted emission band, for example, for FITC with a band-pass filter at 485 nm (20 nm) in combination with one at 520–560 nm, allows one to reduce the observed autofluorescence quite decidely, e.g., with blue, yellow-green or red colours and non-specific conjugate deposition on connective tissue.

For routine use as well as research purposes excitation with reflected light is the most convenient. Thus the simultaneous, or alternating, use of a variety of methods is possible. The type of fluorescence excitation used will depend on the type of fluorescence equipment available and a compromise will have to be achieved between the various microscopic techniques available.

In general, apochromatic objectives are recommended for fluorescence work because of their higher aperture (lower 'f' number). The quadratic increase in sample illumination with increasing lens aperture should be noted.

Table 13. Filter combinations for FITC and TRITC fluorescence.

Fluorochrome	Excitation filter	Beam-splitter	Blocking filter
FITC	LP 450, KP 490, KP 500	TS 500–510	LP 520
	or	or	
	BP 470–495, 450–490	TS 500–510	BP 520–560
	BP 485/20 etc.		
TRITC	BP 515–550	TS 580	LP 590
	or		
	BP 546 etc.		

KP = short-pass filter (Kurzpaßfilter)
BP = band-pass filter
LP = long-pass filter
TS = beam-splitting (dichroic) mirror (Teilerspiegel)

Using reflectance excitation, this is proportional to the fourth power, since the excitation intensity is also dependent quadratically on the aperture.

Objectives with other forms of correction should only be used in exceptional cases. As a result of the development of immersion objectives with the highest possible aperture specially for fluorescence work, the advantages of reflected light over transmitted light can be utilised even at low magnifications, i.e., higher contrast, increased brilliance and intensity of the specific fluorescence, optimisation of the illumination and image sharpness, as well as possibilities for photometric measurements.

5.2 Additional equipment

Of the many recommended additional pieces of apparatus such as microphotographic equipment, apparatus for quantitative immunofluorescence, image amplification using closed-circuit TV, fluorescence polarisation, etc., we will only consider those for photomicrography and confocal laser-scanning microscopy. The latter technique permits automation more easily and the introduction of numerous technical novelties such as the zoom lens, image-processing and storage of the images by computer, as well as colour printing and the possibility of manipulating the scale of any part of the image at will.

5.2.1 Setting up for photomicrography

Because image brightness shows a loss of intensity during irradiation with light (bleaching-out, fading or photo-decomposition), a photomicrographic record of the findings is often made. Because of the low light levels a miniature camera is used. The exposure time should be kept as short as possible because of bleaching-out (fading) and the associated changes in colour. Modern fully automatic microscope cameras are ideal, making the photography itself almost without problems. Since, however, these are not always available, one is sometimes forced to determine the most appropriate exposure time by making a series of exposures with all other parameters held constant. The image distribution in fluorescence microscopy is similar to dark-field microscopy so that moderate over-exposure is feasible. The exposure time calculated automatically by a simple microscope camera, based on the sensitivity of the film used, will be too long because of the extensive dark background of a fluorescence image. The really interesting fluorescent structures are in consequence more or less over-exposed and the background appears lighter in the photograph than actually observed visually through the microscope. With such a camera, therefore, one or two shorter exposures must be tried depending on whether there are many or only few fluorescent structures in the preparation. When there is no possibility available for adjusting the camera, then the exposure meter can be set to a two-fold or four-fold higher film sensitivity. (On the other hand, the situation with normal light-field microphotography is the exact opposite, for example, in pathology or haematology. Here one must then use longer exposures, according to the characteristics of the object field, so that interesting structures are not under-exposed and so that the white background does actually appear white). High sensitivity film is to be preferred for black-and-white photography. Double or multiple fluorescence can be documented microphotographically using modern filter combinations for double or multiple fluorescence. When the different fluorescence intensities are of comparable strengths, then simultaneous photographic recording is not a problem using these filter combinations. If, on the other hand, the fluorescence intensities differ greatly, then the exposure time appropriate for recording with the filter combination for the strongest incident intensity is the determining factor, the weaker intensities not being registered adequately under these circumstances.

The conditions used when making an exposure, especially the filters and objectives used and the final magnification, should always be stated in publications. Details about exposure times are necessary for comparing non-specific fluorescence, e.g., in a negative control, with specific fluorescence if fully automatic cameras are being used. Recording of the exposure time should therefore be carried out. The colour illustrations in this book – unless stated otherwise – were made using an Olympus BX60 microscope and PM–30 photomicrographic equipment.

It is often useful to use broad-band or somewhat limited narrow-band excitation for visual examination of the image, with or without filtering of the emitted light. Restriction of both the excitation and emission spectrum is, however, recommended in order to obtain higher selectivity with FITC on account of using higher illumination intensities for better orientation and for comparing autofluorescence and non-specific staining. Thus in our new colour illustrations we have predominantly used the NIBA set of filters for recording FITC fluorescence: excitation, 470–490 nm; beam-splitting mirror, 505 nm; emission, 515–550 nm. For TRITC and Cy3 fluorescence as well as for recording counter-staining by Evans Blue, we use the SWG set of filters: excitation, 480–55 nm; beam-splitting mirror, 570 nm; emission, 590 nm. The objectives used were UPlanFl 20 × 0.50; UPlanApo 40 × 0.1 Oil (iris=0.5–1.0); UPlanApo 100 × 1.35 Oil (iris=0.5–1.35). The post-magnification used was predominantly × 5. Kodak Ektachrome 400 ASA, Fujichrome 400 ASA and Fujichrome Sensia 200 ASA films were used. The illustrations taken from the 1st edition were obtained mainly using either a transmission microscope type Nf or a reflecting microscope type Fluoval I from Carl Zeiss (Jena).

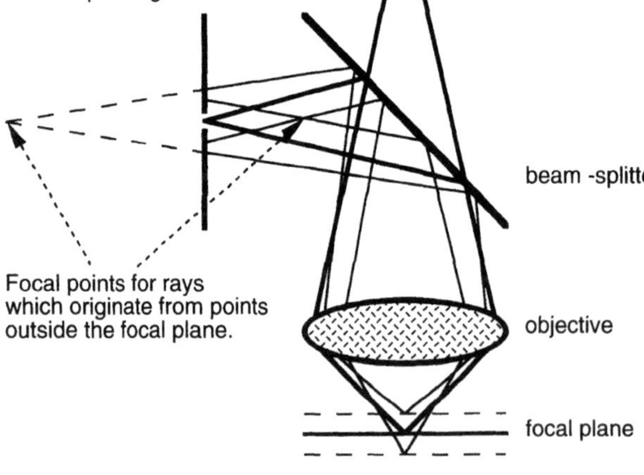

Figure 23. Explanation of the principle of confocal microscopy (according to Engelhardt and Knebel, [1993]). A point source, for example, from a laser, is focused in the plane of the object field as a diffraction-limited image by the objective. This point of light, derived from the laser, then stimulates a point source of fluorescence in the preparation. The emitted fluorescent light enters the objective and, as a result of the dichroic mirror, is then focused exactly on the detector pinhole. The light subsequently passes onto the detector where it is measured. Both the diffraction-limited laser spot focused in the plane of the object and the detector pinhole, where a point image the the spot of light is formed, lie exactly in conjugate planes with respect to one another. Thus fluorescent light which originates from areas of the object outside the focal plane, is not sharply focused at the detector pinhole and is filtered out and, accordingly, not detected. Point sources of light and point detectors are in practice constructed using filters consisting of very small holes.

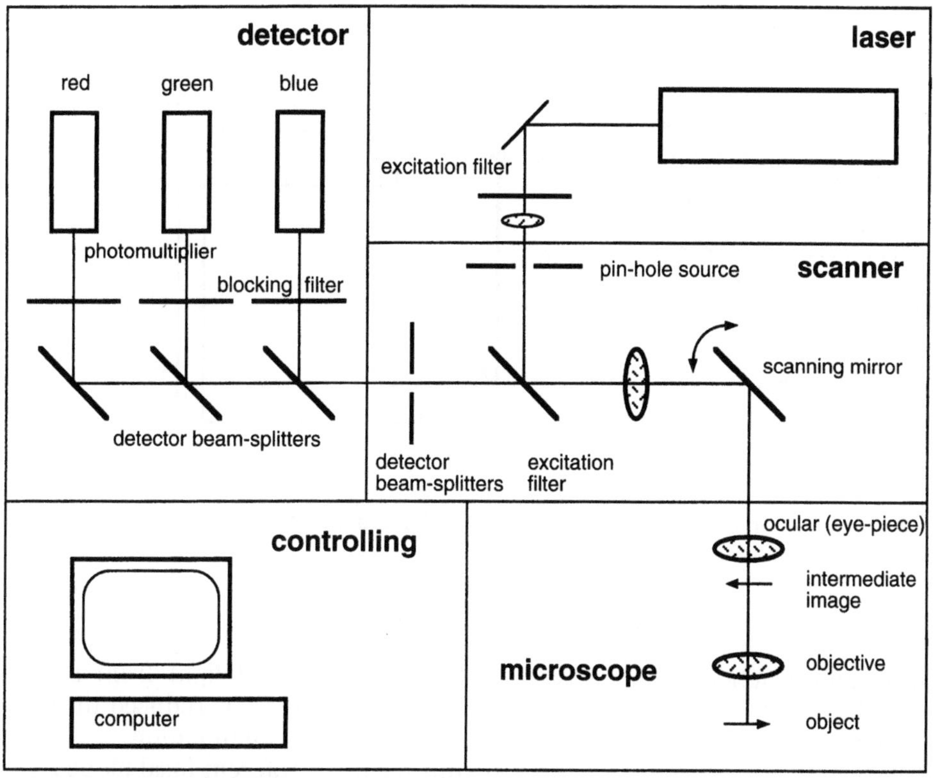

Figure 24. Representation of the five essential components of a confocal laser-scanning microscope (based on Engelhardt & Knebel [1993]). The laser, e.g., a mixed gas argon-krypton laser, acts as a light source capable of generating a number of monochromatic spectral lines for stimulating the fluorescent dye(s). The scanner contains the confocal pin-hole, or point filter arrangement, and a beam-splitter which reflects the excitation wavelengths and transmits the fluorescence to be detected. The excitation beam is moved across the object in the field of view in the form of a raster, using a scanning mirror. Light that is reflected, or emitted, by the object is registered by the detector according to wavelength. The complete system is computer controlled. In this way the beam can be made to traverse all parts of the object in a coordinated manner and the detector signals can be stored in the form of multi-dimensional images. Powerful computers make possible the further processing of the comprehensive image information containing, as a rule, more than a million data points.

5.2.2 Confocal laser-scanning fluorescence microscopy (CLSFM)

The principle is based on the reproduction of only those structures which are directly in the focal plane of the objective (Fig. 23). Thus it is possible to obtain extremely sharp image definition and, by using a combination of a scanner and a computer, an optical image in three dimensions. This technique has been impressively demonstrated by Wijnaendts van Resandt et al. [1985]. A modern confocal microscope can be considered as being composed of five essential parts; the microscope, the laser, the scanner, the detector and the computer which acts as the control unit (Fig. 24). A 'section' is possible in both the horizontal or focal plane (the so-called xy-plane), or in the vertical plane parallel to the optic axis (the so-called xz-plane) – see Figure 25.

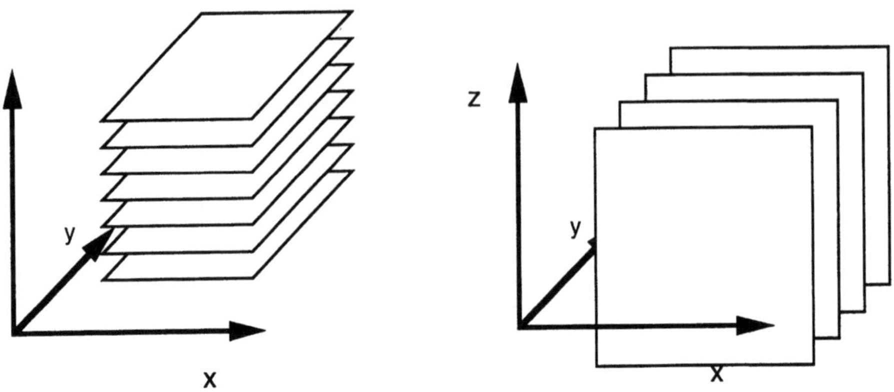

Figure 25. Representation of optical sections in the focal plane (horizontal or xy-sections) and along the optic axis (vertical or xz-sections).

Engelhardt and Knebel [1993] have summarised the most important consequences arising from confocal microscopy as follows:

1. Suppression of light from outside the focal plane.
2. Defocussing does not produce lack of sharpness but rather an exclusion of parts of the object which are not in the focal plane. The object image becomes dark on defocussing before it loses its sharpness. Hence, one speaks of an 'optical' section.
3. True three-dimensional sets of data can be recorded.
4. Side-views, i.e., vertical or xz-sections, are possible by means of continuously moving the object or preparation along the optic axis.
5. Stray or scattered light from neighbouring structures is reduced through the use of point illumination.
6. Images from different focal planes can be combined, using image processing, to give a new image with extensive depth of focus. In conventional microscopy an increase in the depth of field requires a reduction in the aperture, which leads to a loss of resolution.
7. Quantitative measurements no longer depend on the adjustment of the focus or the user because the whole object can be recorded in three dimensions.

In contrast to conventional microscopy the brilliance of the image, even with fluorescence of low intensity and especially with thick preparations and at high magnifications, is particularly conspicu-

Figure 26. Selection from a series of horizontal images representing horizontal and one vertical section obtained by confocal laser-scanning fluorescence microscopy. Incubation with antibodies against cytoplasmic myosin (provided by Prof. Stewart, Darmstadt) produced fluorescence in the cytoplasm and also, questionably, the nucleus. The vertical section (f) proves that the antibody really does bind to the nucleus. The interpretation of this finding, whether the binding indicates the presence of 'cytoplasmic' myosin in the nucleus, or whether this is a cross-reaction with additional antibodies present which produce an artefact, is difficult or impossible to assess. Plane of the sections: (a) 468 nm; (b) 0 nm; (c) 492 nm; (d) 1484 nm; (e) summation image ("extended focus"); (f) vertical section through the lower nucleus (based on Storch [1989]).

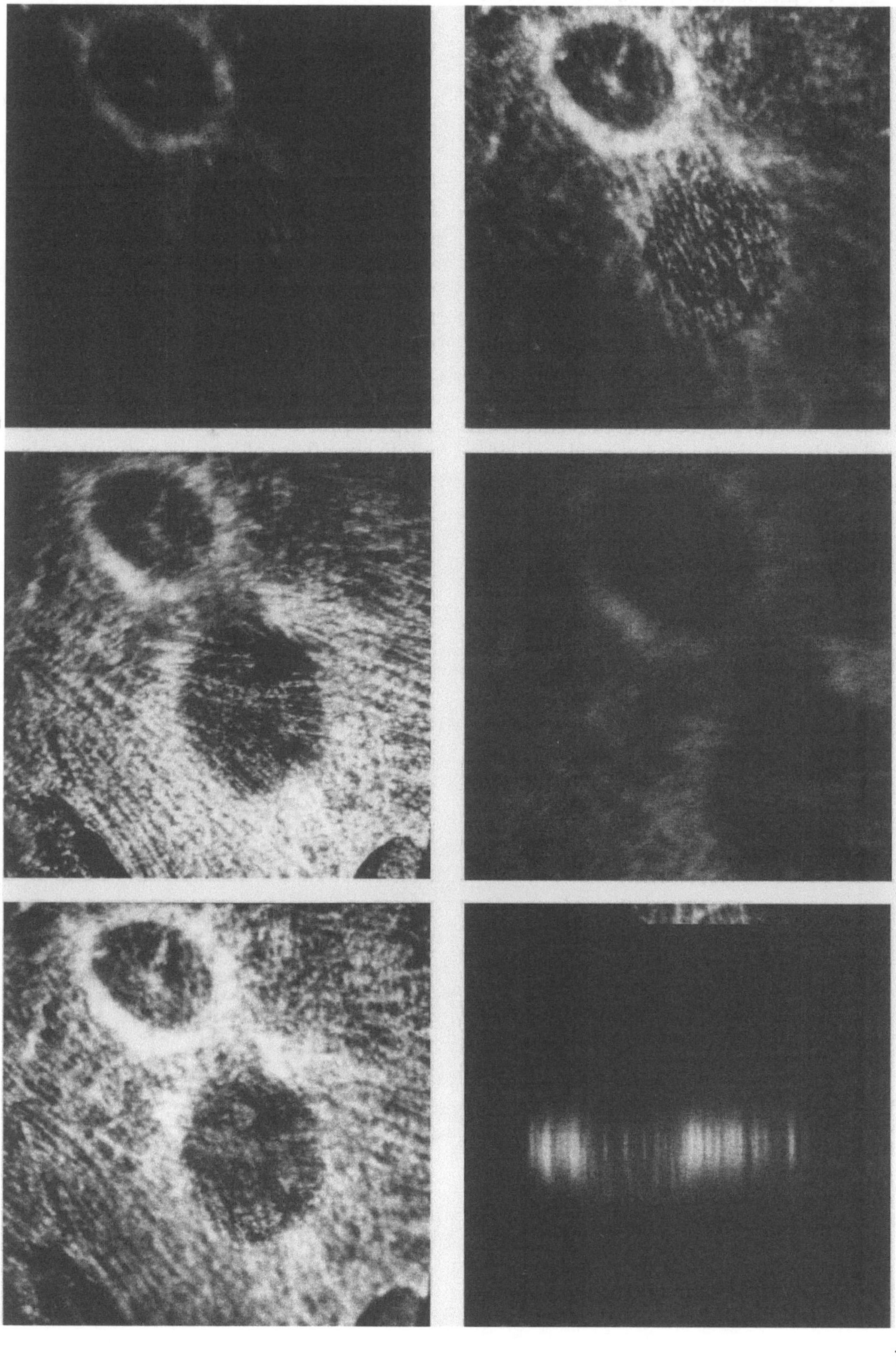

a b c d e f

ous. A 4 µm section suitable for conventional fluorescence microscopy can be subdivided into ten serial 0.04 µm thick sections. Additional, so-called 'maximum projections' (the representation of the brightest image points of all focal planes) as well as so-called 'extended focus images' (summation images) can be created[7]. By comparing horizontal and vertical images an unambiguous structural assignment for the fluorescence may be achieved, for example, as a projection onto or within the nucleus of a cell (Storch [1989], Fig. 26), for a glomerulus (Fig. 56) or for Auerbach's plexus (Fig. 65). Furthermore, three-dimensional views of the object can be calculated using the computer and reproduced as a stereo image. A range of new applications, both in research and routine work, becomes possible in the context of the simultaneous demonstration of multiple antigens and antibodies, which are labelled with different fluorescent dyes, above all for transmission microscopy when additional contrast methods are incorporated. New insights are possible not only into cell structure but also into cell dynamics.

Finally, in spite of all the known advantages, two disadvantages should not be forgotten. The inherent disadvantages of image-processing are the dangers of unwitting false representation associated with misinterpretation, and deliberate falsification or conscious manipulation of data – compare Fig. 115 and 116. The opportunity or possibility for suppressing fluorescence should not be used, for example, in order to 'eliminate' the inadequate specificity of a result without explanation of the grounds for doing so.

For an overview of confocal laser-scanning fluorescence microscopy refer to Kriete [1992], Stevens et al. [1994] or Pawley [1995].

[7] ***maximum projections***: take a stack of images with different focal planes, then create a new image which contains at any x,y position the brightest intensity of all focal planes:

$$mproj[x,y] = \mathbf{max}(I[x,y,z]\,\forall z)$$

extended focus: create a new image by summing the intensity of all focal planes at each x,y position:

$$\mathrm{extfoc}[x,y] = \sum_{\forall z} I[x,y,z]$$

6 General advice

6.1 Sera, antibodies, conjugates and antigenic substrates

We strongly recommend the establishment of a system of record cards containing the identification information and amount in stock for all standard sera, as well as for labelled and unlabelled sera. A record card is produced for each preparation, on which the storage location should also be written.

One should freeze the necessary reference sera and recommended conjugates for specific tests in appropriately small aliquots, so that repetitive freezing and thawing can be avoided and the materials themselves conserved (see 4.1). A microtitration device or a piston-type dispensing pipette can be used for the preparation of serial dilutions. It is an advantage, when using human antigenic substrates, if the donor is blood-group Od positive, since this will reduce interfering fluorescence from blood-group antibodies. When using animal material, it is advisable where possible to starve the animal for 24 hours beforehand, thus reducing autofluorescence. In-bred strains are to be preferred in order to ensure the consistency of results. Young male animals with weights between 100–150 grams are especially suitable for the differentiation of the proximal tubules of rat kidney, particularly for the detection of microsomal antibodies.

Two serial dilutions, e.g., 1:20 and 1:100, should be examined in the detection of antibodies – depending on the antibody – in order to recognise 'prozone' phenomena, a 'specific' or 'non-specific' inhibition of immune-complex formation (Barthelmes & Sönnichsen [1970]). Serial dilutions of serum less than 1:10 are only sensible in exceptional cases because of the risk of unwanted staining. In order to detect multiple antibodies simultaneously, we recommend using an anti-human-IgG preparation as the conjugate with a low F/P-ratio (around 1.5) and an antibody concentration of 50 µg/ml at the working dilution.

Serum titre can differ with different antigen substrates. In view of the positive controls, one should fix on one substrate because of this. The experimental conditions should be kept constant after working out reproducible methods which can be compared with those of a reference laboratory.

6.2 Analysis of results

The use of microscope slides of constant thickness and quality, as well as the cover-slips and embedding medium used, are especially important in transmission microscopy. Microscopic examination should take place immediately after production of the preparation, especially if photographic or fluorimetric recording are to be carried out. Storage of preparations overnight in a refrigerator is acceptable but acclimatisation to room temperature before use for approx. ten minutes is then needed.

The controls, provided with code numbers, should be assessed first. If there are differences observed on examination at this stage, then clarification is necessary before examination of the remaining preparations.

The following basic rules are suitable for microscopy:

1. A solid, firm location should be chosen for the microscope which is free from vibration and shocks and is in a dust-free room which can be darkened.
2. Do not 'put out' (alter) the centering or alignment of the lamp and microscope, if at all possible. The smallest change in the condenser, or a light-source that is not accurately centred, can make assessment by transmission microscopy impossible. The condenser aperture diaphragm should be fully open. The preparation will be better illuminated if a drop of fluorescence-free immersion oil is placed between the first lens of the condenser, with an aperture of >0.9, and the underside of the microscope slide (Holz [1975]).
3. The same filter combinations should be used. The expensive interference filter should be placed as far as possible from the light-source to avoid damage due to heat. The optical path between the lamp and the preparation should in any case be as short and as straight as possible.
4. A constant adaptation period should be used by the investigator.
5. Records should be taken after any changes to the apparatus.

High-pressure mercury-vapour discharge lamps should be allowed to "burn-in" for at least four hours on being brought into service, in order to reduce flickering later. The lamp should be run for approx. ten minutes before microscopy is carried out. Running times and the frequency of striking the arc should be noted. The lamp should be changed for another in good time. For this reason photometric measurement of the light intensity for permanent (stable) preparation should be obtained.

Comparison with a phase-contrast or interference-contrast image is useful, and frequently even a *sine qua non* for membrane immunofluorescence, in order to prevent invalid assessments being made for the structural location of fluorescence.

Finally, it is obvious that one ought to take meticulous care of the cleanliness and unspoilt (unscratched) condition of the objective (polystyrene is, for example, recommended for cleaning objectives).

7 Applications of immunofluorescence

7.1 Overview of the field of application

Immunofluorescence is being used increasingly in all areas of biology. It is used in microbiology in preference to other methods for the detection of viruses, bacteria, rickettsia, protozoa and fungi as well as for demonstrating antibodies against different microorganisms (for reviews see the following: Fraser [1976]; Jentsch [1967]; Mayr et al. [1977]; McEntegart [1976]; Wagner [1967]).

The technique of immunofluorescence serves, amongst others, in clinical and experimental pathology and immunology to demonstrate organ-, tissue- or cell-specific antigens in cells or on their surfaces, as well as antigen-antibody complexes and antibodies which are circulating in the blood or have been partly laid down, together with complement, on particular structural elements. It is also easily used as a histochemical "staining method" for making particular proteins 'visible' using specific antibodies, for example, against hormones, enzymes and plasma proteins, as well as tumour or foreign antigens. Within the scope of this primer the possibilities for application in clinical diagnosis are discussed, in particular with respect to the demonstration of antibodies involved in the immune response as 'markers', 'witnesses' or aids to diagnosis in disease. Monographs and textbooks on immunology should be consulted for further areas of application (Beesley [1993]; Celis [1994]; Lachmann et al. [1993]; Peter & Pichler [1995]).

7.2 Detection of immunoglobulins, immune complexes, auto-, allo-, xeno- and neo-antigens in biopsy material

7.2.1 Kidney biopsy

In order to achieve good results, it is useful to examine the biopsy core (cylinder) directly after aspiration with a hand-lens or dissecting microscope, to determine whether there is enough cortex available. The core is then immediately divided up. A part of the cortex is possibly prepared for electron microscopy (EM), the rest frozen in liquid nitrogen-cooled *iso*-pentane prior to immunofluorescence, after embedding in either Tissue Tek compound, gelatine or rat liver (see 4.2.1). The biopsy core should be sectioned on the same or the following day. Cryotome sections are examined under the microscope in order to choose the best sections, those showing the most glomeruli in as good a condition, i.e., state of preservation, as possible. If there is no possibility of being able to prepare cryotome sections, then special paraffin embedding can be carried out (see 4.2.2).

Suitable conjugates should possess an F/P ratio of 2–3 and an antibody protein content of 200 μg/ml at the working dilution. The direct method is to be preferred on account of a saving in time and a reduction in the danger of non-specific staining. It is, however, less sensitive than the indirect method. The incubations required are based on the clinical question being asked. Incubation with a polyvalent anti-human IgG and an anti-C3 antiserum suffices, for the time being, for screening purposes (Schneider [1976]). Investigations involving the following proteins are, at present, usual: IgG, IgM, IgA, (IgD, IgE), C1q, C3, C4, properdin, and fibrinogen, as well as albumin and transferrin. Antisera against antigens of interest such as hepatitis-B antigen, tumour and streptococcal antigens, are all part of the current repertoire.

The pattern of fluorescence. The many possible pictures seen cannot be gone into in great detail here. As a simplification one can distinguish three types of pattern for the glomerulus (Figs 27–34):

1. Linear fluorescence.
2. Granular fluorescence – coarse or fine.
3. Speckled or patchy fluorescence – also coarse or fine.

The deposition can either be diffuse, in over 80% of all glomeruli examined, or localised (focal or 'hearth'-like), in less than 50% of glomeruli. One speaks of 'global staining' if the whole glomerulus is involved and of 'segmental staining' if only part is involved (Churg et al. [1995]). The strength or intensity of the fluorescence should, in any case, be recorded. The deposition is described, based on the identification of anatomical structures, either as:

- subepithelial or subendothelial;
- intra- or extra-membranal (peri or epi-);
- or as mesangial or podocytic.

It can, of course, also be intracapillary. Deposition outside the glomeruli should be described in an analogous way, by reference to anatomical structures.

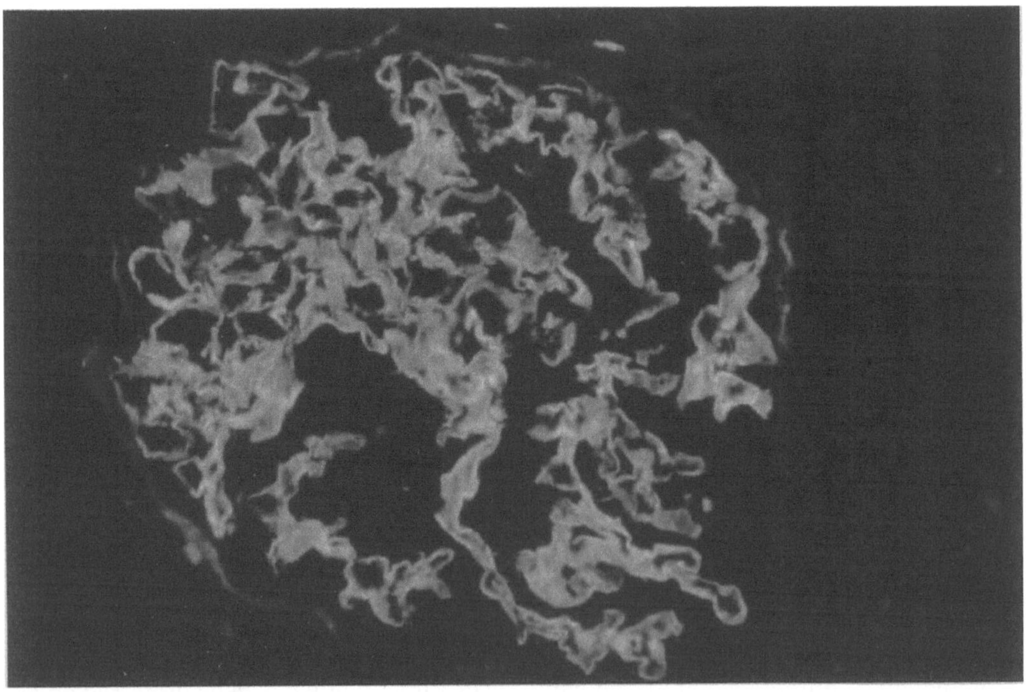

Figure 27. Kidney biopsy: conjugate, anti-human IgG. Anti-basement membrane glomerulonephritis (objective ×25, ×330; Prof. Waldherr, Heidelberg).

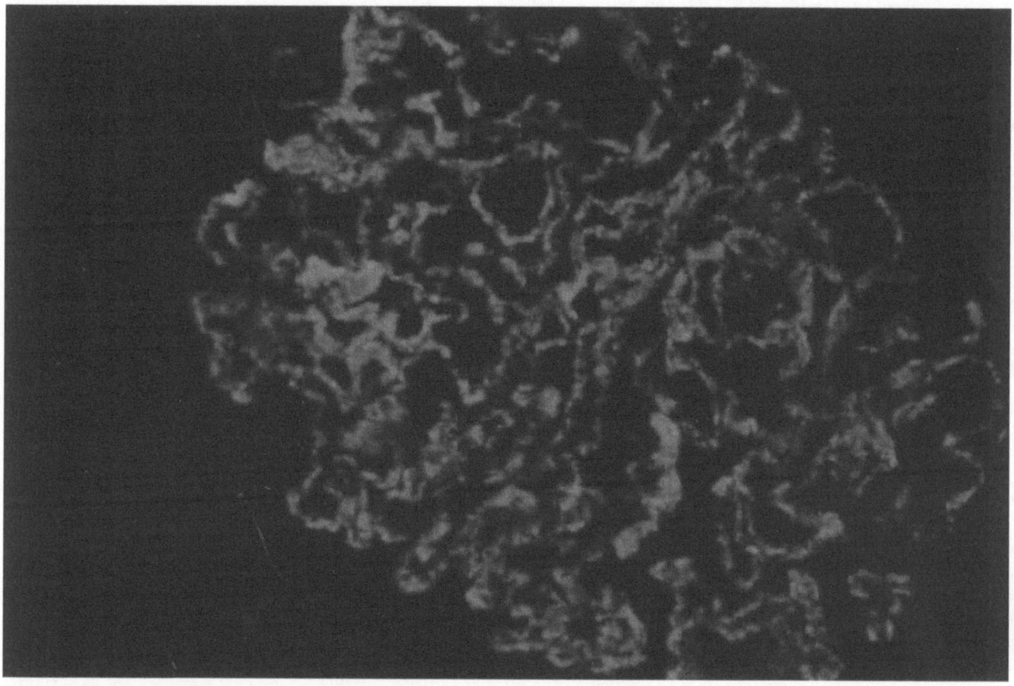

Figure 28. Kidney biopsy: conjugate anti-human IgG. Membranous glomerulonephritis (objective ×25, ×330; Prof. Waldherr, Heidelberg).

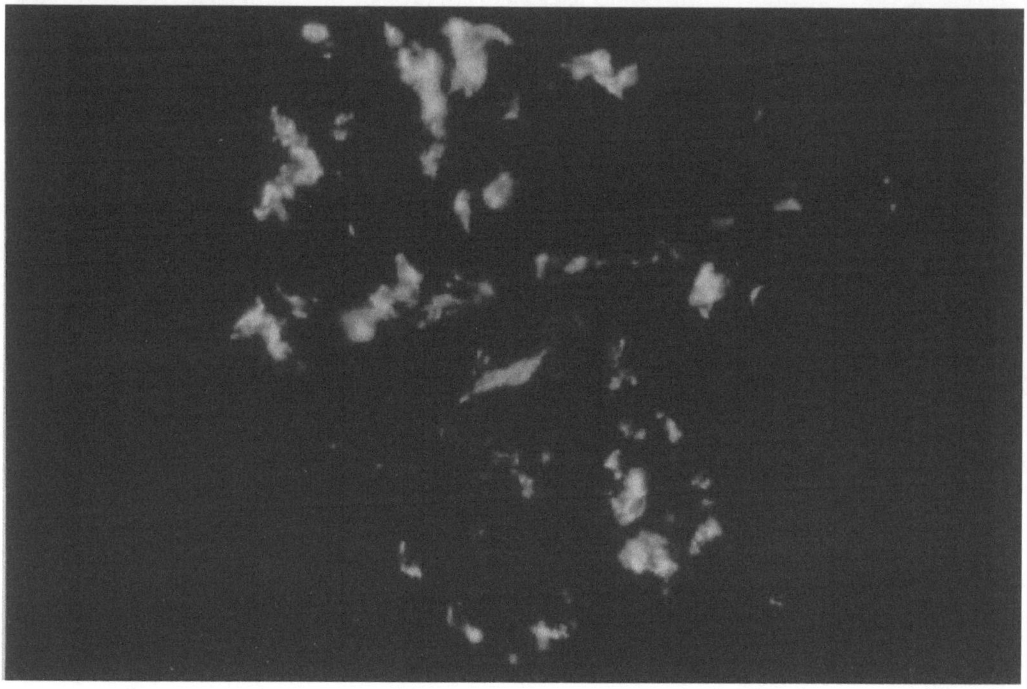

Figure 29. Kidney biopsy: conjugate anti-human IgA. Mesangial IgA-glomerulonephritis (objective ×25, ×330; Prof. Waldherr, Heidelberg).

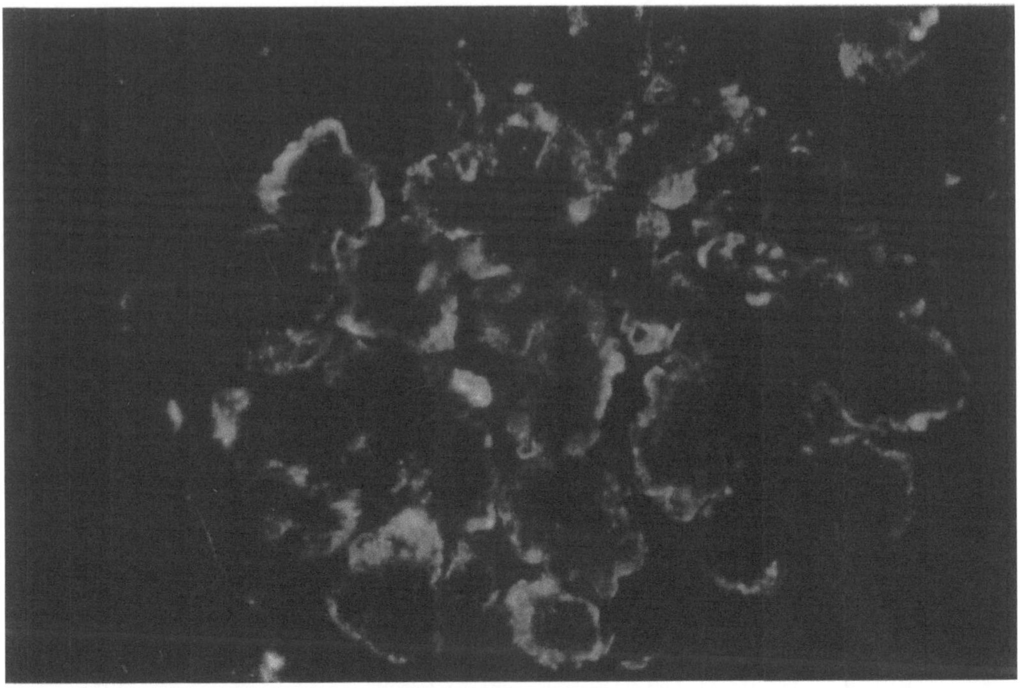

Figure 30. Kidney biopsy: conjugate, anti-human IgG. Membrane-proliferative (mesangio-capillary) glomerulonephritis type I (objective ×25, ×330; Prof. Waldherr, Heidelberg).

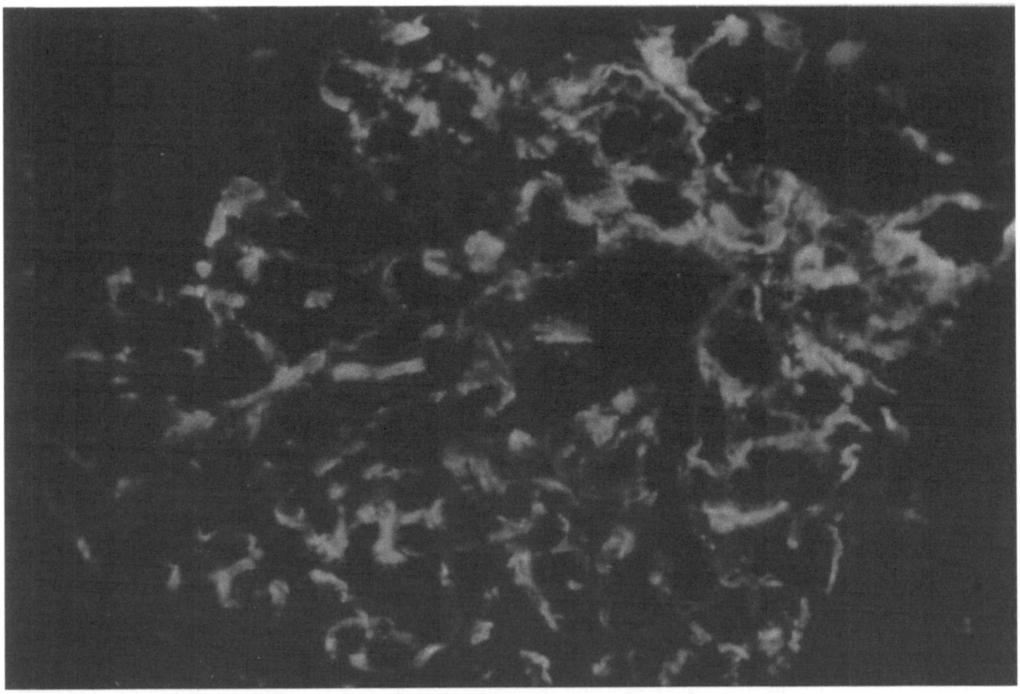

Figure 31. Kidney biopsy: conjugate, anti-human IgG. Proliferative lupus-glomerulonephritis (objective ×25, ×330; Prof. Waldherr, Heidelberg).

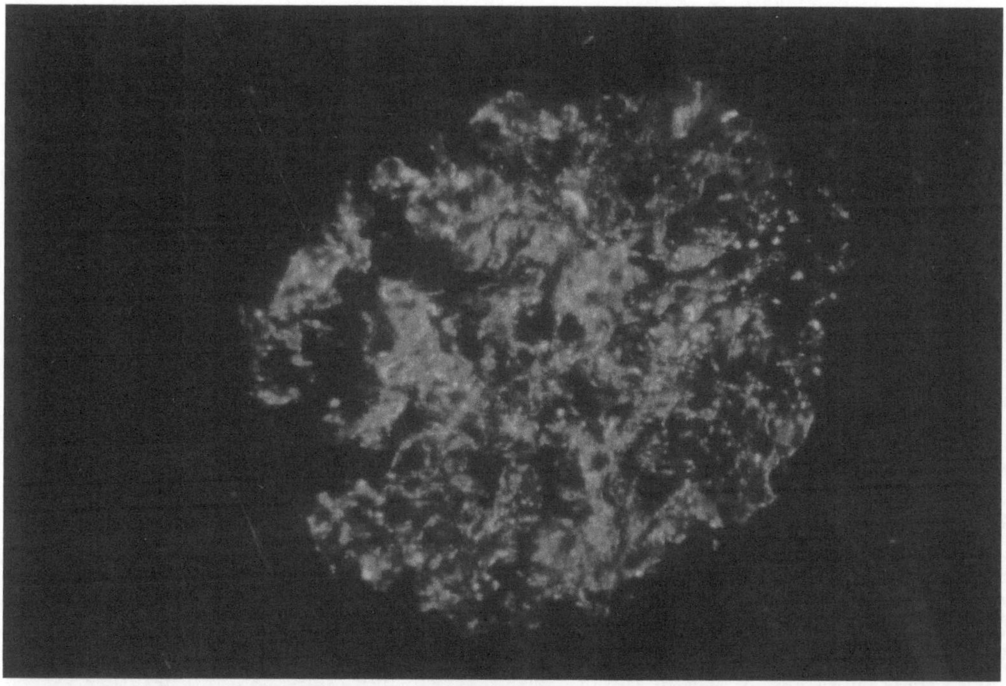

Figure 32. Kidney biopsy: conjugate, anti-human C3c. Acute post-infectious (endocapillary) glomerulonephritis (objective ×25, ×330; Prof. Waldherr, Heidelberg).

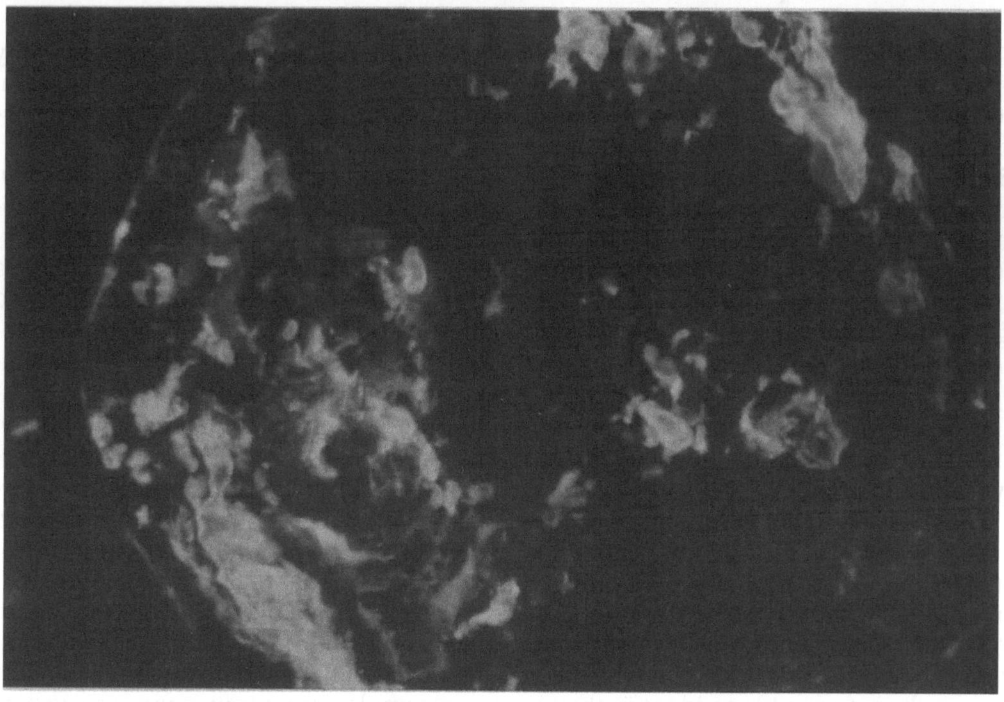

Figure 33. Kidney biopsy: conjugate anti-human fibrin/fibrinogen. Pauci-immune extra-capillary necrotising glomerulonephritis (objective ×25, ×330; Prof. Waldherr, Heidelberg).

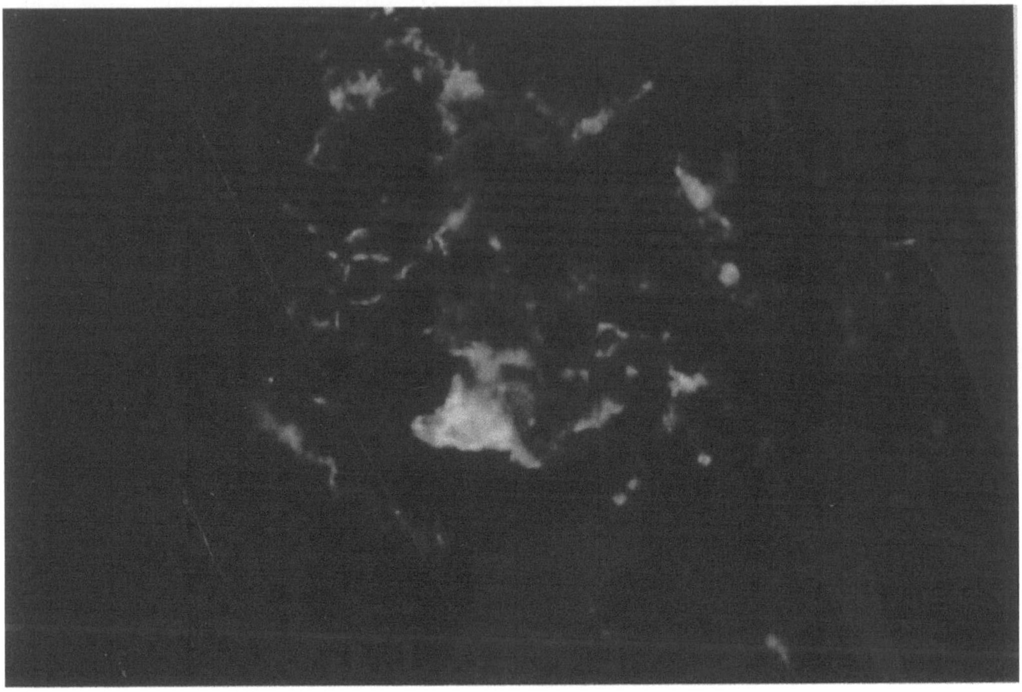

Figure 34. Kidney biopsy: conjugate, anti-human IgM (non-specific). Focal glomerulosclerosis in nephrotic syndrome (objective ×25, ×330; Prof. Waldherr, Heidelberg).

Clinical and diagnostic relevance. Different types of glomerulonephritis are currently to the fore in terms of clinical immunofluorescence. Interpretation is simple only in exceptional cases because the assignment to known structures within the cell is not always easy. Thus the very rare *linear* type of pattern with a uniform, diffuse and hazy deposition of IgG and C3 indicates a *basement membrane antibody-induced nephritis* caused by complement-binding cytotoxic antibodies (prototype: Masugi nephritis seen in animal experiments, or the Goodpasture syndrome in man).

The *granular* type, with a more or less even deposition of IgG and small amounts of IgM and C3, is evidence generally for an *immune-complex nephritis* (prototype in man: 'lupoid' kidney seen in *systemic lupus erythematosus* (SLE)), in which it is frequently not possible to identify the responsible antigen. The finding that the majority of immune-complex nephritides arise from deposition of antigen-antibody complexes circulating in the bloodstream, is important in the context of the glomerulus acting simply as a filter.

Depending on their size, complexes can be deposited either subepithelially, subendothelially or mesangially, so that Germuth & Rodriguez [1973] distinguish a trans-membrane, endo-membrane and mesangiopathic form of glomerulonephritis. The drawing of conclusions about a pathological process based on the identification of immunoglobulins alone is not permissible. It can, as with any other serum proteins such as albumin or transferrin, simply be the result of increased membrane permeability, e.g., due to inflammatory reactions.

On the other hand it is not clear why, in nephrotic syndromes with so-called 'minimal changes', neither immunoglobulins nor complement components are generally found, although the disease responds to immuno-suppressive therapy. An assessment of the immuno-histological findings is frequently only possible if one has used a broad spectrum of antisera and if, at the same time, one has taken into consideration, in the form of an information 'mosaic', all the available clinical, laboratory, chemical, light- and EM-microscopical findings. A table such as that which appeared in the 1st edition is not appropriate on account of the numerous new findings which have been generated explosively in this area (for a review of the literature, see Churg et al. [1995]).

7.2.2 Liver biopsy

The aspirate is divided into three as with the kidney biopsy – for electron microscopy, light microscopy and immuno-histology – especially if research questions are being asked, and where possible immediately frozen in liquid nitrogen for immunofluorescence or prepared for membrane immunofluorescence.

Suitable conjugates should possess an antibody content of 200 µg/ml and a low F/P molar ratio (not greater than 1:1!). In order to reduce non-specific staining, these should be further purified over DEAE-cellulose, as well as by absorption with human liver, and be available as the F(ab'$_2$) fragment. Incubation with antisera against HB_s-, HB_e-, HB_eAg and HCAg, as well as with C3, C4, total complement, IgG, IgA, IgM and fibrinogen, appears to be both necessary and important.

Patterns of fluorescence: (Figs 35 to 39)

1. Cellular fluorescence arising from macrophages, von Kupffer star-cells, plasma cells, lymphocytes, hepatocytes and from mononuclear cells, without any clear distinction.
2. Extracellular fluorescence in the lumen and vessel walls of capillaries, other blood vessels, sinusoids and in the space of Disse, as well as associated with necrosis and, above all, in the periportal area in the region of the bile ducts and in relation to connective tissue.

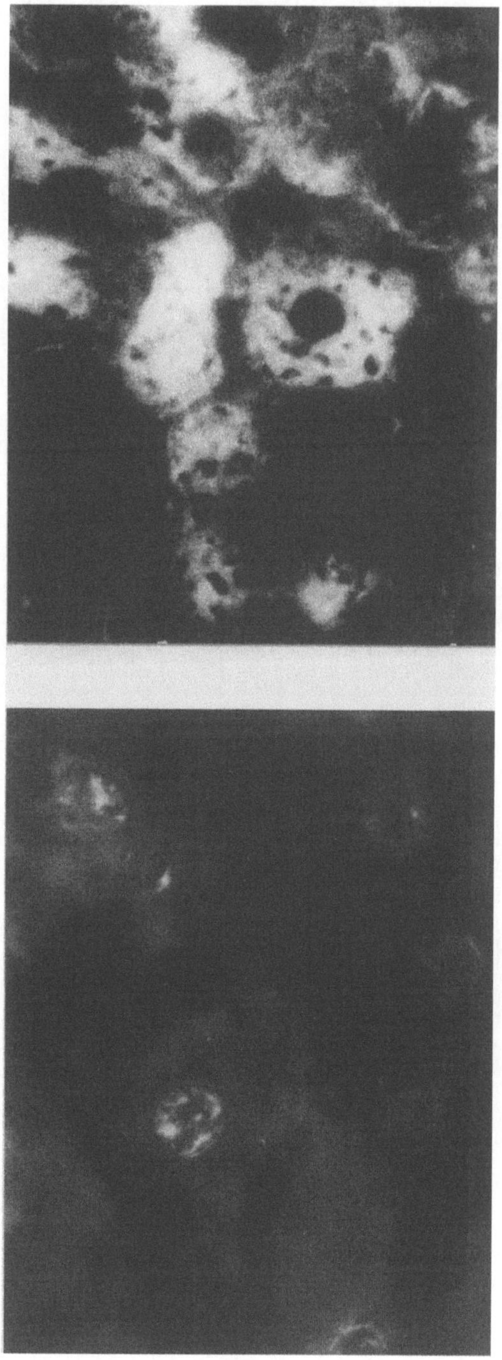

a

b

Figure 35. Liver biopsy: (a) conjugate, anti-HB_sAg. Cytoplasmic fluorescence of the liver cells (objective ×40, ×530); (b) conjugate, anti-HB_cAg. Nuclear fluorescence of the liver cells (objective ×100, ×1320).

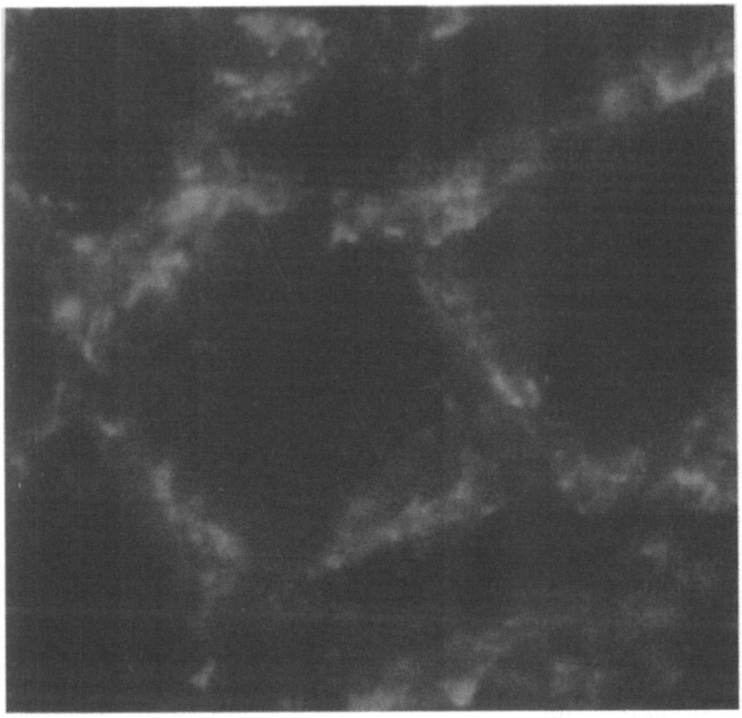

Figure 36. Liver biopsy: conjugate anti-HB_sAg; (a) fluorescence of the cell periphery and cell membrane (objective ×40, ×800); (b) cytoplasmic fluorescence of individual mononuclear cells – lymphocytes? – (objective ×100, ×1250).

b

3. Fluorescence of the *linear* and *granular* form for cell membranes in the liver, as well as along the sinusoids.

Tip: comparison of observations under phase contrast and fluorescence conditions, with the use of different filter combinations, allows assignment to cells or other known structures, respectively, in most cases.

Clinical significance: independently of whether or not the more frequently used PAP technique is chosen in preference to the more unusual liver immunofluorescence technique, immunofluorescence has advantages in the hands of the experienced user. Indications for using immuno-histological methods are currently the same as previously described (compare Storch [1986]):

1. The differential diagnosis of viral hepatitis, especially hepatitis B and C with an assessment of the likely prognosis;
2. Indications for immuno-suppressive therapy in chronic aggressive hepatitis

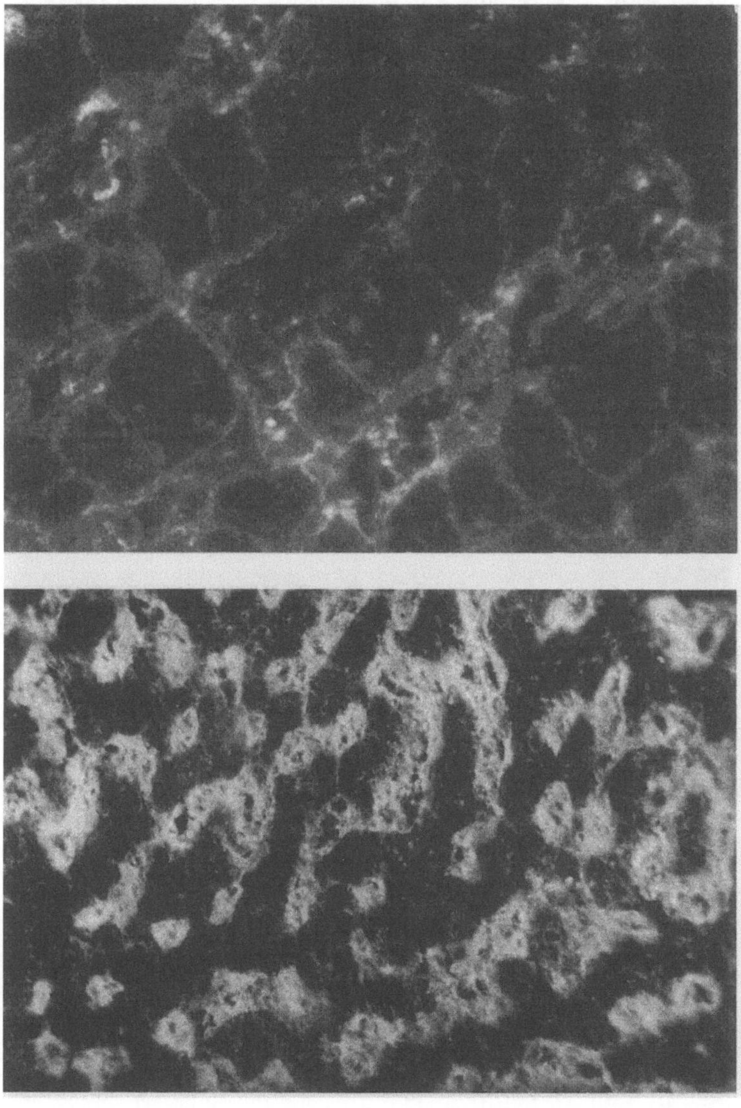

a

b

Figure 37. Liver biopsy: (a) conjugate, anti-HB_sAg; weak, partly linear, partly granular fluorescence along the cell membrane; (b) conjugate, anti-human IgG (the same as in Fig. 37a) – massive deposition of IgG along the sinusoids (objective ×40, ×530).

3. The establishment of the diagnosis for metabolic disorders, especially α1-antitrypsin deficiency but also complement component storage in PAS-negative hyaline bodies (Storch et al. [1982]; Storch [1985]);
4. Classification and differential diagnosis of liver tumours using, amongst other methods, antibodies against intermediate filaments (Bannasch et al. [1981]).

The identification and localisation of viral antigens is of interest in viral hepatitis, in hepatitis B and D, for example, the antigens HB_sAg, HB_cAg, HB_eAg and $HB_\delta Ag$. The distribution pattern of hepatitis B antigens inside and outside of liver cells appears to be associated with particular diseases or the activity of the disease (Tab. 14). Immunofluorescence has advantages over the PAP-technique in the diagnosis of non-A/non-B or C hepatitis,

a

b

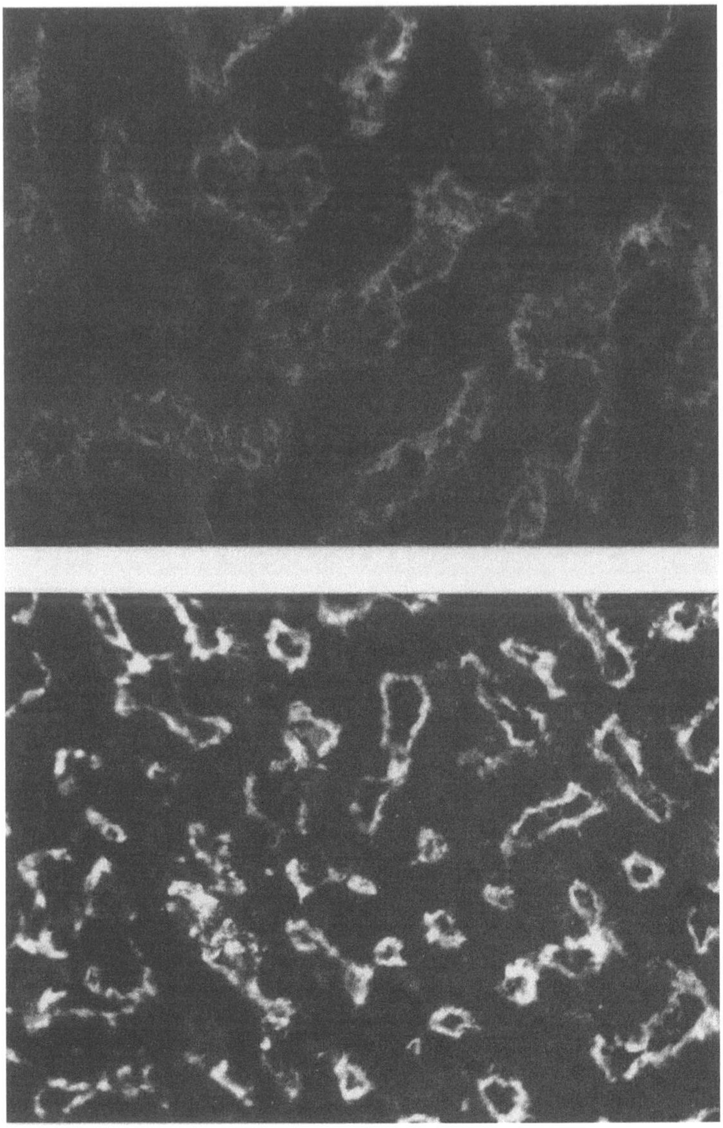

Figure 38. Liver biopsy: (a) conjugate, anti-human IgG – weak fluorescence along the sinusoids (binding via the F_c-receptor); (b) conjugate, anti-human fibrinogen – intense fluorescence along the sinusoids (objective ×40, ×530).

as described previously, because it is more sensitive (Storch et al. [1983]; Ballardini et al. [1995]). The diagnosis of virus-induced or virus-associated autoimmune hepatitis (Storch [1978, 1982]) is made using immuno-histological methods on the liver. Solitary or isolated aggregates of HB_cAg – nuclear and cytoplasmic, occasionally membranous – or cytoplasmic HB_s or hepatitis C antigens, are characteristic. IgG present in a diffuse form along the liver cell membrane, or C3, IgG, IgA and IgM focally deposited in the cytoplasm of individual hepatocytes, are also typical. In the serum of such patients hepatitis B and C antigens, and also most viral antibodies, are undetectable; however, high titre antibodies against nuclear material, smooth muscle, endoplasmic reticulum or liver ribosomes,

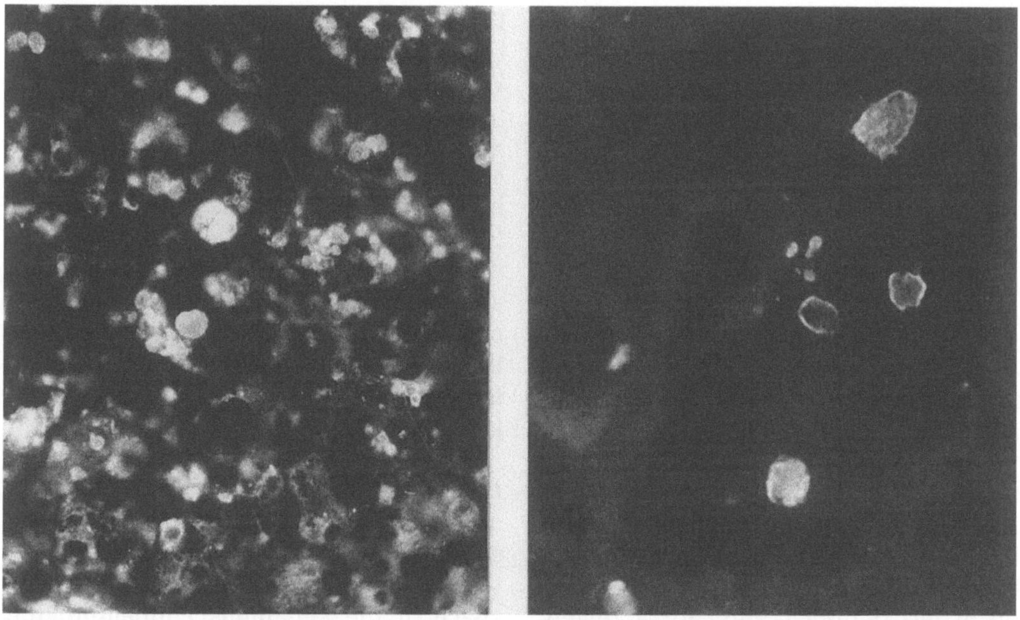

a b

Figure 39. Liver biopsy: (a) pronase pre-treatment of paraffin-embedded tissue; conjugate, anti-alpha–1-antitrypsin – predominantly granular fluorescence of the cytoplasm of hepatocytes with an alpha–1-antitrypsin deficiency, n.b., PAS-positive inclusion bodies (objective ×40, ×480); (b) unfixed tissue; conjugate, antibody against the C4 component of complement – predominantly isolated globules in PAS-negative hyaline inclusion bodies (objective ×100, ×1200).

may be. IgG antibodies against liver ribosomes are frequently deposited in the patient's own liver, i.e., these are true autoantibodies. Whether an isolated massive *linear* deposition of IgG on the liver cell membrane can readily be interpreted as an *autoimmune hepatocyte membrane hepatitis* and a *granular* deposition of IgG and HB_sAg as an *immune-complex hepatitis*, as we have postulated by analogy to glomerulonephritis (Storch [1973]), remains unclear because:

Table 14. Correlation between hepatitis B antigen deposition in the liver and associated disorders (after Gudat & Bianchi [1980], based on Storch [1986]).

Predominant type of hepatitis B Ag deposition	Most important associated disorder
1. generalised cytoplasmic HB_sAg-type (HB_cAg free)	"minimal" hepatitis, healthy carriers in chronic hepatitis B infection
2. generalised membranous, focal cytoplasmic HB_sAg-type	subacute and active chronic hepatitis
3. generalised nuclear HB_cAg-type	chronic persistent hepatitis, during effective immunosuppression
4. focal HB_sAg and HB_cAg-type	strongly active chronic aggressive hepatitis and regressing acute hepatitis
5. solitary granular HB_sAg or HB_cAg-type	waxing or waning acute hepatitis, extremely active chronic aggressive hepatitis, virus-associated autoimmune hepatitis

1. IgG and IgA are also seen in completely normal liver and only minimal changes are observed by light microscopy (Storch [1975];
2. Exact attribution to specific cell structures is rarely achieved with cryotome sections;
3. Simultaneous deposition of complement factors is not always detectable. Nowoslawski et al. [1972] found, however, complexes made up of IgG, HB Ag and complement.

More informative, therefore, are investigations on isolated hepatocytes. One must consider though that liver cell membranes possess receptors for C3 and the F_c portion of immunoglobulins (Hopf et al. [1976]), so that F(ab'$_2$) fragments should be especially used for the detection of IgG (Storch [1986]; for a review see MacSween et al. [1994]).

7.2.3 Skin biopsy

Processing of the biopsy material takes place as described in section 4.2.1. Suitable conjugates should have a molar F/P ratio of between 1.8 and 2.5 at the working dilution and the following antibody-protein concentrations (Beutner et al. [1973]): anti-IgG (unabsorbed, polyvalent), anti-C3 and anti-C4, 50–100 µg/ml; anti-IgA and anti-IgM (absorbed, H-chain specific), 100 µg/ml. With the exception of the direct methods, sections are incubated indirectly with the autologous patient's serum – at dilutions of 1:10, 1:20 and 1:40 – followed by FITC-labelled anti-human IgG. There should be sufficient sections available for possible repetitions, controls and further incubations, for example, for anti-properdin, anti-HBAg, fibrinogen, etc.

Pattern of fluorescence: the localisation and intensity of fluorescent deposits should be described, with particular attention being paid to the inter-cellular space of the squamous epithelium, the transition from epidermis to corium and the blood vessels (Figs 40–43).

Clinical significance: the greatest relevance is for the differential diagnosis of bullous dermatitis. The essential findings are collected together in Table 15 (for overview see Beutner et al. [1987a]; Ueki & Yaoita [1989]; Wojnarowska & Briggaman [1990]; Laaff et al. [1991]).

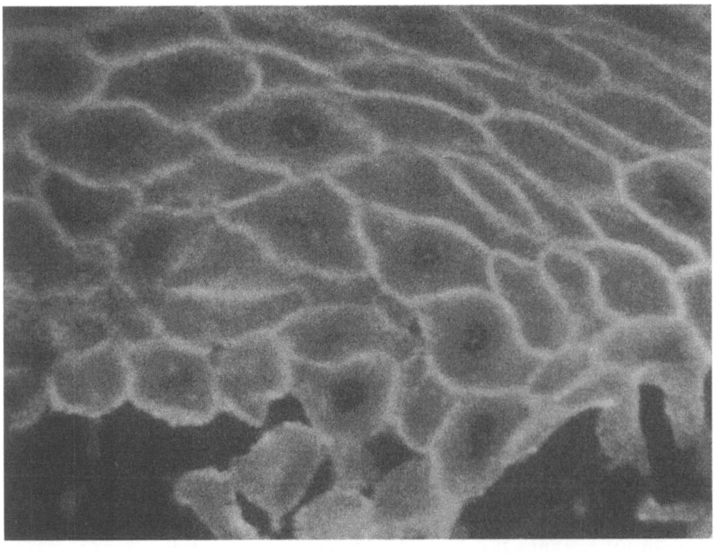

Figure 40. Skin biopsy. Chronic pemphigus (acantholytic) from the top of a blister; conjugate, anti-human IgG – fluorescence of the cell surface brought about by *in vivo* deposition of IgG antibodies (objective ×63, ×800; Dr. Barthelmes, Berlin).

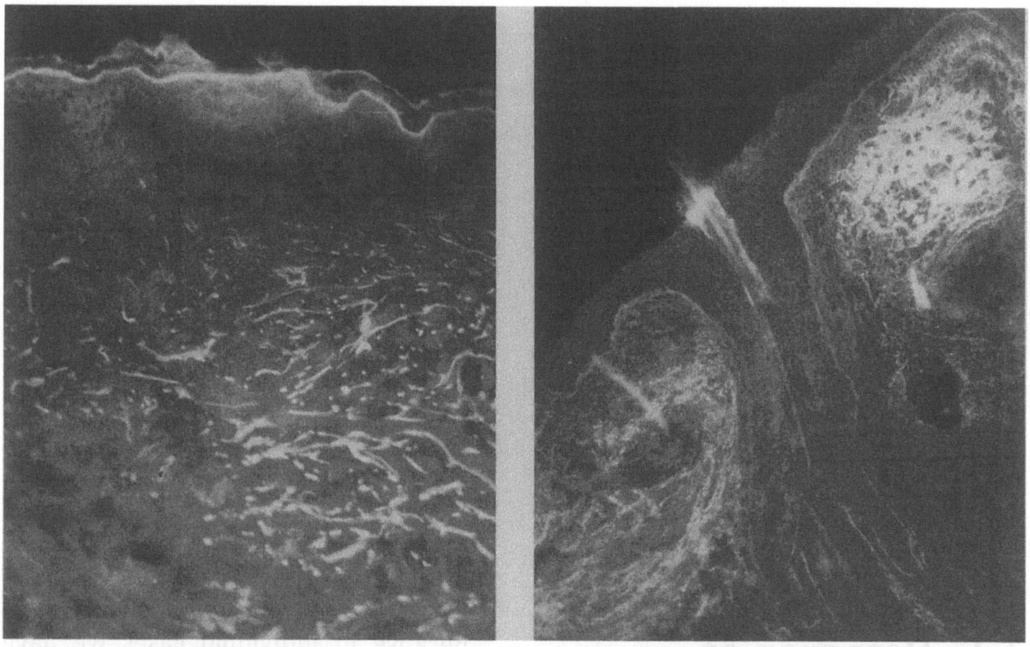

a b

Figure 41. Skin biopsy: (a) normal skin, untreated preparation, showing intense autofluorescence especially of the connective tissue; (b) chronic discoid *Lupus erythematosus* – conjugate, anti-human IgG; sub-epidermal, band-like deposition of IgG in the epidermis and, in the associated inset, collagenous degeneration of connective tissue in the *stratum papillare* (objective ×20, ×140; Dr. Barthelmes, Berlin).

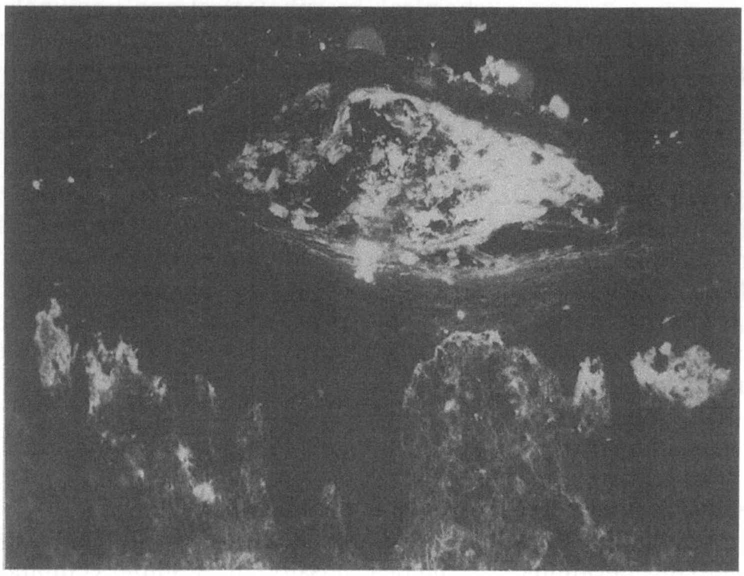

Figure 42. Duhring's herpetiform dermatitis: conjugate, anti-human fibrinogen; deposition of fibrinogen at the tips of the papillae and in an old intra-dermal blister (objective ×20, ×140; Dr. Barthelmes, Berlin).

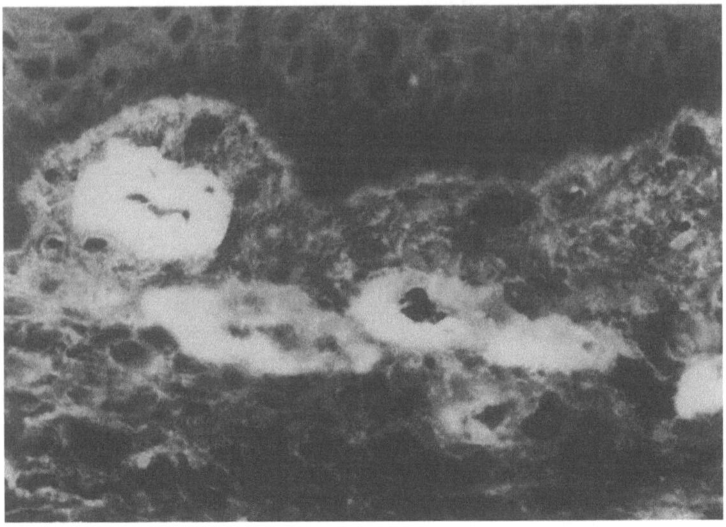

Figure 43. Skin biopsy. Allergic vasculitis of the skin: conjugate, anti-human IgG – massive deposition of IgG, and other immunoglobulins as well as complement, in the walls of the arterioles of the *stratum papillare* (objective ×40, ×300; Dr. Barthelmes, Berlin).

7.3 Detection of humoral antibodies

The sub-division of those antibodies that can be detected can be carried out from a number of points of view, for example, based on:

- the pathogenicity
- the organ, tissue and cell-specificity
- the disease-specificity
- the cell components which are likely to be target antigens, or also based on
- the clinical significance.

Their diagnostic value as indicators, markers or evidence for the presence of an immune pathology in many illnesses, has been proven. Antibodies against auto-antigens are of particular interest to clinicians; cross-reactions with allo-antigens and xeno-antigens must, however, also be recognised. Antibodies against neo-antigens are likely to achieve importance in the future since these are, in general, mostly induced by viruses (Tevethia et al. [1965]). We subdivide these antibodies based on cyto-topological considerations (Whittingham & MacKay [1977]) as shown in Table 30.

Because in individual cases we have little direct evidence as to whether the antibody detected is pathogenic, protective or of no significance to the organism, a clinically appropriate opinion can only be given if a résumé is carried out of all the retrievable details including the age of the patient as well as the clinical, biochemical, morphological and immunological findings. The following parameters should be checked when detecting antibodies (Storch et al. [1976]):

1. the type of antibody (precise differentiation of the different types);
2. the antibody titre;
3. reactions with autogenous material or that already deposited *in vivo*;
4. antibody consistently detectable – constantly or transiently detected, possible dependency on the course of the disease or on therapy, amongst others;
5. immunoglobulin class, possibly antibody sub-class;
6. complement binding;
7. simultaneous occurrence of other antibodies, possibly in a stable combination.

Table 15. Immunological findings in diseases of the skin (based on Gately & Nesbit [1994]).

Disease	Preferred biopsy location	Findings
1 Pemphigus	peri-lesional "normal" skin	Epidermis – intercellular IgG and C3
2 Para-neoplastic Pemphigus	peri-lesional skin	Epidermis – intercellular IgG, cells of the basement membrane (bmz) – linear granular C3 and IgG
3 Intradermal IgA neutrophil dermatosis	peri-lesional skin	Epidermis – intercellular IgA
4 Bullous pemphigoid	erythematous peri-lesional skin	bmz cells – linear C3, IgG
5 Scar pemphigoid	peri-lesional skin	bmz cells – linear C3, IgG
6 Acquired bullous epidermolysis	peri-lesional skin	bmz cells – linear C3, IgG
7 Linear IgA bullous dermatosis	peri-lesional skin	bmz cells – IgA, IgG
8 Herpetiform dermatitis	"normal" skin close to the lesions	Dermal papillae – IgA, C3
9 Herpes of pregnancy	peri-lesional skin	bmz cells – linear C3, occasionally IgG
10 Lichen planus pemphigoid	peri-bullous skin and lichenoid lesions	bmz cells – linear IgG, C3; IgG, IgM, IgA, C3 – cell inclusion (cytoid) bodies
11 Lupus erythematosus	lesional and normal skin (SLE)	Dermal-epidermal junction – granular IgG, IgM and IgA
12 Subacute lupus erythematosus	lesional skin	epidermal IgG (to some extent)
13 Bullous eruptions in lupus erythematosus	peri-lesional skin	bmz cells – linear IgG, IgA (50% cases); dermal-epidermal junction – IgG, IgM, IgA, C3 (50% cases)
14 Dermato-myositis	lesional skin	Dermal-epidermal junction – focal IgG, IgM, cell inclusion (cytoid) bodies; perivascular in skin/muscle – IgM
15 Scleroderma	lesional skin	Usually negative; epidermal nuclei – occasional IgG
16 Graft-hose disease	lesional skin	Dermal-epidermal junction – IgM, C3
17 Lichen planus	lesional skin	Dermal-epidermal junction – linear fibrin, IgM, IgG, IgA, C3, cell inclusion (cytoid) bodies
18 Vasculitis	early lesion or from a NaCl/histamine injection site	Perivascular – IgM, IgG, C3, fibrin
19 Schönlein-Henoch purpura	as for vasculitis	Perivascular – IgA, C3, fibrin
20 Livedo vasculitis	non-ulcerated sites	Vessel walls – IgM, C3, fibrin (superficial vessels)
21 Erythema multiformis	very early lesions	Perivascular – IgM, IgG, C3; cell inclusion bodies
22 Late cutaneous porphyria	skin exposed to sunlight	Dermal-epidermal junction and vessel walls – IgG, hyalin
23 Recurrent polychondritis	lesional skin	Chondrofibrosis transitional region – IgG, C3
24 Psoriasis	lesional skin	Stratum corneum – IgG, C3
25 Drug-induced exanthemata	lesional skin	Dermal-epidermal junction (occasionally) – IgG, IgM; IgM-containing cell inclusion (cytoid) bodies
26 Non-specific inflammation	lesional skin	Dermal-epidermal junction (occasionally) and superficially IgM, C3 and fibrin (non-specific pattern of reaction to skin injury)

bmz: basal membrane zone

Diagnostic value of antibodies. Some of the problems associated with deciding on the diagnostic value of auto-antibodies are listed in Table 16.

1. Different descriptions have been published in the past and exist in the literature for similar or identical disease scenarios.
2. The insufficient standardisation, which especially concerns "new" methods or those in a state of development, for example, the immuno-blotting procedure.
3. Difficulties in nomenclature also arise if one thinks, for example, that mitochondria also contain DNA and ribosomes.
4. Antibodies differ with regards to avidity, affinity and specificity.
5. The results obtained with a method depend on various factors; for example, with immunofluorescence using an antigen substrate (species, sex, age, weight, nutritional status, tissue treatment, etc.).
6. Knowledge of the stage of the disease, its activity and the therapy being given are often pre-requisites for being able to discriminate between clinically relevant and irrelevant findings.
7. The prevalence of an illness is rarely known. One defines *prevalence* as the *a priori* probability of the illness occurring in the population at risk; this means the frequency of the illness at a particular point in time, for example, at the time of carrying out the test. In contrast, one understands *incidence* to mean the annual number of new cases, expressed per 100,000 of the population. *Sensitivity* is defined as the proportion of correct positives (C^+) in relation to the false negatives (F^-), resulting from investigating patients definitely known to be ill.

$$\text{sensitivity} = \frac{C^+}{C^+ + F^-} \quad 0 \leq S \leq 1$$

Specificity is the proportion of negative results for people known to be healthy, thus:

$$\text{sensitivity} = \frac{C^-}{C^- + F^+} \quad 0 \leq Sp \leq 1$$

The *predictive value*, or forecast value, is a direct function of prevalence. One can distinguish a *positive predicted value* (an *a posteriori* probability of a positive test) from a *negative predicted value* (the probability of excluding disease with a negative test result). One can define these parameters in greatly simplified form, as follows:

$$\text{positive predicted value} = \frac{C^+}{C^+ + F^+}$$

$$\text{Negative predicted value} = \frac{C^-}{C^- + F^-}$$

It has proven to be a disadvantage that the value is only valid for groups with equal prevalence and that it depends on

Table 16. Problems of assessing the diagnostic value of auto-antibodies.

1. there is inadequate definition of the diseases in which auto-antibodies may be involved;
2. there is insufficient standardisation of the methods used for detecting the auto-antibodies;
3. the antibodies and antigens are heterogeneous, insufficiently characterised and therefore inadequately classified;
4. multiple antibodies are generally present;
5. the various methods used record to some extent different antibodies;
6. deciding between a clinically relevant and a clinically irrelevant result depends frequently on unknown 'factors';
7. there are problems with the statistics: the prevalence of the disease is rarely known; the prevalence, sensitivity, specificity and predictive values are not, in any case, independent of one another.

the quality of the test procedure. In the case of very rare disease patterns only an extremely low positive predictive value is obtained, in spite of higher sensitivity and specificity.

One defines the *diagnostic efficiency* as the ratio of correct laboratory results to the total number of results for the group:

$$\text{efficiency} = \frac{C^+ + C^-}{C^+ + C^- + F^+ + F^-}$$

This is also dependent on the prevalence, as well as additionally on the sensitivity and specificity of the test. Therefore, one must be more precise.

$$\begin{aligned}\text{efficiency} = {} & \text{prevalence} \times \text{sensitivity} \\ & + (1 - \text{prevalence}) \\ & \times \text{specificity}\end{aligned}$$

For an explanation of other important parameters the reader is recommended to consult Zöfel [1985] and Thomas [1992].

Subsequent investigations of the progression of individual diseases have, furthermore, yielded the result that autoantibodies may be present – for example, against mitochondria in primary biliary cirrhosis – before a diagnosis can be made by other means. This shows that antibodies can and should be used as markers even in the pre-clinical or patent stage. On the other hand, a test can be negative, especially in the rising phase, because the available antibodies do not circulate in the serum but are stored in the primed organ.

7.3.1 Antibodies against cell membrane antigens

7.3.1.1 Antibodies against liver cell membranes

Synonym/abbreviation: liver membrane antibodies (LMA) – in the first instance, a 36 kDa protein.
Target antigen: the liver cell membrane.
Properties of the antibodies: largely organ-specific, not species-specific, typical for the disease. Predominantly IgG.
Antigen substrate: isolated liver cells (see 4.3.2.1, p. 44).
Conjugate: anti-human IgG (where possible use the F(ab'$_2$) fragment).

Pattern of fluorescence: linear or granular fluorescence along the liver cell membranes (Fig. 44).
Tip: Cryotome sections yield clear results only in exceptional cases. Non-specific binding by F_c receptors must be excluded.
Clinical significance: still not clear, possibly a marker for acute inflammatory or chronic HB_sAg-negative hepatitis.
Literature: Hopf et al. [1976a, 1990]; Storch [1975b]

7.3.1.2 Antibodies against thyroid cell membranes

These are discussed fully in section 7.3.2.1.

7.3.1.3 Antibodies against membranes of pancreatic islet (Langerhans) cells

These were first demonstrated using cultured insulinoma cells (MacLaren et al. [1975]) but cannot be used successfully for routine purposes. For an overview of this topic see Scherbaum et al. [1986].

7.3.1.4 Antibodies against lymphocyte membranes

The target antigens and the properties of the antibodies have not been sufficiently investigated.
Antigen substrate: isolated lymphocytes (see p. 52).
Conjugate: anti-human IgG (if possible use the F(ab'$_2$) fragment).
Pattern of fluorescence: linear fluorescence.

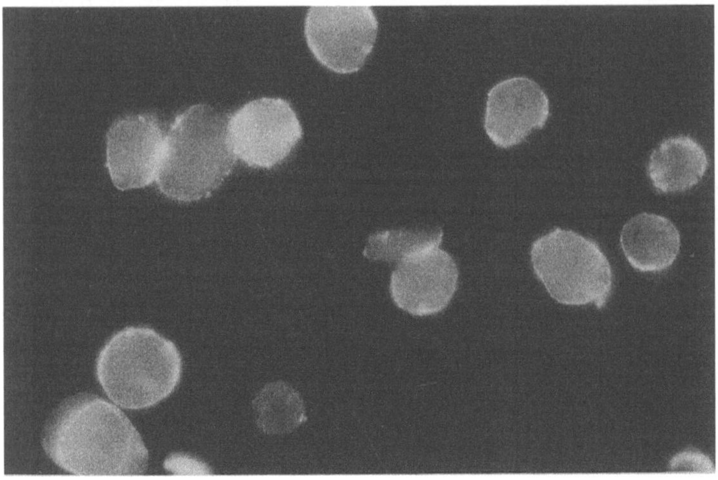

Figure 44. Antibodies against the hepatocyte cell membrane; antigen, isolated rabbit liver cells (objective ×63, ×900; Prof. Hopf, Berlin).

Clinical significance: still not clear.
Literature: Thomas [1973].

7.3.1.5 Antibodies against tumour membrane antigens

The multitude of different antibodies cannot be gone into in detail here; the literature should be referred to if required (Götz & Bücherl [1975]; Nairn et al. [1975]; Pasternak & Schneeweiss [1973]).

7.3.2 Antibodies against components of the cytoplasm

7.3.2.1 Antibodies against thyroid antigens

Other antigens, not only cytoplasmically-directed ones, will be discussed in this summary.
Target antigens:
1. Microsomes from thyroid epithelial cells (peroxidase).
2. Thyroglobulin (so-called *primary* colloid antigen).
3 So-called *secondary* colloid antigen.
4. Membranes from the thyroid epithelium.

Properties of the antibodies: the antibodies are organ-specific and generally species-specific (cross-reactions occur with primate material), not disease-specific but typical of the disease. They belong predominantly to the IgG fraction, more rarely to the IgA and IgM fractions. The microsomal antibodies bind complement and are cytotoxic towards cells in culture.
Antigen substrate: human thyroid, if possible obtained at operation for hyperthyroiditis, blood-group Od. For the detection of membrane-antibodies, a suspension of thyroid cells should be used. For the detection of peroxidase antibodies unfixed sections are used, whereas for antibodies against colloid antigen fixed sections are appropriate (three min in methanol at 56°C) – an organ 'block' together with stomach, kidney and liver is recommended.
Conjugate: anti-human IgG; F/P molar ratio, 1 to 2; antibody content at the working dilution, 50 µg/ml, possibly absorbed beforehand using liver homogenate in order to reduce non-specific staining.
Pattern of fluorescence: microsomal peroxidase antibodies give rise to a diffuse cytoplasmic fluorescence, especially of the apical portions of thyroid epithelial cells (Fig. 45); antibody against the thyro-

globulin produces a speckled or cloudy fluorescence delineating the periphery (Fig. 46); and antibody against secondary colloid Ag leads to a broad, uniformly homogeneous staining of the colloid. A coarse granular or speckled fluorescence is typical for the membrane immunofluorescence.

Tip: in order to avoid errors of interpretation with existing mitochondrial, ribosomal and other antibodies, we recommend the use of combined blocks containing rat kidney and liver. In case interpretation is not possible, an ELISA or a thyroglobulin and turkey cell haemagglutination should be carried out (this is easily done, for example, with test equipment available from Wellcome Laboratories, Beckenham, Kent ,UK).

Clinical significance: the detection of anti-thyroid antibodies has value in the diagnosis, differential diagnosis, therapy and prognosis above all for diseases of the thyroid, especially if one considers the titre and total antibody picture in the context of the clinical findings. One can at present express this as follows (Tab. 17a):

1. Membrane antibodies occur predominantly in thyroiditis and thyrotoxicosis.
2. Low titres of an antibody have little practical significance because these are detectable in many other diseases as well as in healthy patients especially in elderly women – microsomal antibodies are typical though for Grave's disease, exophthalmic goitre (Tab. 17b).
3. Moderate to high titres (over 1:80) occur primarily in Hashimoto's thyroiditis and myxoedema. They are found more rarely in nodular colloid goitre, thyroid carcinoma and granular thyroiditis.
4. The simultaneous occurrence of microsomal and thyroglobulin-directed antibodies at high titres clearly indicates an autoimmune thyroiditis. Colloid goitre, thyroid carcinoma and subacute thyroiditis can be excluded to a large extent.
5. Hyperthyroidism with a high antibody titre should not go to operation because of the danger of a post-operative myxoedema developing but rather it should be treated with anti-thyroid drugs. It can also simply be a multifocal thyroiditis, which need not be progressive and in which the titre drops during the healing phase.
6. A diffuse goitre with a antibody high titre should arouse suspicion of a malignant tumour.

Literature: White [1957]; Balfour et al. [1961]; Scherbaum & Paschke [1995].

Table 17a. Frequency of antibodies against thyroid tissue in % – modified based on Scherbaum & Paschke [1995].

Disease	Microsomal Antibody (MAB)	Thyroglobulin antibody	TSH receptor antibody
Hashimoto's thyroiditis	80	35	rare
Grave's disease exophthalmic goitre	65	20	~90
Primary myxoedema	60	25	15
Post-partum thyroiditis	15	10	rare
Healthy patients	15	10	negative

TSH: Thyroid Stimulating Hormone

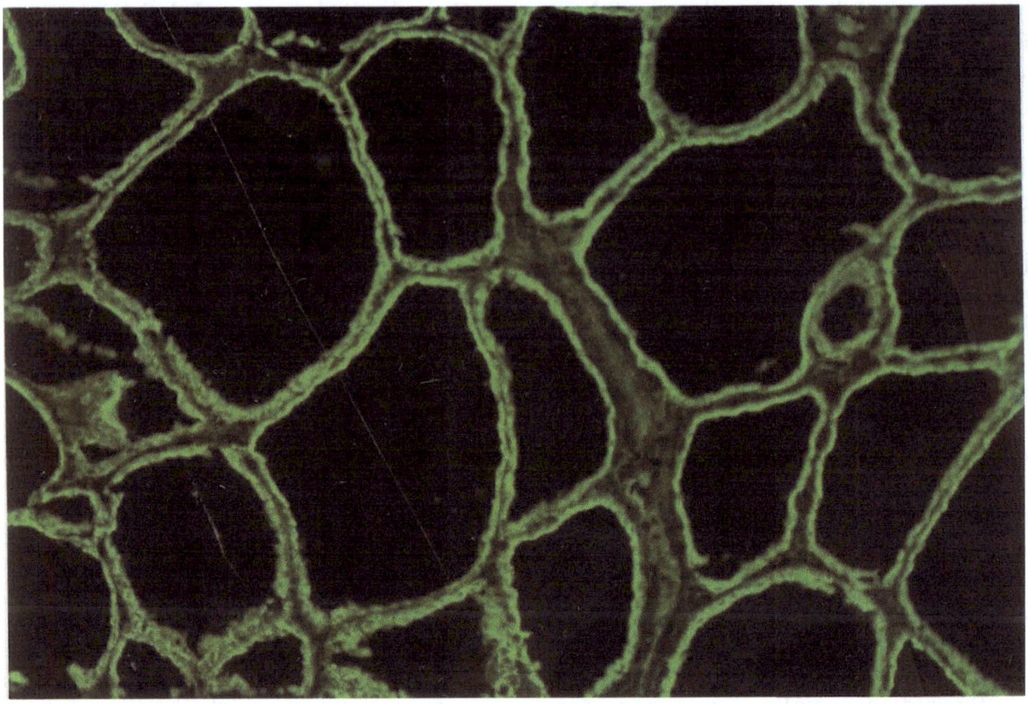

Figure 45. Antibodies against microsomes (peroxidase) from thyroid cells; antigen, monkey thyroid – cytoplasmic fluorescence especially of the apical portions of the epithelial cells (objective ×20, ×330).

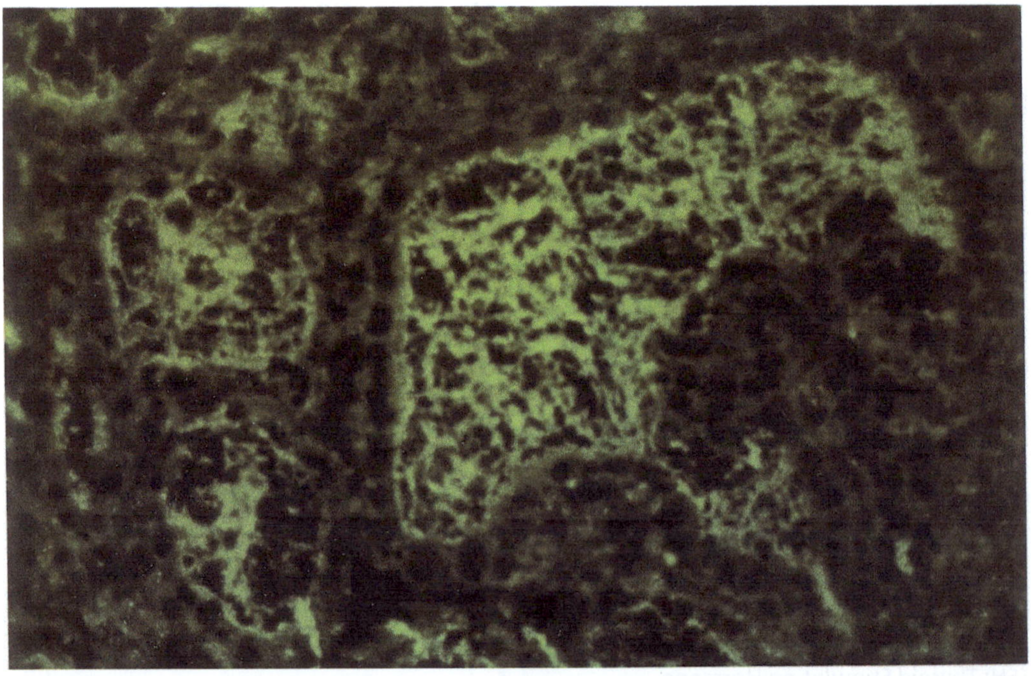

Figure 46. Antibodies against the colloid (thyroglobulin) in thyroid cells; antigen, monkey thyroid (objective ×40, ×500).

Table 17b. Frequency of antibodies against parietal cells and thyroid gland in % (modified based on Chanarin [1975]).

Diagnosis	Parietal cell antibodies	Thyroid antibodies
Controls	0–16	0–15
Healthy patients < 60 yrs.	5–8	3–10
Healthy patients > 60 yrs.	12–20	8
Pernicious anaemia (PA)	84	55
Relatives of patients with pernicious anaemia	36	50
Atrophic gastritis and atrophy of the stomach	10–50	–
Myxoedema	32	87
Hyperthyroiditis	28	53
Relatives of patients with thyroid disease	20	46
Addison's disease	23	31
Insulin-dependent Diabetes mellitus	18	14

7.3.2.2 Antibodies against cells of the stomach

7.3.2.2.1 Antibodies against parietal cells of the stomach

Synonyms/abbreviations: parietal cell antibodies (PCA), acid cell antibodies, border cell antibodies, gastric parietal cell antibodies (GPCA).

Target antigen(s): Ba fragment of the α-subunit of the H^+/K^+-ATPase (the proton pump) – Song et al. [1994]; Karlsson et al. [1988].

Properties of the antibodies: organ-specific, not species-specific – cross react with parietal cells from rat, monkey, cat, mouse, guinea pig and dog – predominantly IgG but also IgA in gastric secretion, binds complement, titre rarely over 1:160.

Antigen substrate: fundus of the stomach from patients who have had elective operations for duodenal ulcer, if possible blood-group Od, mouse stomach.

Conjugate: anti-human IgG.

Pattern of fluorescence: typically this is a fine, granular fluorescence exclusively of the cytoplasm of the parietal cells (Fig. 47).

Tip: in order to avoid confusion with mitochondrial antibodies, a comparison with kidney sections is advantageous. Brush-border antibodies can only be interpreted using rat stomach (see Tab. 19). If the simultaneous occurrence of mitochondrial and parietal cell antibodies is suspected, then absorption with mitochondria should be carried out.

Clinical significance: the detection of parietal cell antibodies makes a distinction between relevant differential diagnoses possible, such as between pernicious anaemia and non-immune megaloblastic anaemia. The detection of these antibodies can also be useful in genetic investigations. The increase seen with age is significant (see Tab. 17b).

Literature: Taylor et al. [1962]; review by Chanarin [1975].

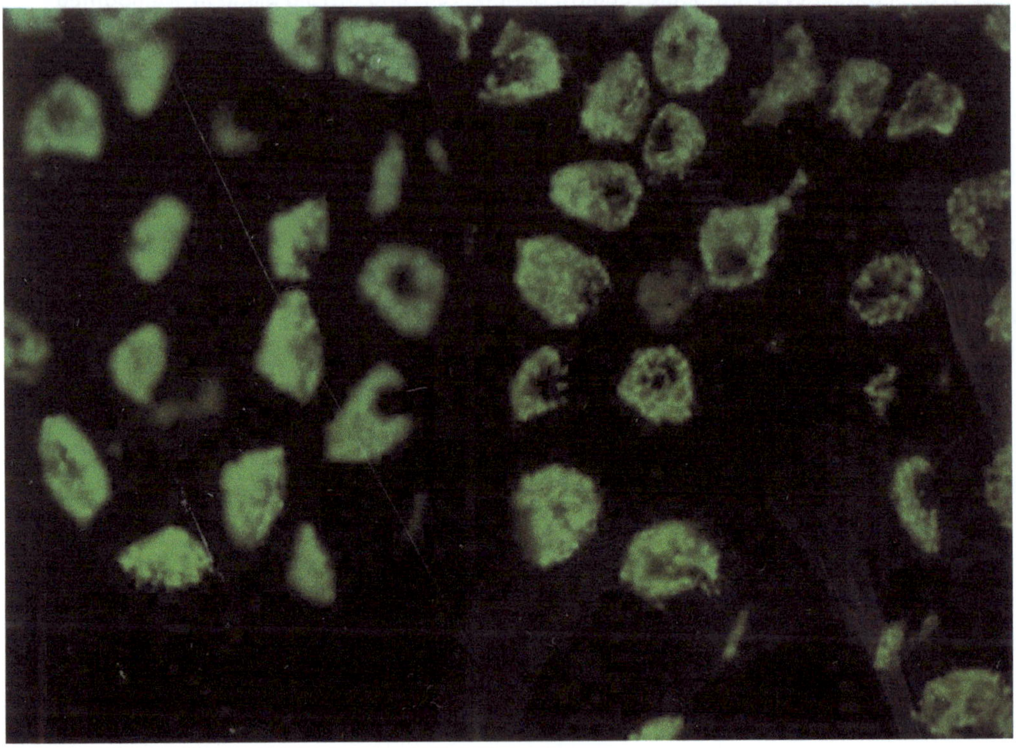

Figure 47. Antibodies against stomach parietal cells; antigen, mouse stomach (longitudinal through glandular crypts) – cytoplasmic fluorescence of the parietal cells. The *submucosa* and *muscularis* are negative (objective ×40, ×500).

7.3.2.2.2 Antibodies against gastrin-producing cells of the stomach

Target antigen(s): the cytoplasm of gastrin-producing cells of the stomach.

Properties of the antibodies: IgG, binds complement.

Antigen substrate: mucosa of the antrum from human stomach, blood-group O.

Conjugate: anti-human IgG, double immunofluorescence with antibody against gastrin.

Pattern of fluorescence: cytoplasmic fluorescence of gastrin-producing cells in the antrum.

Tip: the suitability of the antigen substrate must first of all be tested by screening for the presence of gastrin-producing cells using incubation with antibody against gastrin alone.

Clinical significance: marker for a rare autoimmune form of chronic antral gastritis (type B).

Literature: Vandelilet al. [1979].

7.3.2.2.3 Antibodies against the 'chief' cells of the stomach

At least two different antibodies must be distinguished, an antibody with broad reactivity towards ribosomes (see 7.3.2.19) and an antibody which only reacts with the chief cells, especially those in the region of the lumen (Fig. 48). These antibodies have to date not be described in the literature. They are rare, occurring together with antibody against the brush-border. Some antibodies react, in addition, with individual perinuclear granules in liver cells.

Figure 48. Antibodies against the chief cells of the stomach; antigen, mouse stomach (longitudinal section through the glandular crypts) – granular cytoplasmic fluorescence especially of the apical portions of the chief cells (objective ×40, ×500).

7.3.2.3 Antibodies against pancreas

7.3.2.3.1 Antibodies against the pancreatic islet cells (Langerhans)

Synonyms/abbreviations: islet cell antibodies (ICA).

Target antigen(s): predominantly the glutamine decarboxylase II (Baekkeskov et al. [1989]).

Properties of the antibodies: broadly organ-specific but not species-specific and not disease specific, although typical of the disease. They are mainly of the IgG type and frequently bind complement.

Antigen substrate: fresh human pancreatic tissue, if possible blood-group Od.

Conjugate: anti-human IgG.

Pattern of fluorescence: cytoplasm of almost all islet cells, mostly staining with different intensities (Fig. 49). The typical pattern of this so-called "non-restricted" ICA, because it also reacts with mouse tissue, must be distinguished from the so-called "restricted" ICA. These react only with the ß-cells of human pancreas. Furthermore, it must be distinguished from atypical patterns which can be produced, for example, by antibodies against glucagon, somatostatin and pancreatic polypeptide-producing cells (Bottazzo & Lendrum [1976]; Larsson [1980]; Ravazzola & Orci [1980]; Winter et al. [1984]).

Tip: on account of the mostly low titres (1:2 to 1:40), special value should be placed on suitable conjugates and antigen substrates. Conversion to "Juvenile Diabetes Foundation" units (JDFU) is possible using a reference serum. The reference serum can be obtained either from Dr. Riley or Dr. MacLaren, Box J–275 JHMHC, Gainesville, Florida 32610, USA. They are also the coordinators for the quality controls, which are carried out at approximately yearly intervals.

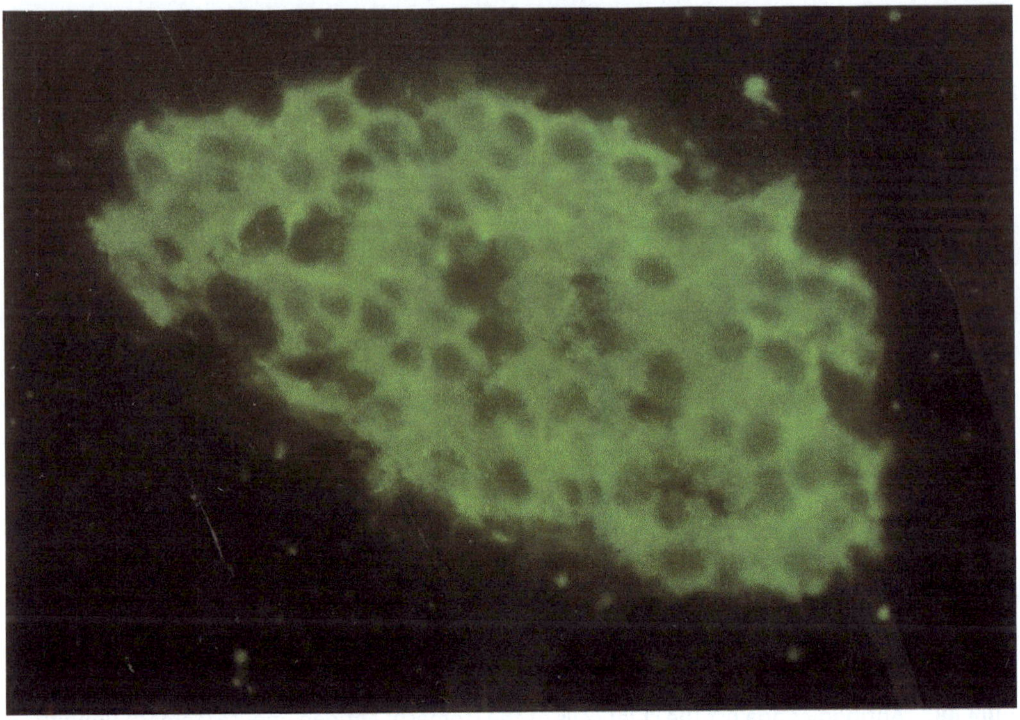

Figure 49. Antibodies directed against pancreatic islet cells; antigen, monkey pancreas (objective ×40, ×500).

Literature: Bottazzo et al. [1974, 1980]; MacCuish et al. [1974]; Bingley et al. [1994].

7.3.2.3.2 Antibodies against pancreatic acinus cells

Target antigen(s): microsomal antigen in the cytoplasm of pancreatic acinus cells.

Properties of the antibodies: they appear to be broadly organ- and species-specific as well as typical of the disease. IgG and IgA predominate; binding of complement is rare.

Antigen substrate: where possible fresh pancreatic tissue from a child, blood-group Od. Initial dilution of the patient's serum, 1:5.

Conjugate: anti-human IgG, IgA and IgM.

Pattern of fluorescence: coarse granular fluorescence of the cytoplasm of the pancreatic acinus cells (Fig. 50).

Clinical significance: their status as a marker for a special type of acute pancreatitis, occurring in 1/3 of all cases where the antibody was found, remains unclear and has not been confirmed. Since the antibody is frequently also seen in healthy patients, the proposed association with Crohn's disease is questionable.

Literature: Lendrum & Walker [1975]; Stöcker et al. [1984].

Figure 50. Antibodies directed against pancreatic acinar cells; antigen, monkey pancreas – granular fluorescence especially of the apical regions of the acinar cells, with the islet cells being negative (objective ×40, ×500).

7.3.2.4 Antibodies against steroid-producing cells in the adrenal, testis, ovary and placenta

Target antigen(s): cytoplasm – predominantly the microsomes but also the mitochondrial fraction of steroid-producing cells, primarily from the adrenal *Zona glomerulosa*, or the testis, ovary and placenta.
Properties of the antibodies: not organ-specific, presumably species-specific, bind complement to some extent.
Antigen substrate: fresh organs; where possible the adrenals should be from patients with Cushing's syndrome.
Conjugate: anti-human IgG.
Pattern of fluorescence: cytoplasmic fluorescence of hormonally active cells. Undiluted titre up to 1:64, mostly 1:4 to 1:8 – see Table 18 and Figure 51.
Clinical significance: a positive result is of greatest significance in the differential diagnosis of a tubercular adrenal insufficiency. The simultaneous occurrence of a thyroiditis, primary myxoedema, pernicious anaemia, atrophic gastritis or diabetes mellitus, is conspicuous. The significance of the antibody for the diagnosis of premature ovarian insufficiency and infertility remains, as previously, unclear.
Literature: Blizzard et al. [1962, 1967]; Anderson et al. [1968]; Sotsiou et al. [1980]; Betterle et al. [1983]; Belvisi et al. [1993]; for a review see Gleicher [1994].

7.3.2.5 Antibodies against cells of the parathyroid

Target antigen(s): oxyphil and principal cells of the parathyroid.

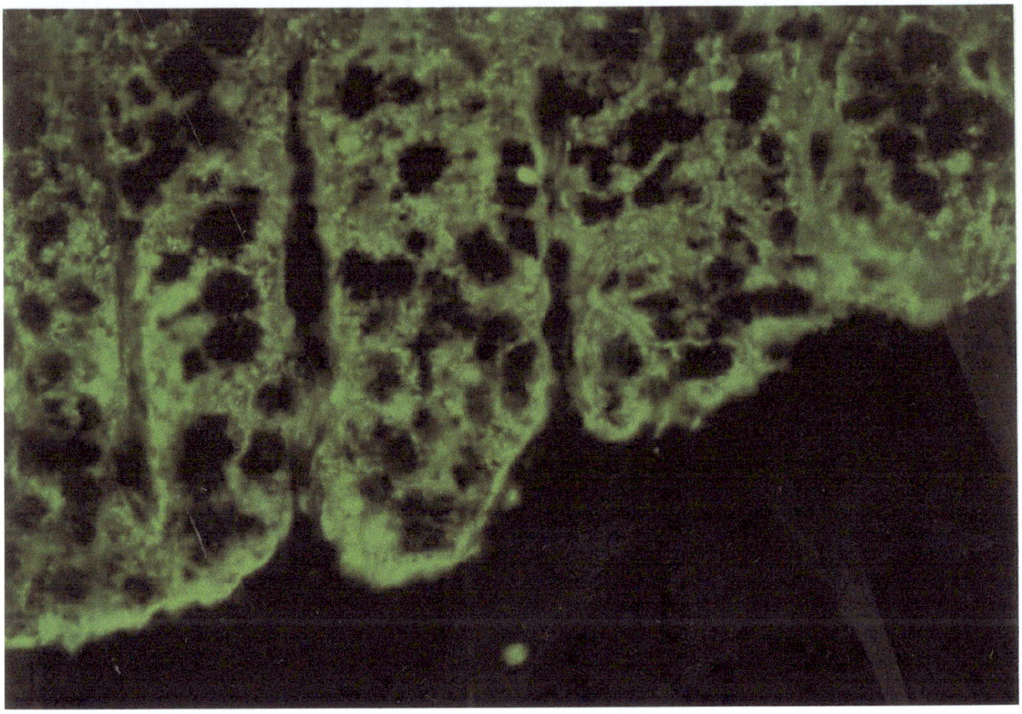

Figure 51. Antibodies against cells of the adrenal cortex; antigen, monkey adrenal – cytoplasmic fluorescence of the cells of the adrenal cortex, with the medullary cells (lower part) being negative (objective ×40, ×500).

Table 18. Frequency of antibodies against the steroid-producing cells of the adrenals, the gonads and the placenta (based on Irvine & Barnes [1975]).

Diagnosis	Total number of sera	Number of cases with antibodies against steroid-producing cells	
		Adrenals	Ovary, testis or placenta
Idiopathic Addison's (female)	175	127	36
Idiopathic Addison's (male)	71	34	3
Idiopathic hypoparathyroidism	20	6	–
Tuberculous adrenal insufficiency	62	0	–
Diabetes mellitus	93	1	–
Pernicious anaemia	102	0	–
Thyrotoxicosis	50	0	–

Properties of the antibodies: IgG, broadly organ-specific and species-specific but not disease-specific – do not cross-react with adrenal, thyroid, stomach, kidney or liver; bind complement to some extent at low titres.
Antigen substrate: fresh normal parathyroid glands.
Conjugate: anti-human IgG.
Pattern of fluorescence: cytoplasmic fluorescence of oxyphil and principal cells.
Occurrence: Blizzard et al. [1966] found the following frequencies – (i) 38% in idiopathic hypo-parathyroidism; (ii) 26% in idiopathic Addison's disease; (iii) 12% in Hashimoto's thyroiditis; and (iv) 6% in control patients.
Clinical significance: not clear because other authors have found antibodies less frequently, or not all.
Literature: Blizzard et al. [1966]; for a review see Irvine & Barnes [1970].

7.3.2.6 Antibodies against cells of the hypophysis

Target antigen(s): cytoplasm of cells producing different hormones, for example, prolactin, ACTH, somatotropin, luteinising hormone (LH) or follicle-stimulating hormone (FSH).
Properties of the antibodies: IgG, low titre.
Antigen substrate: human foetal hypophysis, since this does not contain F_c-receptors on, for example, ACTH-producing cells.
Conjugate: anti-human IgG; double immunofluorescence with specific anti-hormone sera.
Pattern of fluorescence: cytoplasmic fluorescence of hypophyseal cells which produce, for example, ACTH (Fig. 52).
Clinical significance: not clear, possibly a marker for an autoimmune variant, for example, of idiopathic pan-hypopituitar-

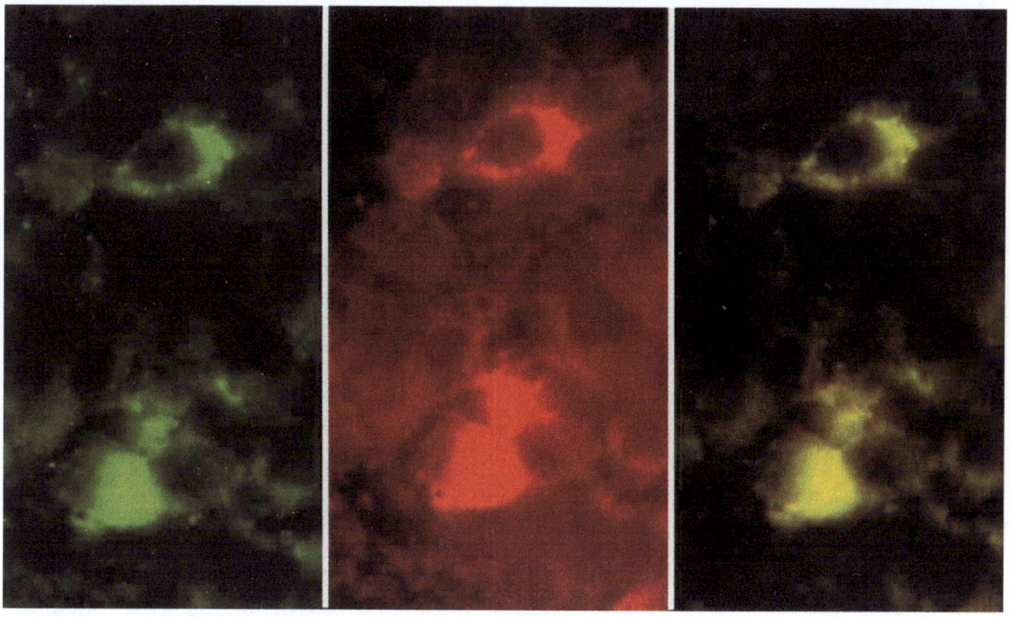

a b c

Figure 52. Antibodies against ACTH-producing cells in the hypophysis – cytoplasmic fluorescence of individual cells by double immunofluorescence; (a) incubated with human serum which contains the antibodies, detected as usual using FITC-labelled anti-human IgG; (b) the binding of anti-bodies directed against ACTH made visible using a TRITC-labelled antiserum; (c) on double illumination (excitation of both FITC and TRITC) the yellow colour points to an identical localisation (objective ×40, ×500; Dr. Glück, Ulm).

ism, Cushing's syndrome or poly-endocrinopathy.
Literature: Engelberth & Jezkova [1965]; Nerup et al. [1969]; Bottazzo et al. [1975]; Ludwig & Schernthaner [1978]; Bottazzo et al. [1980]; Pouplard-Barthelaix et al. [1984]; Scherbaum et al. [1987]; Glück et al. [1993]; for a review see Mauerhoff et al. [1987].

7.3.2.7 Antibodies against cells of the hypothalamus

Target antigen(s): cytoplasm of vasopressin-producing cells.
Properties of the antibodies: mainly IgG and IgA but also IgM; partial binding of complement: low titre, maximally 1:32.
Antigen substrate: best is human foetal hypothalamus, for example, obtained as a result of a therapeutic hysterectomy.
Conjugate: mainly anti-human IgG, double immunofluorescence with anti-vasopressin antisera.
Pattern of fluorescence: cytoplasmic fluorescence of vasopressin-producing cells of the hypothalamus.
Occurrence: in around 1/3 of cases of idiopathic diabetes insipidus and in 6% of symptomatic cases.
Clinical significance: not clear, but possibly a marker for an autoimmune variant of diabetes insipidus.
Literature: Scherbaum & Bottazzo [1983].

7.3.2.8 Antibodies against cells of the salivary gland excretory ducts

Synonyms/abbreviations: salivary gland antibodies.
Target antigen(s): cytoplasm of the epithelial cells of the excretory ducts of

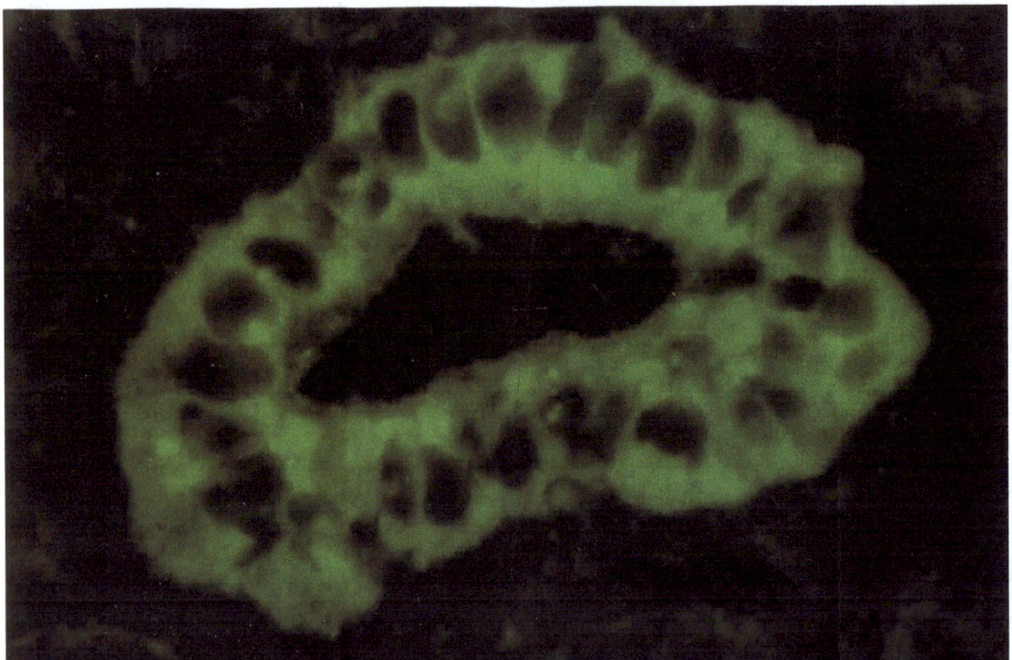

Figure 53. Antibodies directed against the efferent ducts of the salivary glands; antigen, monkey parotid – cytoplasmic fluorescence of the epithelial cells lining the ducts (objective ×100, ×900).

the salivary glands, especially those of the parotid.
Properties of the antibodies: IgG, broadly organ-specific, not species-specific, not disease-specific.
Antigen substrate: fresh salivary gland.
Conjugate: anti-human IgG.
Pattern of fluorescence: cytoplasmic fluorescence of the epithelial cells lining the excretory ducts, only associated with the glandular epithelium (Fig. 53).
Tip: there is a range of atypical patterns of fluorescence, one of which is an uncommon fluorescence of the basal cells (Nozicka [1989]).
Occurrence: 60% approx. in Sjögren's disease, 25% approx. in SLE (systemic lupus erythematosus); 20% approx. in rheumatoid arthritis; 5% in healthy patients under 60 years old, 10% in those over 60 years old.
Clinical significance: relatively little; however, makes the differential diagnosis of parotid tumours easier.
Literature: Bertram & Halberg [1964].

7.3.2.9 Antibodies against kidney antigens

Numerous antibodies are known. Antibodies, for example, against mitochondria, endoplasmatic reticulum (liver and kidney microsomes) and ribosomes, which also react with kidney tissue, as well as antibodies against the basement membrane of the glomerulus and tubules, are dealt with separately. Reference should be made to the appropriate section. In order to make orientation easier, the customary nomenclature and subdivision of kidney sections (Fig. 54), as well as the characteristic features of proximal tubules (Fig. 55), need to be explained. If a microscope with the facility for contrast illumination is not available, then orientation must be achieved using autofluorescence. For this purpose broad-band excitation, without restriction of the fluorescence wavelengths produced, is recommended.

7.3.2.9.1 Antibodies against podocytes (APA)

Target antigen(s): cytoplasm of podocytes.
Properties of the antibodies: IgG, organ-specific, not species-specific, not disease-specific.
Antigen substrate: block containing rat kidney, stomach and liver (also human material).
Conjugate: anti-human IgG
Pattern of fluorescence: coarse granular or speckled fluorescence exclusively of

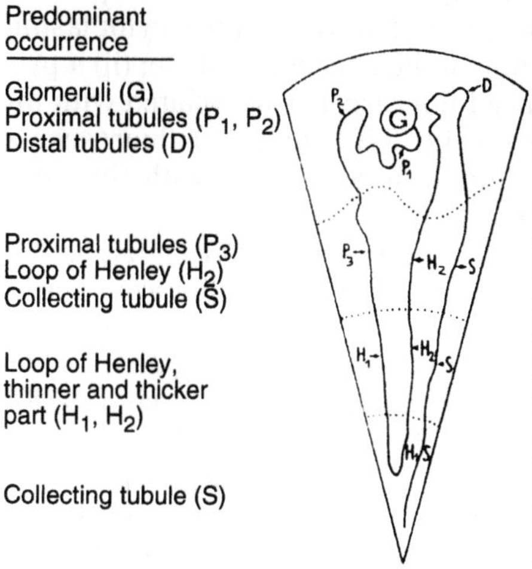

Nomenclature

after Maunsbach, Beard and Novikoff	after Sternberg
cortex	surface cortex
outer medulla	deep cortex
inner medulla	surface medulla
papillae	deep medulla

Figure 54. Nomenclature and sub-division of kidney sections (according to Homberg et al. [1974]).

Schematic representation of the proximal tubules	Predominant occurrence of the tubules	Size of the granules	Number of granules
1. Segment P1  brush-border of irregular width; few apical vacuoles; continuous transition to -	directly connected to the glomerulus	small	many
2. Segment P2  brush-border irregularly narrower than for P1; many apical vacuoles; continuous transition to -	superficial cortex	medium to large	very many
3. Segment P3 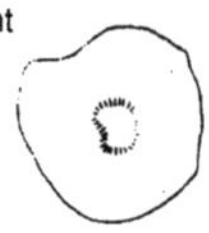 brush-border regular; almost no apical vacuoles; many fewer mitochondria as for P_1 or P_2	Cortico-medullary border	very small	few

Figure 55. Differentiation of the proximal tubules of the kidney in male rats, based on the number and size of the autofluorescent granules.

the glomeruli, the so-called podocyte pattern (Fig. 56).
Tip: this pattern also turns up in the presence of cold agglutinins, especially Pr1 and PRM.
Occurrence: a relationship exists primarily for diseases of the kidney, liver and those of the 'rheumatoid' type, as indicated by the few existing known cases.
Clinical significance: not clear.
Literature: Storch [1995].

7.3.2.9.2 Antibodies against cells of Henle's loop in the kidney

Target antigen(s): cytoplasm of the epithelial cells lining Henle's loop.
Properties of the antibodies: IgG.
Antigen substrate: rat kidney.
Pattern of fluorescence: cytoplasmic fluorescence (Fig. 57a).
Clinical significance: not clear, occur naturally in healthy patients, also in chronic liver disease.
Literature: Ford [1973]; Storch [1973, 1975, 1978]; Chanarin et al. [1974].

7.3.2.9.3 Antibodies against the collecting tubules

Target antigen(s): unknown, but primarily the junctional cells of the collecting tubule.
Properties of the antibodies: predominantly IgG, in 40% of cases IgM and in ~17% of cases IgA; generally organ-specific, not species-specific, bind complement, not disease-specific.
Antigen substrate: block containing rat or mouse kidney, stomach and liver.
Conjugate: anti-human IgG.
Pattern of fluorescence: three types can be distinguished; (i) a diffuse cytoplasmic staining of the collecting tubule; (ii) a pronounced fluorescence associated with the lumen (Fig. 57b); and (iii) fluorescence predominantly associated with the junctional cells (Fig. 58).
Tip: the fluorescence picture must be distinguished from those which are caused, for example, by antibodies against the cells of Henle's loop and against individual distal tubules (see Storch [1978]). Ford [1973] has presumably depicted fluorescence of the junctional cells, although the picture was interpreted as antibodies against the loop of Henle.

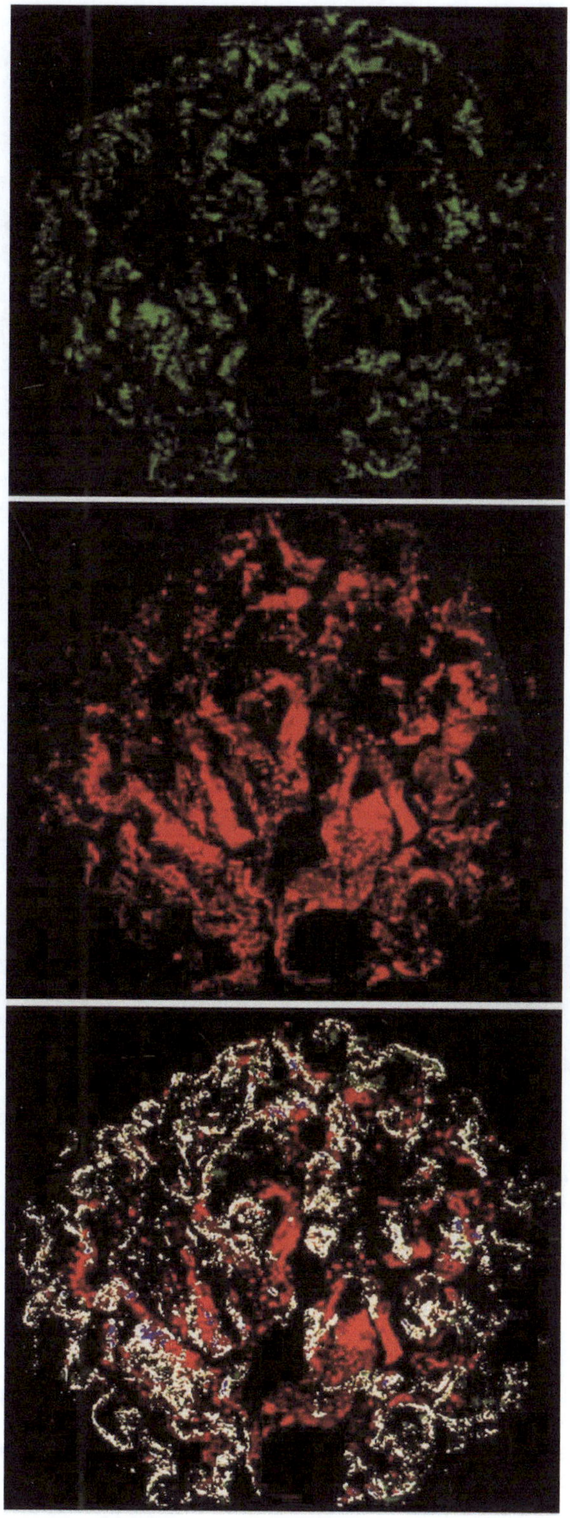

a

b

c

Figure 56. Antibodies directed against podocytes; antigen, rat kidney – using double immunofluorescence:

(i) first incubation with a mixture containing the patient's serum diluted 1:40 and monoclonal antibodies against the 44 kDa protein from podocytes, diluted 1:20;

(ii) second incubation with a mixture of FITC-labelled anti-human IgG (Viramed, München) at its working dilution, and Cy 3-labelled anti-mouse immunoglobulin (Dianova, Hamburg) diluted 1:40. Photomicrographic recording from the monitor of a Leitz TCS (Wetzlar) confocal laser-scanning fluorescence microscope, with excitation using an argon-krypton laser at 488 nm and 568 nm, a double dichroic filter 488/568 nm, a blocking filter for FITC-fluorescence BP 535 nm, and a blocking filter for Cy 3-fluorescence LP 590 nm: (a) excitation at 488 nm – the human antibody appears in green; (b) excitation at 568 nm – the monoclonal antibody against the 44 kDa protein from podocytes appears in red; (c) excitation at both 488 and 568 nm – simultaneous demonstration of the human antibody and the monoclonal antibody against the podocyte 44 kDa protein, with the white colour indicating the broadly equivalent localisation for both antibodies (zoom objective ×40, ×800; based on Storch [1995]).

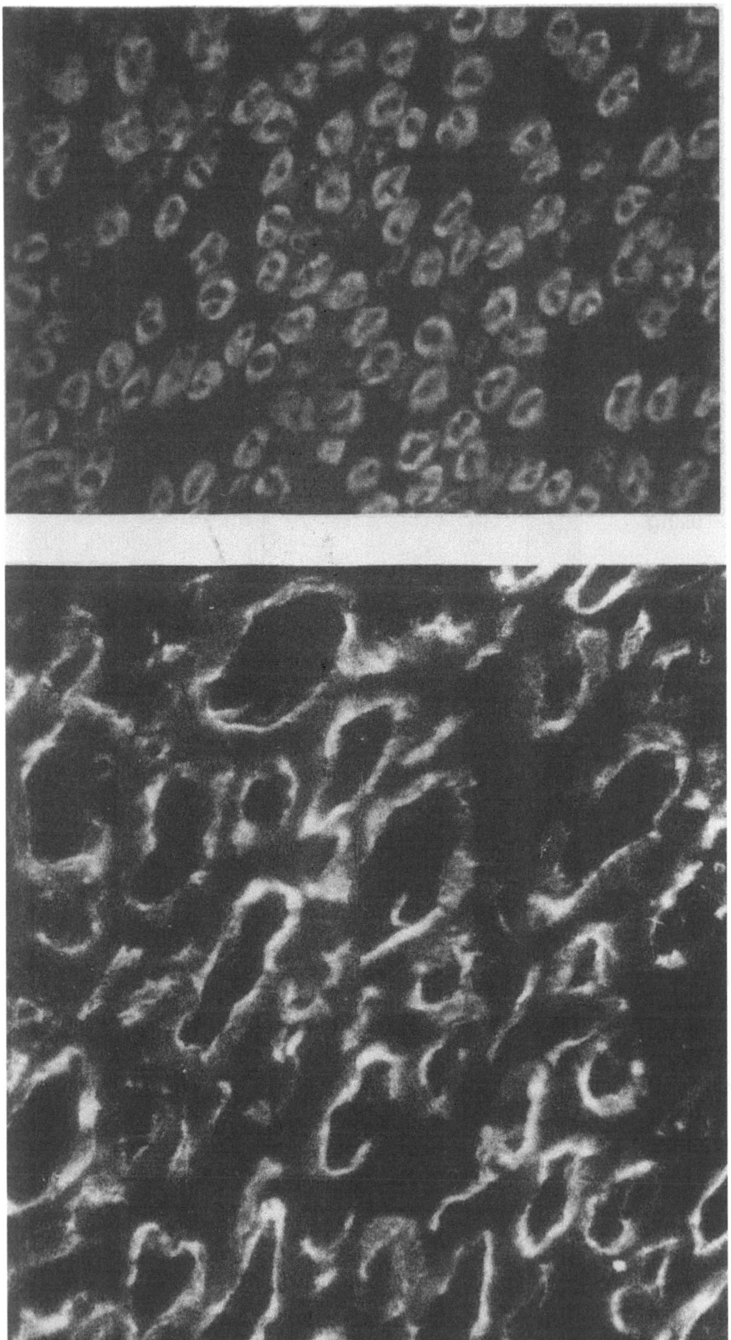

a

b

Figure 57. Antibodies directed against (a) Henle's loop and (b) the collecting tubules; antigen, rat renal medulla – (i) cytoplasmic fluorescence of the cells of Henle's loop and also of the small vessels (objective ×16, ×196), and (ii) fluorescence of the collecting tubules with a preference for segments close to the lumen (objective ×40, ×480).

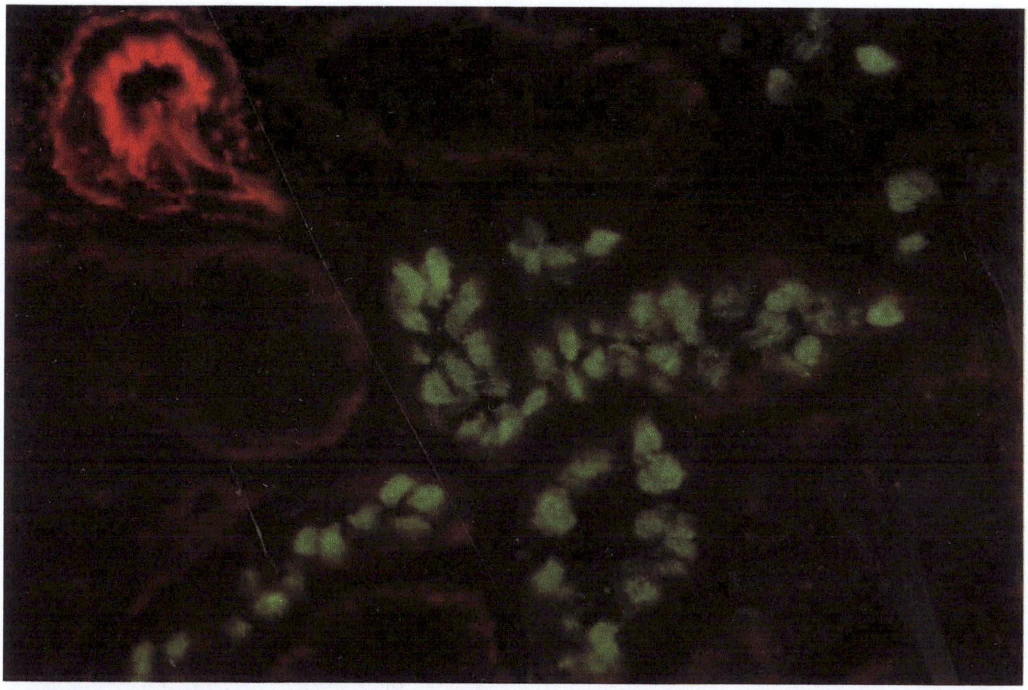

Figure 58. Antibodies directed against junctional cells of the collecting tubules; antigen, rat renal cortex – cytoplasmic fluorescence of the junctional cells of the cortical collecting tubules, with the glomeruli, proximal and remaining distal tubules, as well as blood vessels, all negative (objective ×40, ×500).

Occurrence: Gaarder & Heier [1983] found the junctional cell type of fluorescence, at a titre of 1:8 or higher, in 113 out of 45,000 sera (~0.25%); Storch (unpublished) has found six positive results out of 9,000 sera (~0.07%) with titres of at least 1:20. Forbes et al. [1976] saw fluorescence especially in sections from malignant tumours of liver and kidney. Gaarder & Heier observed these antibodies primarily in thyroid disease and occasionally in renal tubular acidosis. This relationship has been described again more recently by Konishi et al. [1994].
Clinical significance: not clear.
Literature: Ford [1973]; Storch [1975].

7.3.2.9.4 Antibodies against the brush-border of proximal tubules of the kidney (and against parietal cells) in rats

Synonyms/abbreviations: "brush-border" antibodies (BBA).

Target antigen(s): unknown, occurs simultaneously in the brush-border of rat kidney proximal tubules and in the cytoplasm of parietal cells – cross-reacts with blood-group substances?
Properties of the antibodies: IgG, generally organ-specific and species-specific.
Pattern of fluorescence: fluorescence of the brush border and, at high titres of antibody, also of the basement membrane of rat kidney tubules, as well as a fine granular fluorescence of the parietal cells. A special form reacts only very weakly with parietal cells (Fig. 59).
Tip: in order to avoid confusion between mitochondrial and parietal antibodies, human or mouse tissue should be examined at the same time (see Tab. 19).
Clinical significance: not clear; hetero-antibodies occur naturally. It appears, however, to be laid down in autoimmune diseases. One subtype may have a relationship to liver disease.

Table 19. Differentiation of antibodies which react with parietal cells.

Antibody	Fluorescence of parietal cells			Fluorescence after absorption with rat liver mitochondiria	Fluorscence of brush-border proximal renal tubules or stomach		
	Man	Rat	Mouse		Man	Rat	Mouse
"customary" parietal cell antibodies (PCA) Perniziosa	+	+	+	+	∅	∅	∅
PCA-hetero-antibodies (brush-border Ab)	∅	+	∅	+	∅	+	∅
mitochondrial antibodies	+	+	+	∅	∅	∅	∅

∅ = not detected.

Literature: Ireton et al. [1971]; Muller et al. [1971]; Storch [1974]; Strickland & Hooper [1972].

7.3.2.9.5 Antibodies against various proximal and distal renal tubules

Target antigen(s): cytoplasm of proximal and distal kidney tubules.

Properties of the antibodies: predominantly IgG, organ- but not species- or disease-specific.

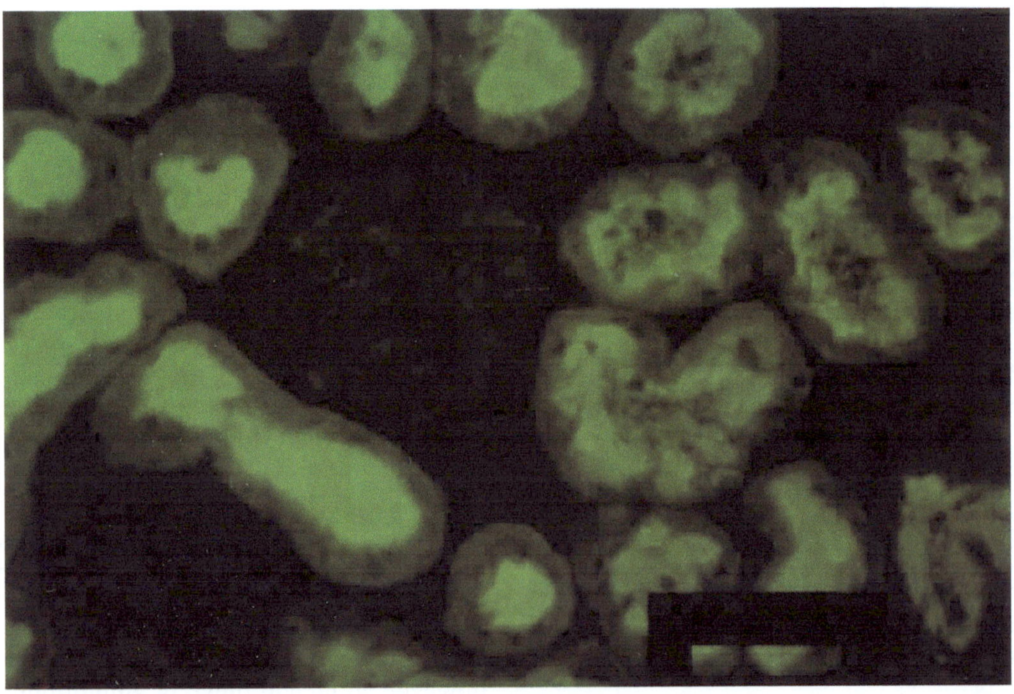

Figure 59. Antibodies against the brush-border of renal proximal tubules in the rat; antigen, rat renal cortex (objective ×40, ×330).

Antigen substrate: rat or mouse kidney.
Conjugate: anti-human IgG; F/P molar ratio, 1.5.
Pattern of fluorescence: different types, for example, antibodies which only react with the primary, secondary or tertiary proximal tubules, or with all of them; there are, furthermore, antibodies which only react with individual distal tubules.
Clinical significance: not clear; blood-group antibodies, atypical mitochondrial antibodies.
Literature: Homberg, personal communication [1975]; Storch [1975] and unpublished observations.

7.3.2.10 Antibodies against cells of the gut

Numerous antibodies have been described in the past, especially in inflammatory diseases of the gut (for a review, see Elson [1989]). These can and ought not to be discussed here, especially as their clinical significance is judged to be variable. Antibodies against goblet cells have achieved a 'renaissance' with the involvement of foetal tissue. The relevance of antibodies against hormone-producing cells, such as those producing the duodenal gastric inhibitory peptide and secretin (Mirakian et al. [1980]), and of antibodies against epithelial cells, for example, in pediatric diarrhoea (McCarthy et al. [1978]; Unsworth et al. [1982]; Martin-Villa et al. [1991]), is not certain. The deposition of antibodies against a 40 kDa protein in combination with complement situated on the apical and basolateral cell membrane of colonic cells, has been emphasised recently as a characteristic finding for ulcerative colitis (Halstensen et al. [1993]).

Antibodies (Ab) against enterocytes
Target antigen: many but especially a 75 kDa protein (Kobayashi et al., 1999)
Properties of the antibodies: in the first instance IgG, less frequently IgA and IgM, to some extent complement fixing, broadly organ-specific, not species-specific.
Antigenic substrate: depending on availability either human or monkey intestine, primarily duodenum or jejunum.
Conjugate: anti-human-IgG: also anti-human-IgA, best done immediately as double immunofluorescence with, for example, FITC-anti-human-IgG and Cy3-anti-human-IgA (see above).
Pattern of fluorescence: primarily fluorescence of the apical portions of enterocytes (Fig. 60a).
Occurrence: rarely during determination of other reactions and antibodies.
Indications: use only blood-group 0 donors for application to human tissue, when available. Confusion with numerous non-specific staining and other reactions must be avoided, primarily with antibodies (Ab) against the cytoskeleton (Figs. 60b and 60c, Fig. 61b). Also beware of confusion with blood-group antibodies!
Clinical relevance: a marker for a so-called autoimmune enteropathy especially in childhood.
Literature: McCarthy et al. [1978]; Mirakian et al. [1986]; Corazza et al. [1997]

Antibodies against goblet cells
Target antigen(s): unknown.
Properties of the antibodies: IgG, IgA, do not bind complement.
Antigen substrate: human foetal intestine.
Conjugate: anti-human IgG.
Pattern of fluorescence: fluorescence of the goblet cells (Figs 60, 61).
Occurrence: predominantly in ulcerative colitis but also in healthy patients and other inflammatory diseases of the bowel.
Clinical significance: not clear.
Literature: Stöcker et al. [1984].

7.3.2.11 Antibodies against the endometrium

Target antigen(s): cytoplasm of cells of the endometrial stroma.
Properties of the antibodies: IgG, also IgA and occasionally IgM, clearly broadly organ- and species-specific.

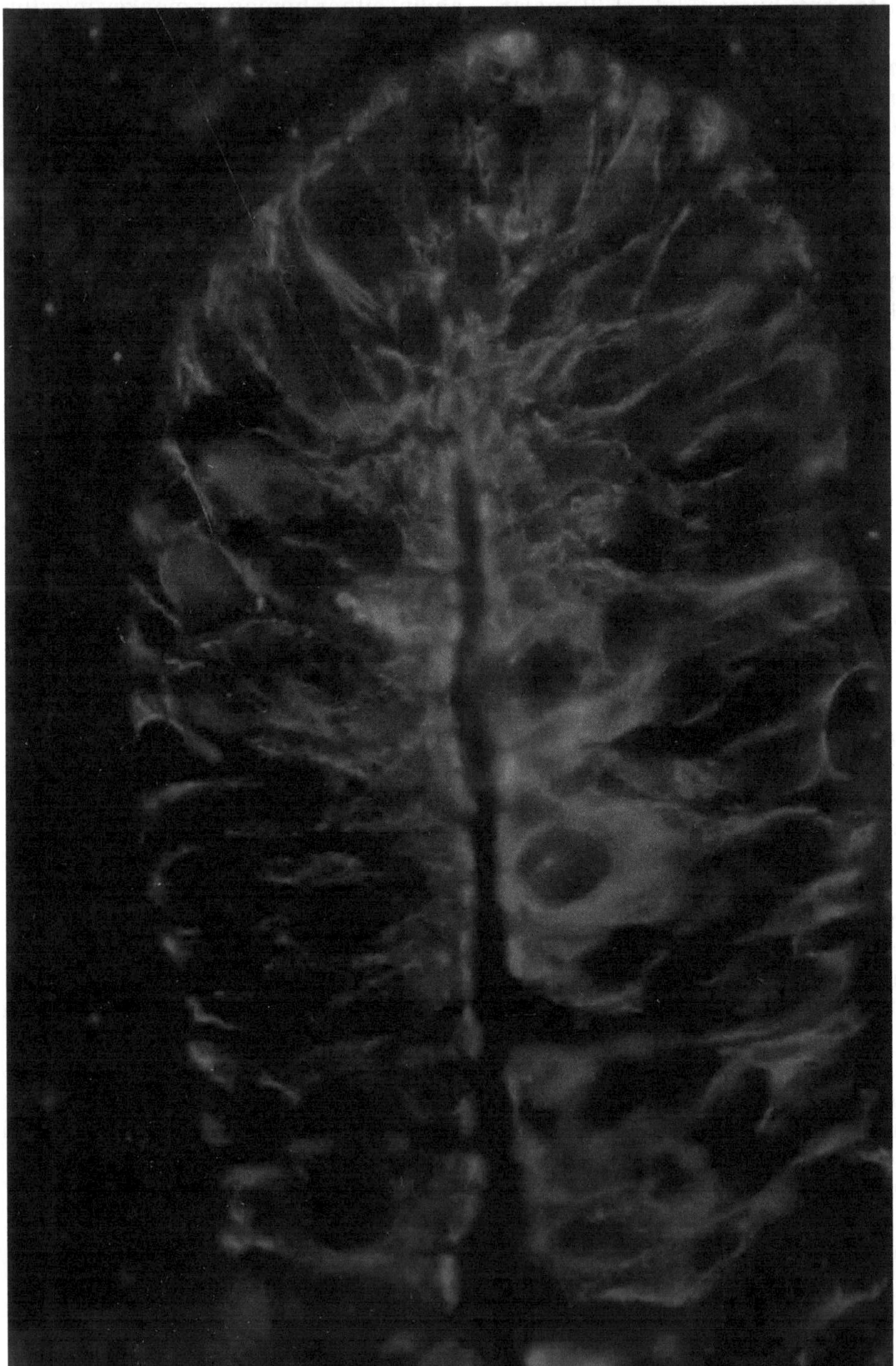

a

Figure 60a. Antibodies against enterocytes. Antigen (Ag): monkey jejunum; fluorescence especially of the spicular cell membrane (objective 100×, × 1250).

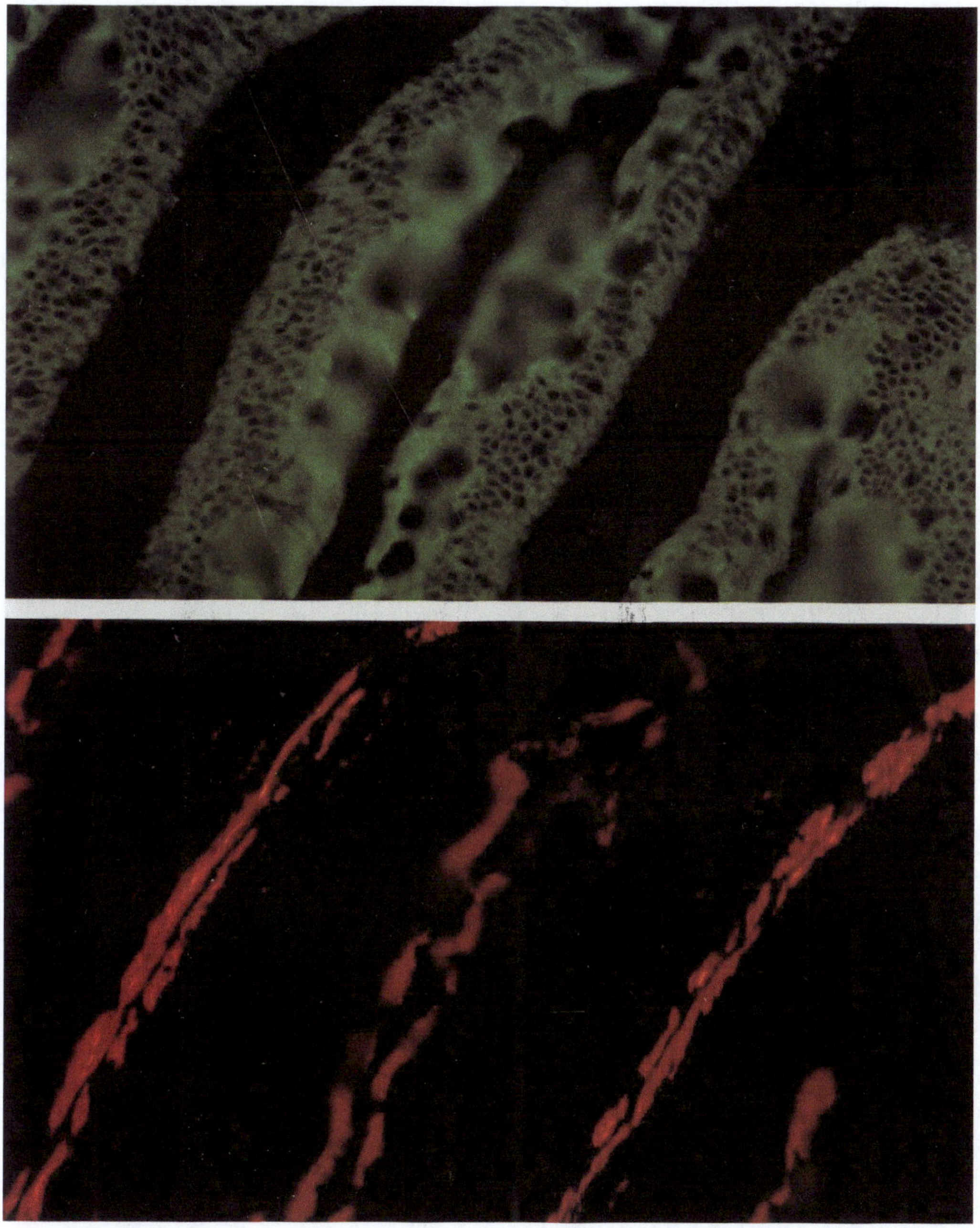

b

c

Figure 60b/c. Antibodies against the endoplasmatic reticulum, probably together with antibodies against enterocytes. Double immunofluorescence using TRITC-labelled Phalloidin: antigen, monkey jejunum. The green fluorescence of the enterocyte cytoplasm (b) demonstrates the reaction of human auto-antibodies (auto-Ab), whereas the red fluorescence (c) indicates the localisation of Phalloidin on the actin filaments, primarily between the glands in the region containing smooth muscle and on the luminal (apical) side of the epithelial cells (objective 40×, × 500).

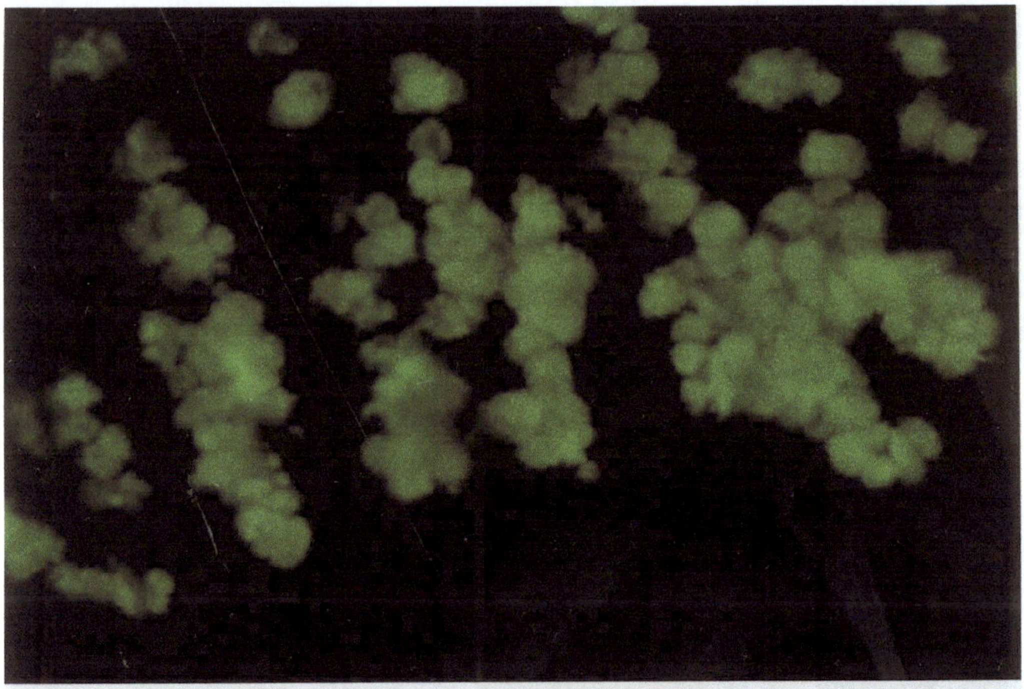

Figure 61a. Antibodies directed against the goblet cells of the gut; antigen, human foetal gut – globular fluorescence of the goblet cells (objective ×40, ×500).

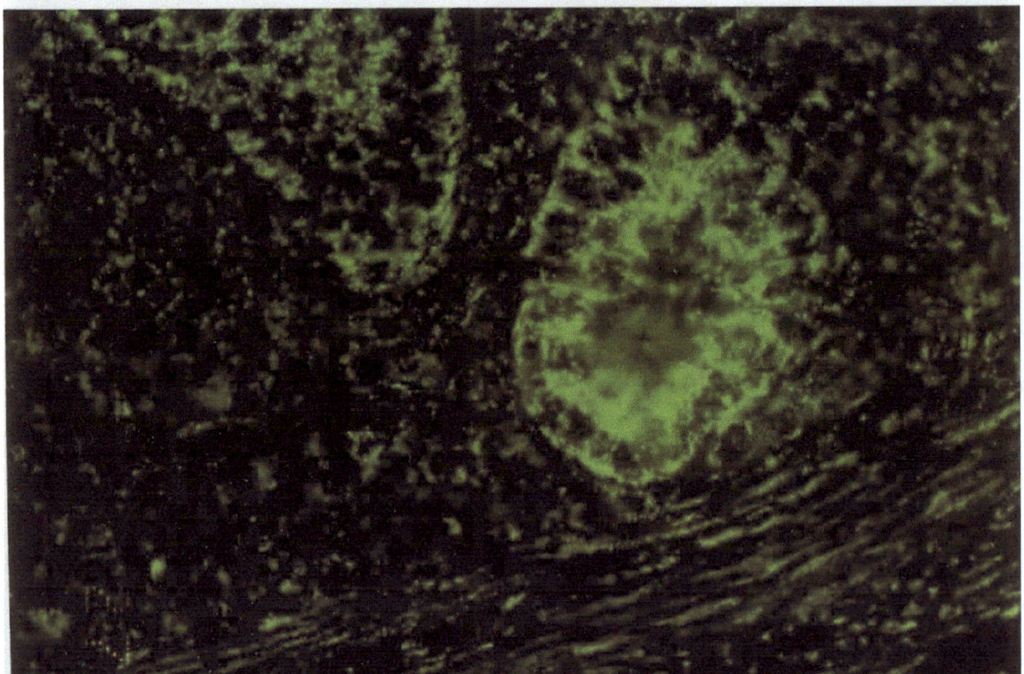

Figure 61b. Antibodies directed against mitochondria in the gut mucosa; antigen, as for Figure 61a – diffuse, finely granular fluorescence especially affecting glandular cells with sparing of the goblet cells (objective ×40, ×500).

Antigen substrate: human endometrial biopsies; sections should be fixed in acetic acid / methanol (3:1 v/v).
Conjugate: anti-human IgG.
Pattern of fluorescence: a colloidal and floccular fluorescence can be distinguished.
Tip: sera should be absorbed beforehand using pooled ABO erythrocytes, in order to avoid false positives caused by blood-group antibodies.
Clinical significance: not clear; the few cases that have been described suggest a relationship to endometriosis.
Literature: Mathur et al. [1982]; Wild et al. [1991].

7.3.2.12 Antibodies against nervous tissue

A multitude of antibodies against nervous tissue have been described since 1965 (Wilkinson & Zeromski [1965]), which can be looked on as markers for particular diseases. It has not, however, been possible, in almost all cases, to use them for routine work.

It is not, nor has it often been considered that IgG from healthy patients is able to bind to nervous tissue via the F_c region (Aarli et al. [1975]). That is why dilutions less than 1:20 are to be viewed with scepticism. Only some antibodies, which may attain relevance in the future, will be discussed here. This may also be true for various cultured cell lines for the detection of, for example, antibodies against neuronal cell surfaces (Hanly et al. [1992]).

7.3.2.12.1 Antibodies against peripheral nerves

Target antigen(s): not known.
Properties of the antibodies: IgG, partly fix complement.

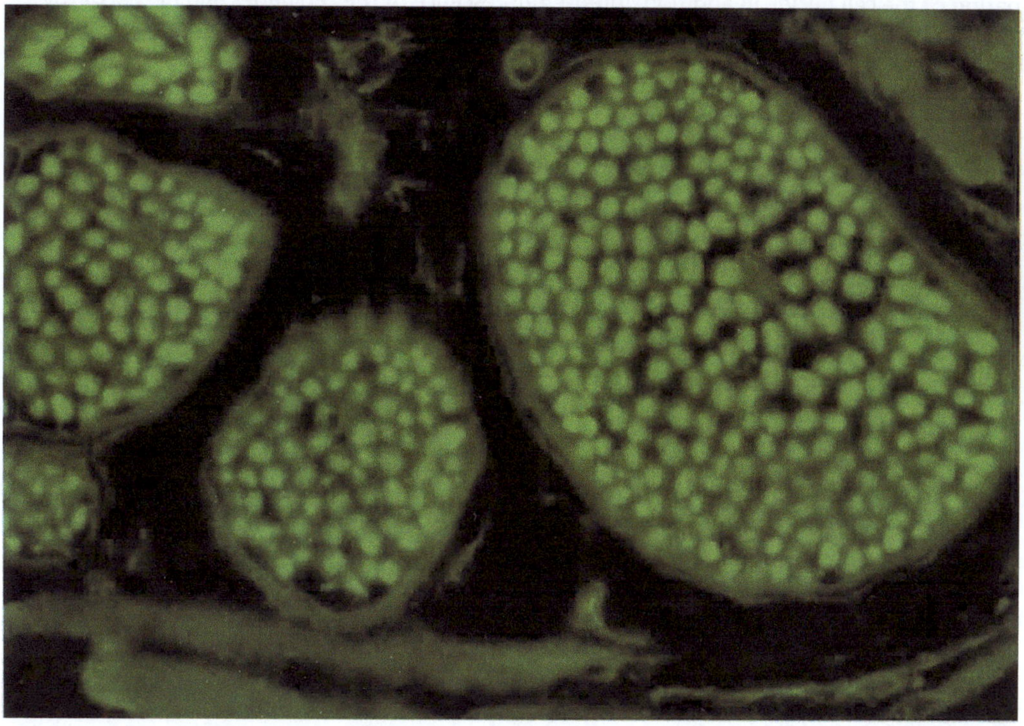

Figure 62. Antibodies against peripheral nerves; antigen, rat tongue – fluorescence predominantly of nerves seen side-on, with the perineurium and surrounding muscle fibres showing non-specific staining (from a preparation kept in a refrigerator for nine years, objective ×40, ×500).

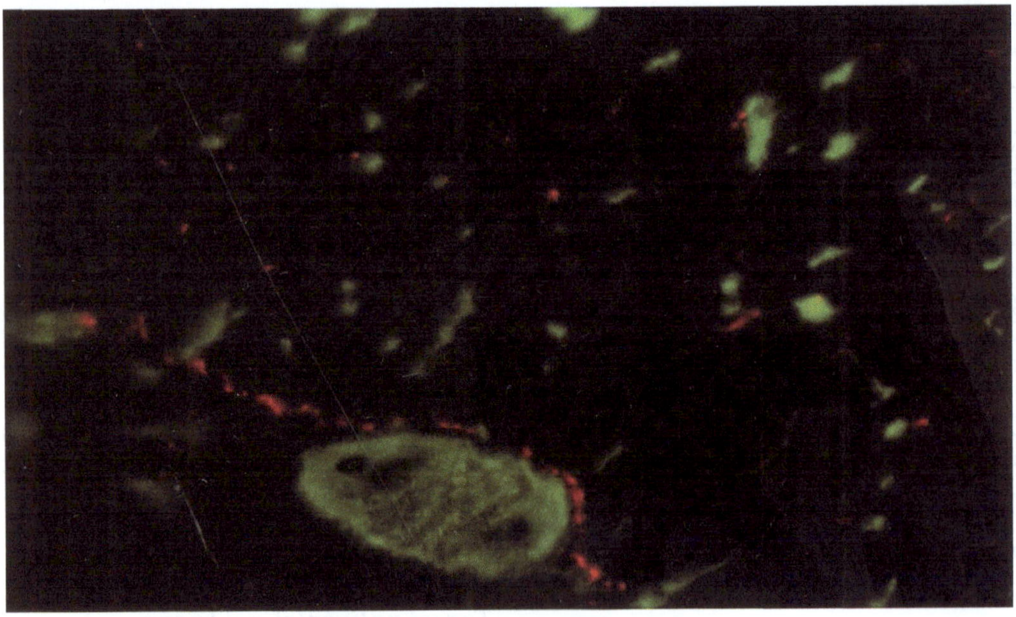

Figure 63. Antibodies against Auerbach's plexus and against vascular endothelium; antigen, rat stomach – finely granular fluorescence of the cytoplasm of the cells of Auerbach's plexus with sparing of the cell nuclei. In addition, intense fluorescence of the capillaries and nerves of the *muscularis* and *submucosa*, red staining of the connective tissue using Evans Blue as counterstain added to the conjugate (objective ×20, ×250).

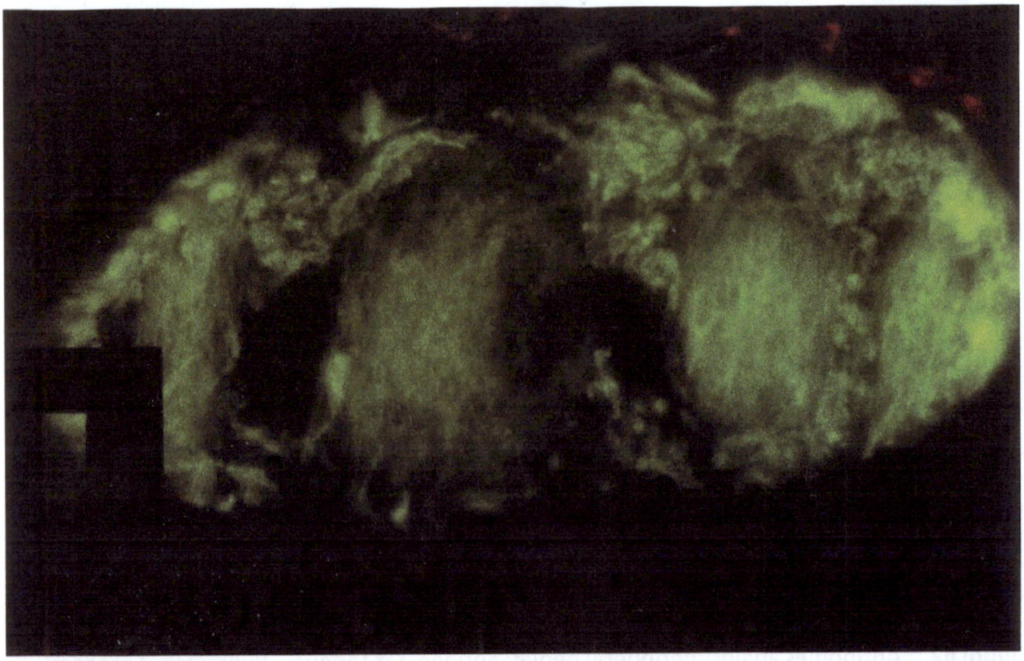

Figure 64. Antibodies directed against Auerbach's plexus, in spite of the filamentous and granular fluorescence, with weak specific staining of the nerves of the *muscularis* (objective ×100, ×1250).

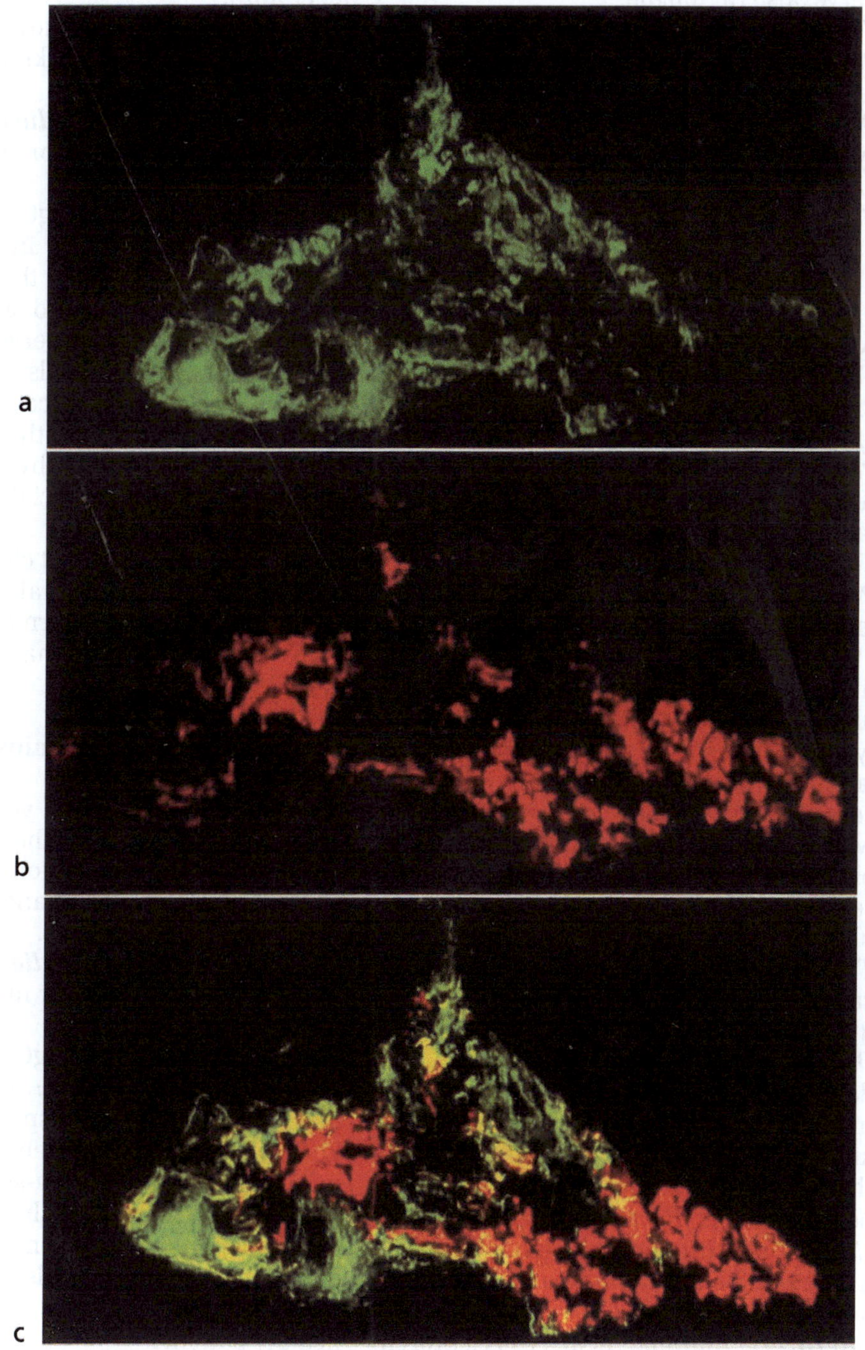

Figure 65. Antibodies against Auerbach's plexus. Double immunofluorescence using monoclonal mouse antibodies against the 68 kDa neurofilament protein and Cy 3-labelled anti-mouse immunoglobulin. Recording using confocal laser-scanning fluorescence microscopy – for the conditions used see the legend to Figure 56. The green fluorescence (a) demonstrates the binding of human auto-antibodies, the red fluorescence (b) the binding of antibodies against the neurofilament protein. Their identical localisation (c) appears yellow on simultaneous display of the images from (a) and (b) (objective ×40, ×800).

Antigen substrate: human peripheral nerve, also rat tongue.
Conjugate: anti-human IgG.
Pattern of fluorescence: numerous patterns can be distinguished (Fig. 62).
Occurrence: myasthenia, Guillain-Barré syndrome, numerous peripheral neuropathies, healthy patients.
Clinical significance: not clear.
Literature: Storch et al. [1985].

7.3.2.12.2 Antibodies against Auerbach's plexus

Target antigen(s): predominantly the cytoplasm of ganglion cells especially those of Auerbach's plexus.
Properties of the antibodies: IgG.
Antigen substrate: an organ block containing rat stomach, liver and kidney.
Conjugate: anti-human IgG.
Pattern of fluorescence: numerous different patterns and combinations can be observed. At least three types can be distinguished: a homogeneous type, a granular or speckled type and a filamentous type of fluorescence (Figs 63–65). In isolation the antibodies stain only ganglion cells and in addition the nerve fibres, as well as endothelial cells and the cytoplasm of other neurones; furthermore, reaction with, for example, renal glomeruli has been observed.
Tip: titres less than 1:20 have no significance because these occur very frequently in healthy patients.
Occurrence: antibodies have been, to date, most frequently observed in rheumatic disease, diseases of the liver and kidneys, as well as especially commonly in the relatively rare disease, achalasia of the oesophagus, as well as in Chagas' disease.
Clinical significance: not clear, possibly a marker for an autoimmune variant of achalasia.
Literature: Altermatt et al. [1991]; Storch [1993, 1994]; Storch et al. [1995].

7.3.2.12.3 Antibodies against neuron cell nuclei

Synonyms/abbreviations: anti-neuronal nuclear antibodies (ANNA), anti-Ri (Luque et al. [1981, 1991]), anti-Hu (Graus et al. [1985]).
Target antigen(s): unknown, suspected to be proteins of 35–40 kDa (Dalmau et al. [1990]).
Properties of the antibodies: IgG.
Antigen substrate: rat or mouse spinal cord dorsal roots.
Conjugate: anti-human IgG.
Pattern of fluorescence: intense, homogeneous fluorescence of the cell nuclei, sparing the nucleoli, and a weak cytoplasmic fluorescence of nerve cell bodies, especially of ganglion cells in the posterior columns of the spinal cord.
Occurrence: predominantly patients with sensorimotor neuropathy, frequently with malignancies, especially carcinoma of the lung.
Clinical significance: not clear.
Literature: Anderson et al. [1988]; Lennon [1989]; reviews, Nairn [1976]; Altermatt et al. [1993]; Dalmau and Posner [1996].

7.3.2.12.4 Antibodies against Purkinje cells

Target antigen(s): not yet sufficiently known, presumed to be the ribosomes of Purkinje and other ganglion cells (Hida et al. [1994]), especially 34 and 62 kDa proteins, so-called Yo and Tr.
Properties of the antibodies: IgG.
Antigen substrate: rat or mouse cerebellum.
Conjugate: anti-human IgG.
Pattern of fluorescence: granular cytoplasmic fluorescence especially of the Purkinje cells. An homogeneous cytoplasmic staining is mostly non-specific.
Occurrence: patients with para-neoplastic cerebellar degeneration appear to predominate; the antibody also occurs in patients without malignant disease and in healthy patients.
Clinical significance: not clear, perhaps associated with cerebellar degeneration.
Literature: Jaeckle et al. [1985]; Fourneaux et al. [1989]; Trotter et al. [1976]; review, see Altermatt et al. [1993] and Dalmau and Posner [1996].

7.3.2.13 Antibodies against cells of the retina

Target antigen(s): still relatively unknown.
Properties of the antibodies: predominantly IgG.
Antigen substrate: fresh human retina.
Conjugate: anti-human IgG.
Pattern of fluorescence: fourteen different patterns of reaction have been described.
Occurrence: predominantly in patients with uveitis and hereditary retinal degeneration, but also in other diseases and in healthy patients.
Clinical significance: not clear; no pattern turns out to be specific for uveitis or any other disease of the eye, even for the group of hereditary retinal degenerations (Nölle [1992]). Although any retinal layer may be involved, there is probably no pattern which can, with safety, be used diagnostically; routine investigations involving this antibody do not appear at present to be sensible.
Literature: Nölle & Müller-Ruchholtz [1987]; Nölle [1994].

7.3.2.14 Antibodies against granulocytes and monocytes

Synonyms/abbreviations: anti-neutrophil cytoplasmic antibodies (ANCA, formerly also ACPA). At least three different groups have been distinguished – c-ANCA (the cytoplasmic or classical pattern); p-ANCA (the perinuclear pattern); and a-ANCA (an atypical pattern) also described as x-ANCA.
Target antigen(s): Proteinase 3 in c-ANCA (Goldschmeding et al. [1989], Niles et al. [1989], Lüdemann et al. [1990]; review, Wieslander & Wiik [1994]), myeloperoxidase in p-ANCA (Falk & Jennette [1988]), as well as lactoferrin, cathepsin G, elastase, lysozyme, ß-glucuronidase (review, Gross et al. [1993]), BPI, bacterial permeability increasing protein (Zhao et al. [1995] and azurocidin (Zhao et al. [1996]).
Properties of the antibodies: predominantly IgG but also IgM and IgA, partly complement fixing.
Antigen substrate: human granulocyte smears.
Conjugate: anti-human IgG with an F/P ratio less than 2.5.
Pattern of fluorescence: at least three different patterns can be distinguished, which are designated as c-ANCA, p-ANCA and a-ANCA (or x-ANCA), as shown in Figures 66 and 67. Cytoplasmic fluorescence is judged to be c-ANCA if the whole of the cytoplasm is involved, often with 'central' emphasis. Perinuclear fluorescence is regarded as p-ANCA. All other currently non-classified or atypical patterns are labelled as a-ANCA.
Clinical significance: c-ANCA counts as a marker for Wegener's syndrome (granulomatosis) with a specificity of approx. 95%. The sensitivity depends on the activity and stage of the disease but is generally around 90%. Higher titres (≥ 1:160) are found predominantly in disease running a highly active course and only exceptionally in any other disorders. p-ANCA is seen above all in microscopic polyangiitis and renal vasculitis (idiopathic rapidly progressive glomerulonephritis), but also in many other diseases, for example, in rheumatoid arthritis, SLE, Sjögren's syndrome, cholangiitis and diseases of the bowel. a-ANCA has been observed to date in, amongst other diseases, ulcerative colitis, primary sclerosing cholangiitis, autoimmune hepatitides and in Crohn's disease.
Literature: Storch et al. [1981a]; Davies et al. [1982]; van der Woude et al. [1985]; Wieslander & Wiik [1994]; Kallenberg et al. [1994]; Gross & Csernok [1995].

The proven procedure of Wiik et al. [1994] is given below, in order to foster standardisation:

Apparatus and reagents

1. Dextran-metrizoate solution
   - 24.7 ml distilled water
   - 89.4 ml 6% Dextran T–500 (Pharmacia, Uppsala, Sweden)
   - 20 ml sodium metrizoate 75% w/v (Nycomed, Oslo, Norway)

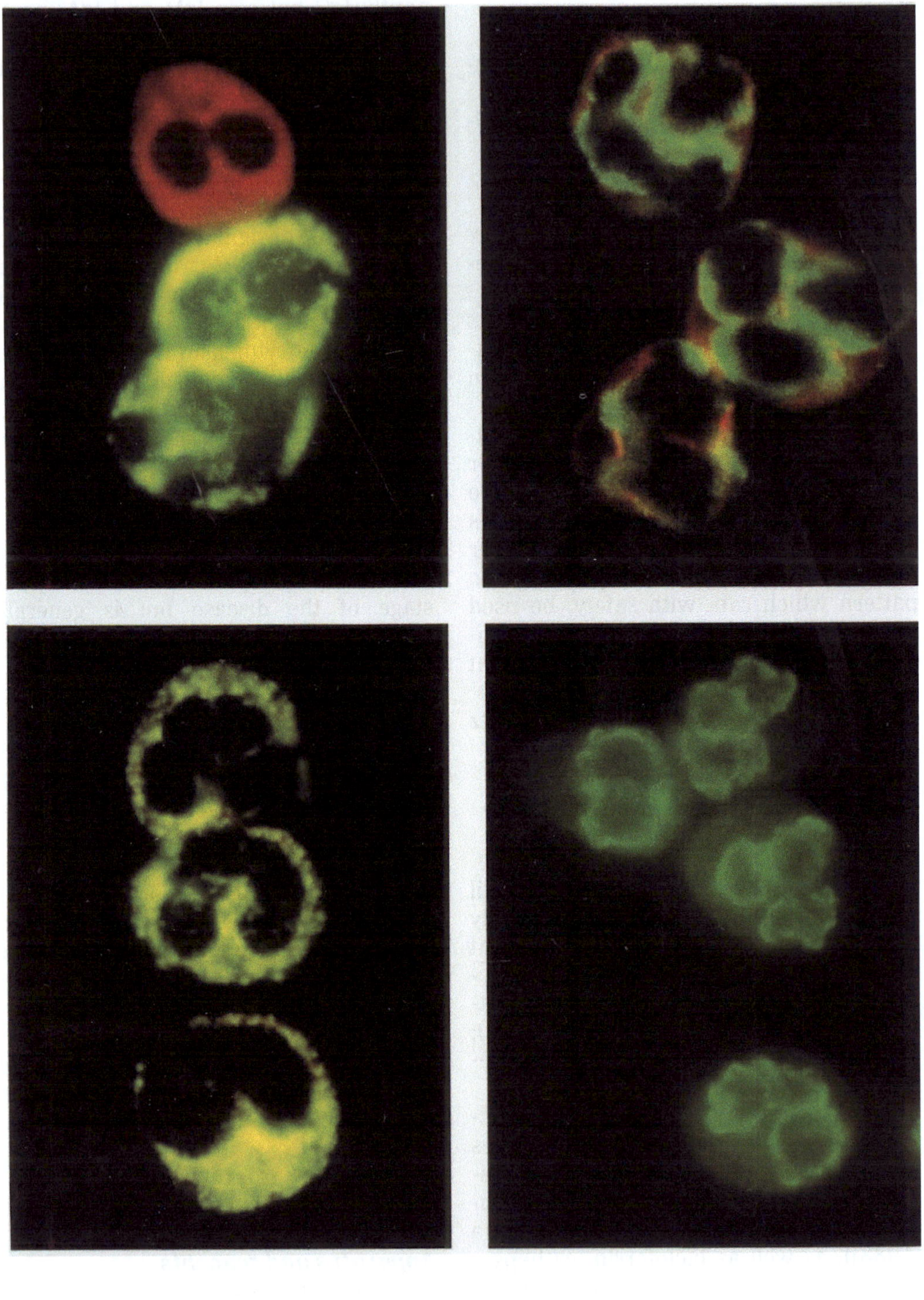

a

c

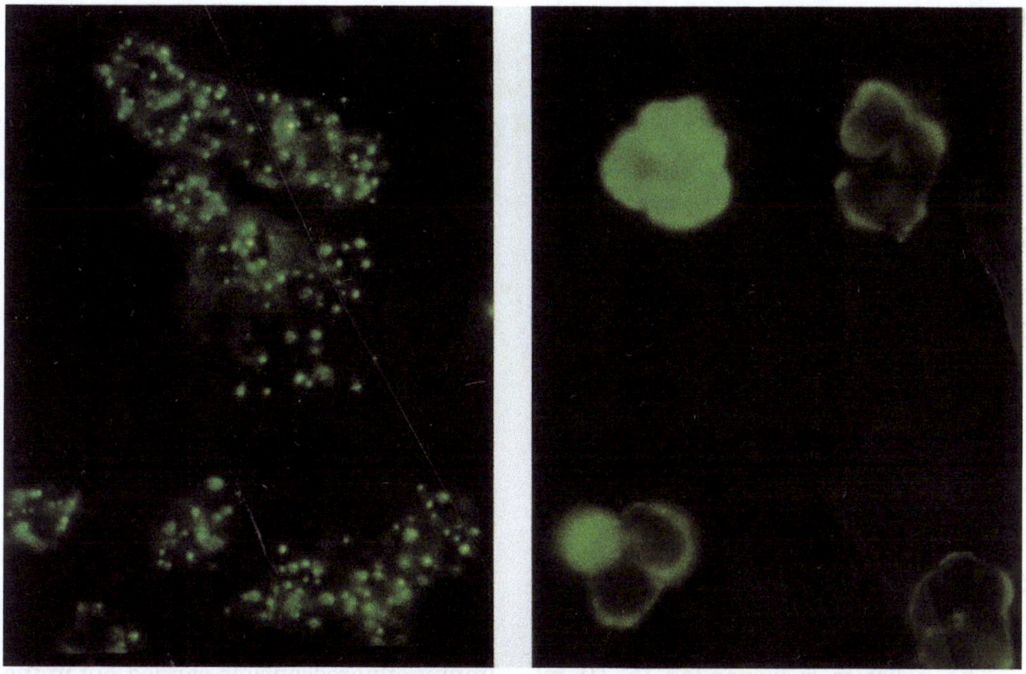

e f

Figure 66. Antibodies directed against neutrophil granulocytes (ANCA); antigen, a granulocyte smear which also contains monocytes and lymphocytes:

(a) *c-ANCA* – using relatively broad-band excitation (450–490 nm) and no blocking of the emission in the red (LP 520), one is able to distinguish a negative lymphoid cell (upper) and two granulocytes showing cytoplasmic fluorescence (lower); counterstaining is with Evans Blue; artefacts are present due to the less than ideal preparation of the antigen substrate;
(b) *p-ANCA* – using narrow-band excitation (470–490 nm) and again no blocking of the emission in the red, one can recognise fluorescence with perinuclear emphasis;
(c) *c-ANCA* – a finely granular fluorescence is easily recognised in ideal antigen preparations;
(d) *p-ANCA* – fluorescence with perinuclear emphasis as in (b); the red is suppressed using narrow-band excitation (470–490 nm) and filtering the emitted light (BP 520–550), whereby a typical apple-green colour for FITC is obtained; the weak staining in the plane of projection for the nuclei is a consequence of the objective's poor depth of focus with intense fluorescence;
(e) *a-ANCA* – granular fluorescence of the cytoplasm.
(f) simultaneous occurrence of p-ANCA and ANF showing a homogeneous pattern; note that overlapping of the pattern only occurs rarely (objective ×100 (a-d and f), ×1250 and ×40 (e), ×800). Figure 66(e), courtesy of Prof. Waldherr, Heidelberg.

2. PBS pH 7.4
   - 8.0 g NaCl
   - 200 mg KCl
   - 1.44 g $Na_2HPO_4.2H_2O$
   - 200 mg KH_2PO_4
   - distilled water to 1000 ml
3. PBS/albumin 1% w/v
   - PBS as in (2) with the addition of 1% w/v human serum albumin (HSA)
4. Ethanol 99%, AR
5. Glycerol-PBS mixture
   - PBS as in (2) with the addition of glycerol so as to give a 66% v/v (i.e., 2:1) mixture
6. Narrow Pasteur pipettes, if necessary pulled out freshly through a gas flame, de-fatted microscope slides, glass marker, syringes, tubes, glass beads.

Preparation of granulocyte smears for the detection of ANCA

1. 10 ml blood is taken from a healthy donor, using a 10 ml syringe without the addition of anticoagulant, and defibrinated immediately by shaking gently with ten glass beads in a conical flask for 5–7 minutes.
2. The defibrinated blood is put into a polystyrene tube together with 250 IU heparin and thoroughly mixed.
3. 2 ml aliquots are added to 5 ml of the dextran-metrizoate solution in 10 ml polystyrene tubes and allowed to stand for 45 minutes at room temperature.
4. The leucocyte-containing plasma is pipetted from each tube into another 10 ml polystyrene tube.
5. These tubes are then centrifuged at 200 g_{av} for ten minutes to sediment the leucocytes.
6. The pellet is resuspended in 10 ml PBS containing 1% w/v human albumin.
7. Stages 5 and 6 are repeated in order to remove serum components.
8. The leucocyte pellet is freed from the wash solution by pipetting, so that only 150–200 µl fluid remains over the pellet.
9. The leucocytes are resuspended in the remaining fluid.
10. The cell suspension is then placed on carefully de-fatted microscope slides, using a fine Pasteur pipette and approx. 1 mm droplets, and a smear made within ten seconds.
11. After air-drying, the microscope slide is fixed in cold 99% ethanol at +4°C for five minutes.
12. After drying, the smear should be labelled using a glass marker and either used for immunofluorescence within 30 minutes, or stored at –20°C in an airtight container.

Immunofluorescence is carried out in the customary way.

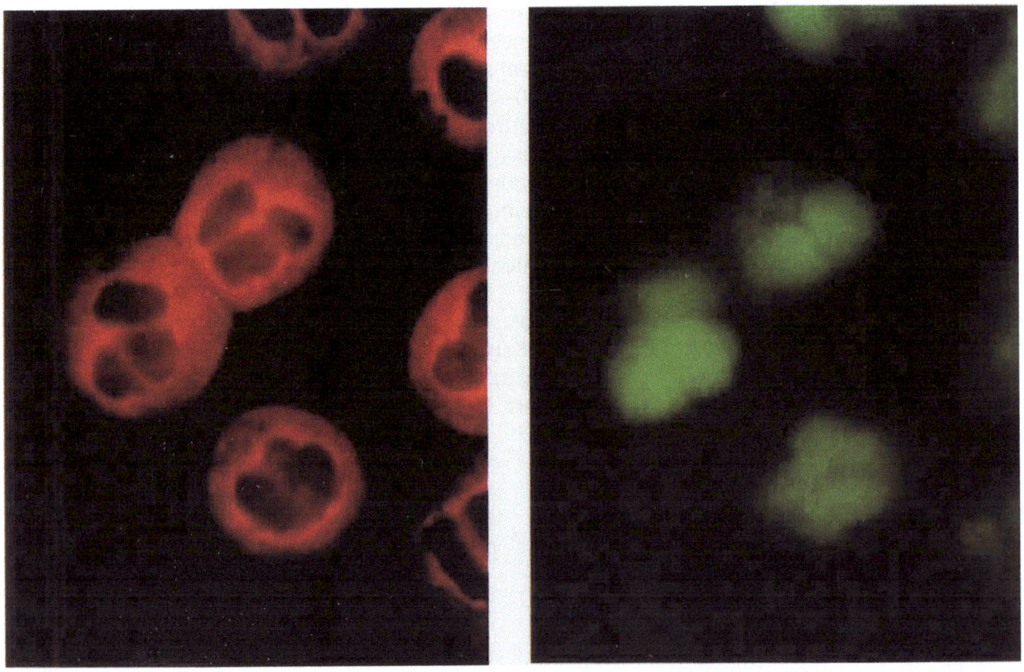

a

Figure 67. Antibodies directed against nuclear material in granulocyte smears; antigen as for Figure 64 – (a) green excitation, appearance of granulocytes using Evans Blue, (b) blue excitation (narrow-band with blocking of the red); in spite of the granulocytes being sharply in focus, there is a diffuse, out-of-focus homogeneous staining affecting the cell nuclei (objective ×100, ×1250).

Tip: the embedding medium should contain an anti-fading agent, either 1 mg/ml *para*-phenylene-diamine (Sigma), or 25 mg/ml DABCO (1,4-diazobicyclo–2,2,-octane: Aldrich) and 10 mg/ml sodium azide, dissolved in the embedding medium.
Pattern of fluorescence: the pattern of fluorescence demonstrates an artificial localisation of granulocyte antigen induced by the ethanol fixation (Pryzwansky et al. [1978]). In order to distinguish between granulocyte-specific anti-nuclear factors (ANF) and p-ANCA we would recommend repeating the procedure using a formalin-fixed smear. ANF retain their fluorescence, whereas p-ANCA show cytoplasmic fluorescence as for c-ANCA.

7.3.2.15 Antibodies against thrombocytes

Target antigen(s): cytoplasmic components of thrombocytes and megakaryocytes.
Properties of the antibodies: broadly cell-specific, partially species-specific and typical of the disease; predominantly IgG, but also IgA and IgM, to some extent able to fix complement.
Antigen substrate: autogenous sternal marrow smear or isolated platelets.
Conjugate: anti-human IgG.
Pattern of fluorescence: cytoplasmic fluorescence affecting megakaryocytes.
Tip: in order to reduce non-specific binding to plasma cells and leucocytes, absorption of the conjugate by bone marrow cells is recommended.
Clinical significance: a positive result using autogenous megakaryocytes is evidence for an idiopathic thrombocytopenic purpura – iso-antibodies must, if necessary, be excluded; detection is successful in approx. 30–40% of cases for untreated patients. Findings are mostly negative in cases of remission or those on prednisone treatment.
Literature: McKenna & Pisciotta [1962]; Storch [1971]; review, Mueller-Eckhardt [1977].

Platelet-adhesion-immunofluorescence test based on Schneider & Schnaidt [1981]

Preparation of the platelets

Platelet-rich plasma, obtained from blood group 0 donors, is prepared from blood containing 0.5% EDTA, and treated three times with 0.3% EDTA-PBS (9 mM sodium EDTA; 26.4 mM $Na_2HPO_4.2H_2O$; 140 mM NaCl; pH 7.1) in polystyrene tubes according to the procedure described by Borne [1978]. The platelet-rich plasma is distributed as 500 µl aliquots in 2 ml plastic tubes for freezing. DMSO is added to give a final concentration of 6%. Freezing should take place slowly at –2°C/min down to –80°C. The frozen platelets can be stored for up to six months at –80°C. Prior to use, the stabilates should be thawed out in a water bath at 37°C, immediately placed in plastic tubes containing 10 ml 0.3% EDTA-PBS, centrifuged and washed three times with 0.3% EDTA-PBS.

Carrying out the test

1. Platelets are washed as described above and resuspended in 0.3% EDTA-PBS at a cell density of approx. 1×10^6 /µl.
2. 20 µl aliquots of this suspension are then placed in the wells of a glass micro-titre plate (Hamax).
3. The plate is covered and allowed to stand for ten minutes at room temperature so that the platelets can adhere to the surface of the glass.
4. The rest of the 20 µl suspension, together with the remaining cells, is sucked off with a micro-pipette. Following this, 20 µl test serum are immediately added. The test serum must also contain 0.3% EDTA-PBS.
5. Incubate for 45 minutes at room temperature.
6. Wash with 0.3% EDTA-PBS, where possible using a suitable dispenser, i.e., an automatic pipette. In this way approx. 1 ml of EDTA-PBS is added repeatedly and sucked off.
7. Cover the cell layer with 20 µl conjugate (working dilution, approx.

1:250). Nine parts of foetal calf serum and one part 0.3% EDTA-PBS should be added to the conjugate in order to minimise non-specific fluorescence.
8. Incubate for 45 minutes at room temperature.
9. Aspirate the conjugate and wash three times by flooding the plate with a total of approx. 5 ml 0.3% EDTA-PBS, gently tilt the plate and decant. The platelets should be covered with 20 µl 0.3% EDTA-PBS after the last decantation.
10. Assess the results using fluorescence microscopy with a magnification of ×400 (Fig. 68).

Clinical significance: the test is suitable as a simple screening test for the detection of antibodies against thrombocytes of different origin, especially in neonatal allo-immune thrombocytopenia (review, Kiefel et al. [1994]).

7.3.2.16 Antibodies against the cytoplasm of liver cells

Target antigen(s): liver cytosol, ribosomes and unknown antigens, especially in the microsomal fraction.

Properties of the antibodies: partly organ-specific, not species-specific (although, in part, typical for a species), typical of the disease, IgG, to some extent able to fix complement.

Antigen substrate: organ block containing rat liver, stomach and kidney.

Conjugate: anti-human IgG.

Pattern of fluorescence: at least four types occur, which can be further sub-divided as follows:

1. Cytoplasmic fluorescence exclusively of the liver cells.

1.1 Antibodies against the hepatic cytosol of type 1 (LC1): fluorescence affecting virtually all hepatocytes, basically omitting 2–3 cell layers around the central vein (Storch [1979]; Martini et al. [1988]; Fig. 69).

Figure 68. Antibodies directed against thrombocytes; antigen, dried thrombocyte suspension – cytoplasmic fluorescence of thrombocytes (objective ×40, ×640; Prof. Mueller-Eckhardt, Gießen).

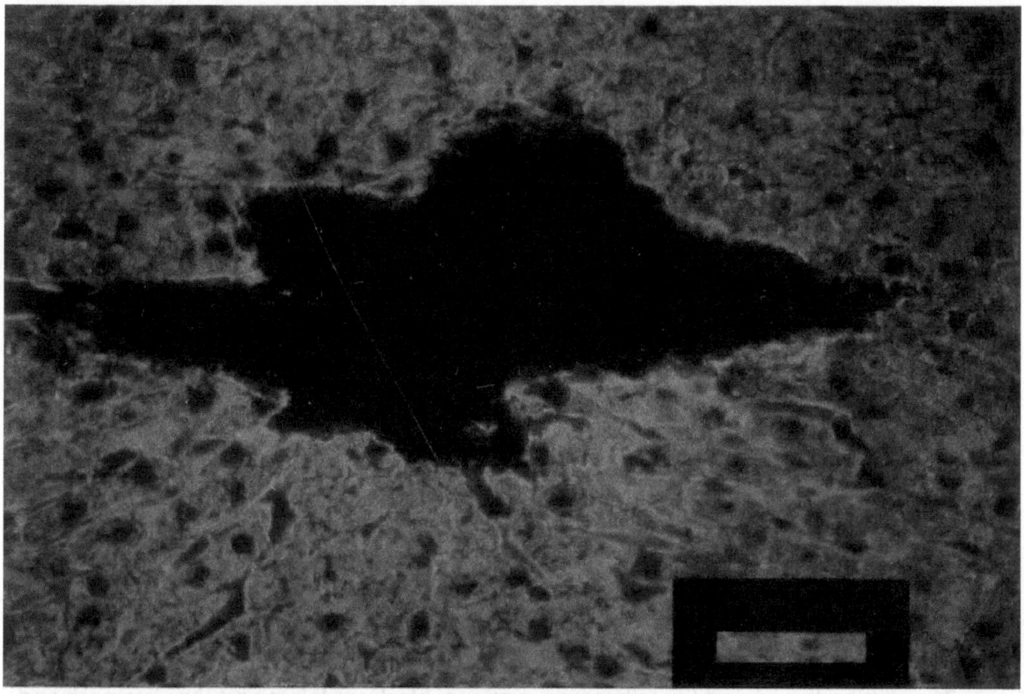

Figure 69. Antibodies directed against hepatocyte cytoplasm type 1 (LC–1); antigen, block containing rat stomach, liver and kidney – cytoplasmic fluorescence of virtually all hepatocytes, with only two to three cell layers remaining unstained (objective ×40, ×330).

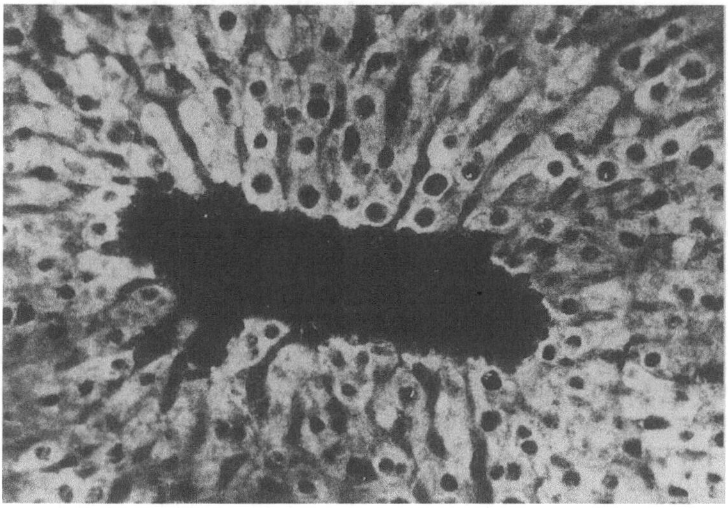

Figure 70. Antibodies directed against hepatocyte cytoplasm type 2 (LC–2); antigen, block containing rat liver, stomach and kidney – cytoplasmic fluorescence of hepatocytes with emphasis of the portal regions (objective ×25, ×510).

1.2 Antibodies affecting hepatocytes especially around the portal regions (Storch [1975, 1978]; Abuaf et al. [1991]; Fig. 70).

1.3 Antibodies against liver ribosomes: fluorescence especially of hepatocytes around the central veins, with the areas round the portal regions being generally negative (Storch & Wohlgemuth [1979]; Storch [1983]; Fig. 71).

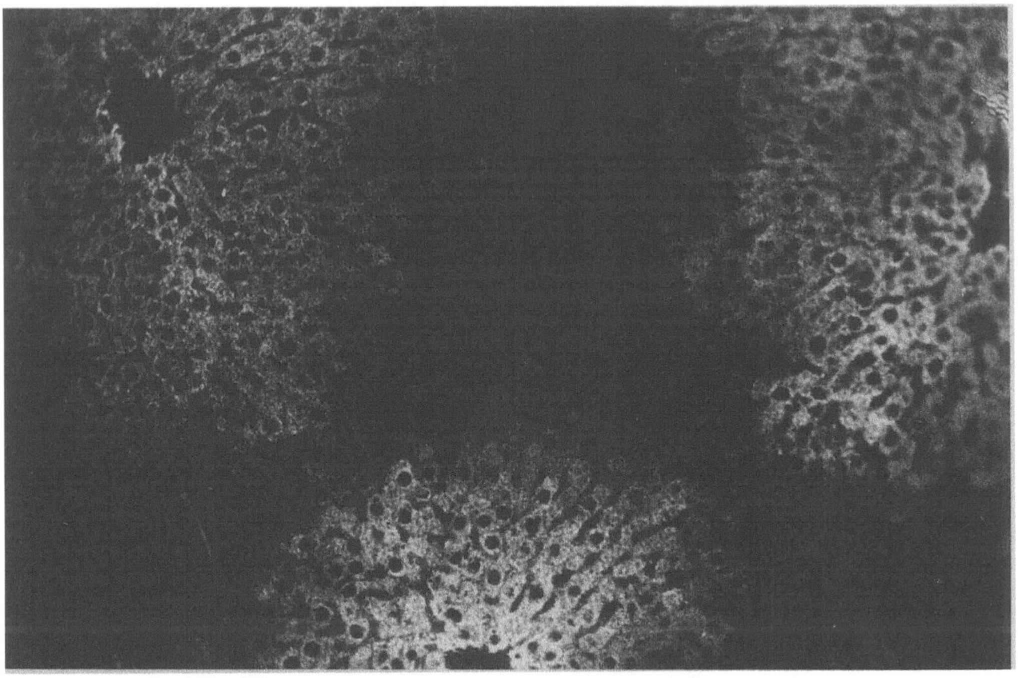

Figure 71. Antibodies against liver ribosomes; antigen, block containing rat liver, stomach and kidney – cytoplasmic fluorescence affecting hepatocytes especially around the central veins, with the region around the portal areas almost completely negative (objective ×25, ×330).

2. Fluorescence affecting liver cells as well as the second and third segments of the proximal tubule in rat kidney.
3. Fluorescence affecting liver cells as well as the parietal and chief cells of the stomach.

Tip: exclusion of mitochondrial as well as other microsomal and ribosomal antibodies is necessary.
Clinical significance: not clear, occur predominantly in acute and chronic hepatitis, with types 1.3 and 4 appearing to be markers for virus-induced autoimmune hepatitis with a good prognosis.
Literature: Storch [1975, 1992]; Abuaf et al. [1993].

7.3.2.17 Antibodies against endoplasmic reticulum – so-called liver and kidney microsomes (LKM)

Synonyms/abbreviations: antibodies against endoplasmic reticulum (AER); liver-kidney-microsomal antibodies (LKMA).
Target antigen(s): see Table 20; especially enzymes of the cytochrome P450 super-gene family (Johnson [1993]).
Properties of the antibodies: partly organ, tissue and species-specific, typical of the disease, predominantly IgG; titres from 1:20 to over 1:2000, mostly 1:40 to 1:80.

Antigen substrate: mainly a block containing male rat liver, kidney and stomach (for types 1 and 2, 4–6) and human tissue for type 3.
Conjugate: anti-human IgG.
Pattern of fluorescence: see Table 21; diffuse cytoplasmic staining of the hepatocytes is typical for type 1 with a com-

Table 20. Antibodies directed against endoplasmic reticulum; target antigens and associated diseases.

Type	Synonym	Disease	Antigen	Literature
1	LKM–1 AER	Special form of auto-immune hepatitis especially in childhood (differential diagnosis – viral hepatitis C)	Cytochrome P–450 II D 6	Storch [1973] Rizzetto et al. [1973] Homberg et al. [1974] Kiffel et al. [1986] Zanger et al. [1988]
2	LKM–2	Drug-induced hepatitis from Tienyl acid	Cytochrome P–450 II C 8–11	Andre et al. [1981] Beaune et al. [1987] Pons et al. [1991]
3	LKM–3 LKM-delta	Hepatitis D	UDP-glucuronyl transferase	Crivelli et al. [1983] Philipp et al. [1994]
4	LM	Drug-induced (Dihydralazine)	Cytochrome P–450 I A2	Nataf et al. [1986] Bourdi et al. [1990]
5	–	Halothane-induced hepatitis	Carboxyesterase Disulphide isomerase	Smith et al. [1993] Satoh et al. [1989] Martin et al. [1993]
6	–	Drug-induced (Carbamazepine)	–	Pirmohamed et al [1992]

Table 21. Fluorescence pattern for antibodies directed against endoplasmic reticulum (liver and kidney microsomes) using various tissues. Comparison with antibodies against mitochondria type 5; modified from Storch [1992], and own further unpublished work.

Organ	Type 1		Type 2		Type 3	Type 5 Ab v. mitochondria		
	Rat / Mouse male	Rat / Mouse female	Rat male	Rat female	Man Rat Ape	Rat male	Mouse male	
Proximal tubules:								
P1	±	++	+++	+++	+	±	+++	±
Kidney:								
P2	+	++	+++	+++	+	±	+++	±
P3	+++	-	+	+	++	±	++	±
Distal tubules:	-	-	-	-	-	-	±	+++
Hepatocytes:								
– centrilobular	+++	++	+++	+++	++	±	++	+
– periportal	+++	++	+	++	++	±	++	+

– = negative   + = weak
± = marginal   ++ = moderate
                       +++ = strong

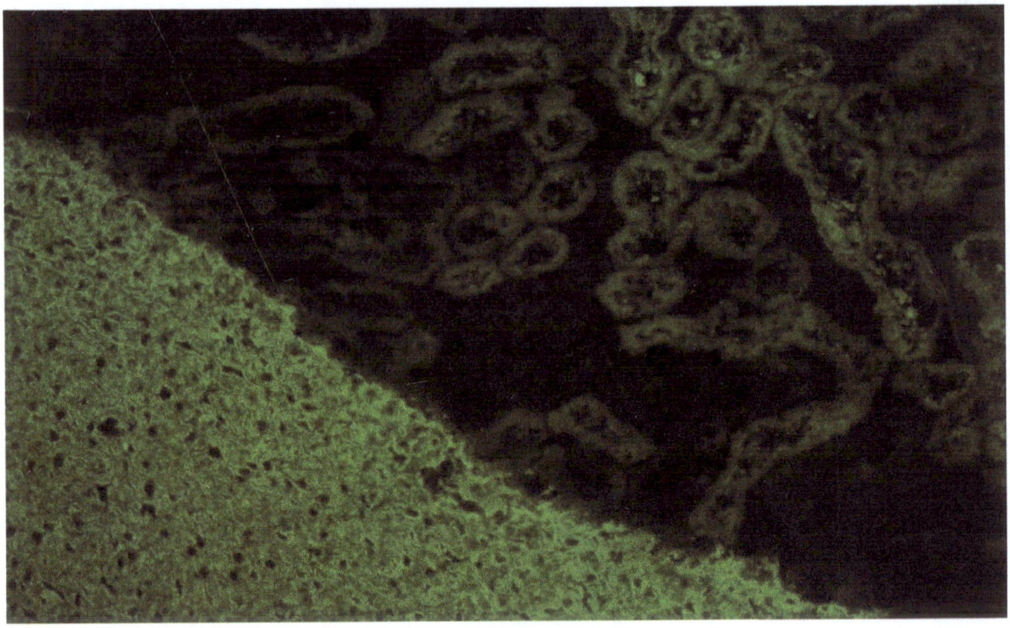

Figure 72. Antibodies directed against endoplasmic reticulum (LKM) type 1; antigen, block containing rat liver, stomach and kidney – intense fluorescence affecting all hepatocytes without preference for any particular region and a comparatively low fluorescence intensity for the kidney, with the third segment of the proximal tubules mainly reacting, and the distal tubules and glomeruli being almost always clearly negative (objective ×16, ×132).

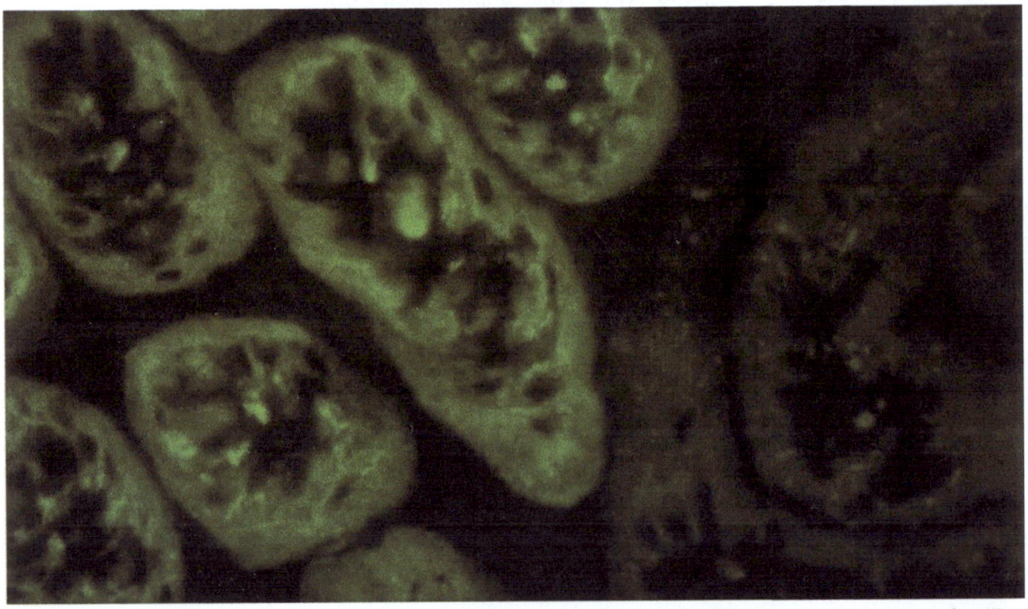

Figure 73. Antibodies directed against endoplasmic reticulum (LKM) type 1; antigen, as for Figure 68 – higher magnification with the appearance of the strongly positive third segment of the proximal tubules and a somewhat weaker fluorescence affecting the first and second segments (right); the distal tubules are clearly negative – compare with Figure 75(b) (objective ×40, ×500).

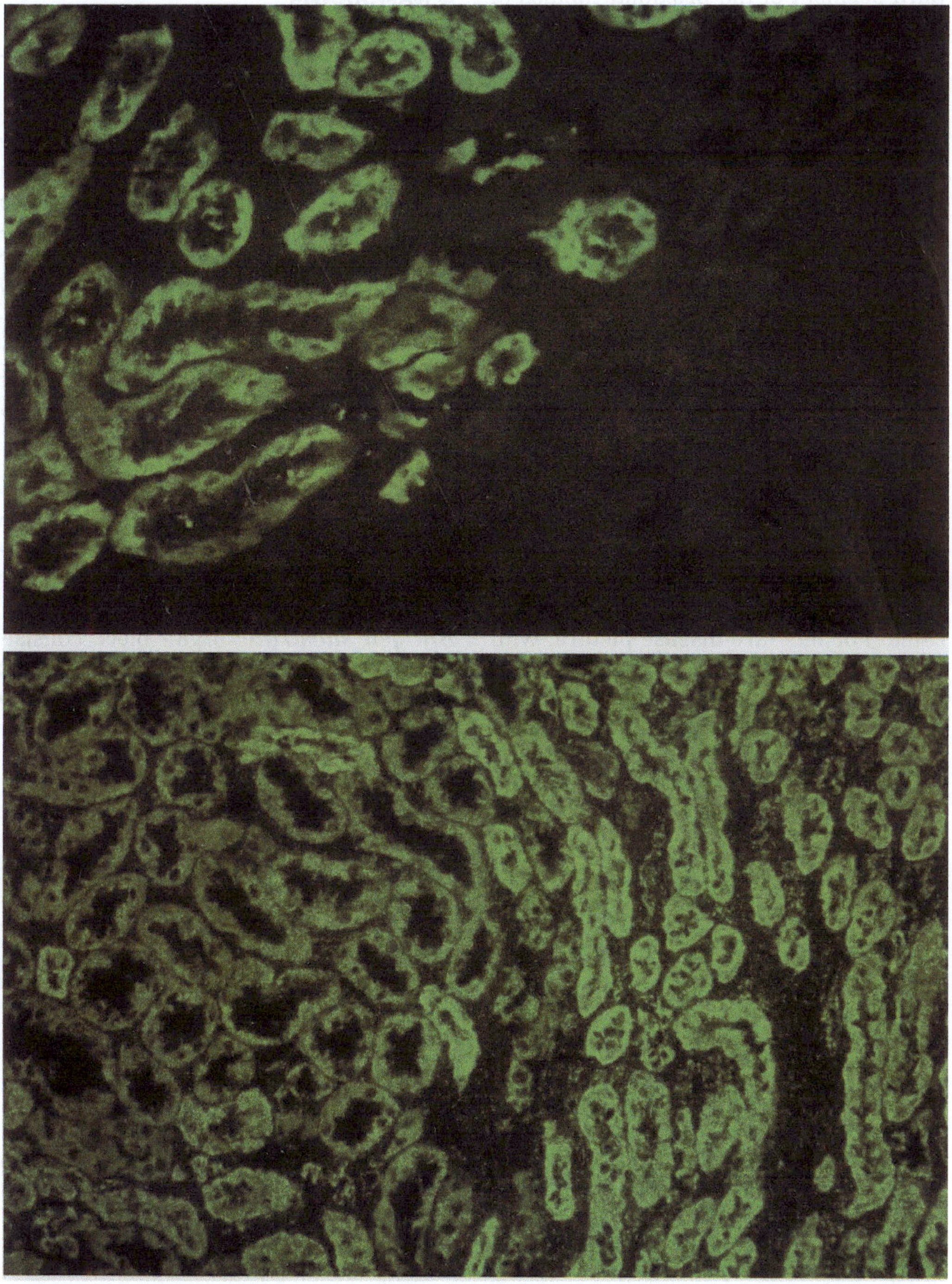

a

b

Figure 74. Antibodies against endoplasmic reticulum compared with those against mitochondria; antigen, as for Figure 68 – a deep cortical section (top left) with crossing over into the renal medulla (lower right); antibodies against the ER (a) react strongly and exclusively with the third segment of the proximal tubules, the distal tubules being negative; antibodies against mitochondria (b) react only weakly with the third segment but strongly with the distal tubules (objective ×16, ×200).

paratively weaker fluorescence of the third segment of the proximal tubules of the kidney; the first and second segments only react weakly (Figs 72–74). Type 2 shows above all fluorescence of the first and second segments of the proximal tubule and a centrilobular fluorescence of the hepatocytes. Type 3 reacts more or less equally with human tissues and scarcely at all with rat tissues. Types 4 and 5 react with both rat and human tissue, with a predilection for the centrilobular regions. Little experience with type 6 is available to date.

Tip: the distal tubules – the preferred reaction site for mitochondrial antibodies – are almost always negative (see Tab. 25).

Clinical significance: the fundamental relationships can be inferred from Table 20; the detection of type 1 antibodies in pediatric autoimmune hepatitis is of particular practical significance, because in this case immuno-suppression is indicated. For this rare disease picture, with an incidence of approx. 0.3/million inhabitants per year and a prevalence of approx. 5/million inhabitants, the specificity is over 90%, with virtually all the titres in excess of 1:640, and a definitive sensitivity of 100%. The relationship of type 1 to viral hepatitis NonA/NonB(C), currently under special discussion, is not new. This had already been pointed out in the 1st edition in 1979.

Literature: Doniach et al. [1971]; Abuaf et al. [1991]; Storch [1992].

7.3.2.18 Antibodies against mitochondria

Synonyms/abbreviations: (anti)-mitochondrial antibodies (AMA).

Target antigen(s): very numerous, antigens for type 2 antibodies are currently the best known; these are associated with enzymes of the pyruvate-dehydrogenase or α-keto-acid-(2-oxo-carboxylic acid)-dehydrogenase complexes of the inner mitochondrial membrane, respectively (Tab. 22). Glycogen phosphorylase, sarcosine dehydrogenase and sulphite-oxidase can, furthermore, also be considered antigens (Klein & Berg [1990, 1991]).

Properties of the antibodies: neither organ nor tissue-specific, nor species-specific; typical of the disease, IgG but also IgA, IgM and IgD, predominantly complement fixing; titres from 1:20 to over 1:10,000.

Antigen substrate: organ block consisting of stomach, liver, kidney and thyroid; where possible human tissue blood group Od, or a block containing rat kidney, rabbit liver and mouse stomach.

Conjugate: anti-human IgG.

Pattern of fluorescence: ten types have been described (Storch [1982]; see Tab. 23). The most important type diagnostically is type 2; it produces a finely granular fluorescence of the cytoplasm, in particular of the distal tubules of the kidney, the parietal cells of the stomach, thyroid

Table 22. Antigens of the 2-oxo-acid dehydrogenase complex versus antibodies against mitochondria in primary biliary cirrhosis (based on Leung et al. [1994]).

Molecular weight	Enzyme	% AMA
74 kDa	E2 acetyl-transferase	> 95%
56 kDa	Protein X	> 95%
52 kDa	E2 acyl-transferase	71%
48 kDa	E2 succinyl-transferase	72%
41 kDa	E1 a-decarboxylase	66%
36 kDa	E1 ß-decarboxylase	2%

Table 23. Antibody sub-types directed against mitochondria. Average fluorescence intensities with rat kidney sections (based on Storch [1982]).

Antibody type	Fluorescence intensity for rat kidney tubules			
	proximal tubules			distal tubules
	P1	P2	P3	
AMA-1	++	++	+	+++
AMA-2	++	++	±	+++
AMA-3	++	++	++	++
AMA-4	++	++	+	+++
AMA-5	++	++	+	±
AMA-6	++	±	-	-
AMA-7	-	-	±	+++
AMA-8	-	-	-	+++
AMA-9	++	++	-	-
AMA-10	++	++	+++	+++

epithelia and in hepatocytes (Figs 75–77, Tab. 25).

Tip: an accurate comparative examination of liver, kidney and stomach sections is necessary in order to avoid confusion with other antibodies, especially those directed against ribosomes and microsomes (see Tab. 25). A positive complement fixation test (CFT) is, furthermore, observed with types 2 and 3 for cardiolipin, and with type 6 for mitochondria. Non-specific staining using low serum dilutions of conjugates which are barely suitable, must also be excluded. One must also realise that two systems of classification are in use, and that these are only identical up to and including type 6 – one based on Storch [1981], AMA 1–10, and the other based on Berg & Klein [1992], AMA 1–9. Types 7,8 and 9 of the Berg classification do not react under the conditions used for

Table 24. Sub-types of anti-mitochondrial antibodies and their associated diseases (based on Storch [1983]).

Type	Disease predominantly associated with sub-type
AMA-1	Secondary syphilis; anti-cardiolipin antibody, Wassermann reaction.
AMA-2	Primary biliary cirrhosis.
AMA-3	Pseudo-lupus erythematosus syndrome.
AMA-4	The mixed form of primary biliary cirrhosis and chronic aggressive hepatitis.
AMA-5	Visceral lupus erythematosus, autoimmune haemolytic anaemia.
AMA-6	Drug-induced hepatitis following Iproniazid (Marsilid) treatment.
AMA-7	Autoimmune hepatitis.
AMA-8	Autoimmune hepatitis, obliterative endangiitis, German measles (rubella) – blood group associated antibodies
AMA-9	Drug-induced hepatitis.
AMA-10	Visceral lupus erythematosus.

Table 25. Localisation and subjective average fluorescence intensity for mitochondrial, microsomal and ribosomal antibodies.

Antigen substrate	Mitochondrial Ab	Microsomal Ab	Ribosomal Ab
Rat kidney:			
Cortex (P1, P2)	+/++	+	+/++
Cortico-medullary junction (P3)	+	++/+++	+/++
Henle's loop	+++	-	+/++
Distal tubules	+++	-	+/++
Rat liver:			
Hepatocyte cytoplasm	(+)/+ diffuse granular	++/+++ diffuse homogeneous	++ perinuclear
Rat stomach:			
Parietal cells	+++	-/+	-
Chief cells	(+)/+	(+)	++/+++

immunofluorescence; they are not considered further.

Occurrence: see Table 24. AMA type 2 are of particular practical importance; they occur primarily in primary biliary cirrhosis with a frequency of 96%. The relationship of AMA to chronic aggressive hepatitis (CAH) and other diseases has been judged to be somewhat variable.

Clinical significance: AMA type 2 counts as a characteristic marker for primary biliary cirrhosis or its early stages, for non-purulent destructive cholangiitis or in the pre-clinical stage when other signs are absent; its diagnostic value must be emphasised. The absence of AMA, and also ANF, in clinically or histologically suspect cases cannot be used to exclude the presence of primary biliary cirrhosis but does, however, make it extremely unlikely. Literature values are available for the incidence, $0.6–1.5/10^5$ per year, the prevalence, $3.7–14.4/10^5$, the specificity, 11–49%, and the sensitivity, 90–100%. In cases of jaundice of unknown cause, high titres of AMA >1:1000 make a diagnosis of an extra-hepatic blockage or secondary biliary cirrhosis very unlikely. These indicate, in the first instance, a primary biliary cirrhosis of currently unknown origin or its early stages. High titres of AMA occur only exceptionally in other diseases. Drugs which can also induce high antibody titres are no longer commercially available. Low titres have been observed in numerous other diseases, for example, in scleroderma, Sjögren's syndrome and SLE in approximately 5–10% of cases, in rheumatoid arthritis and other autoimmune diseases in around 2% of cases, and in less than 1% of healthy patients.

Literature: Walker et al. [1965]; review: Berg & Klein [1992].

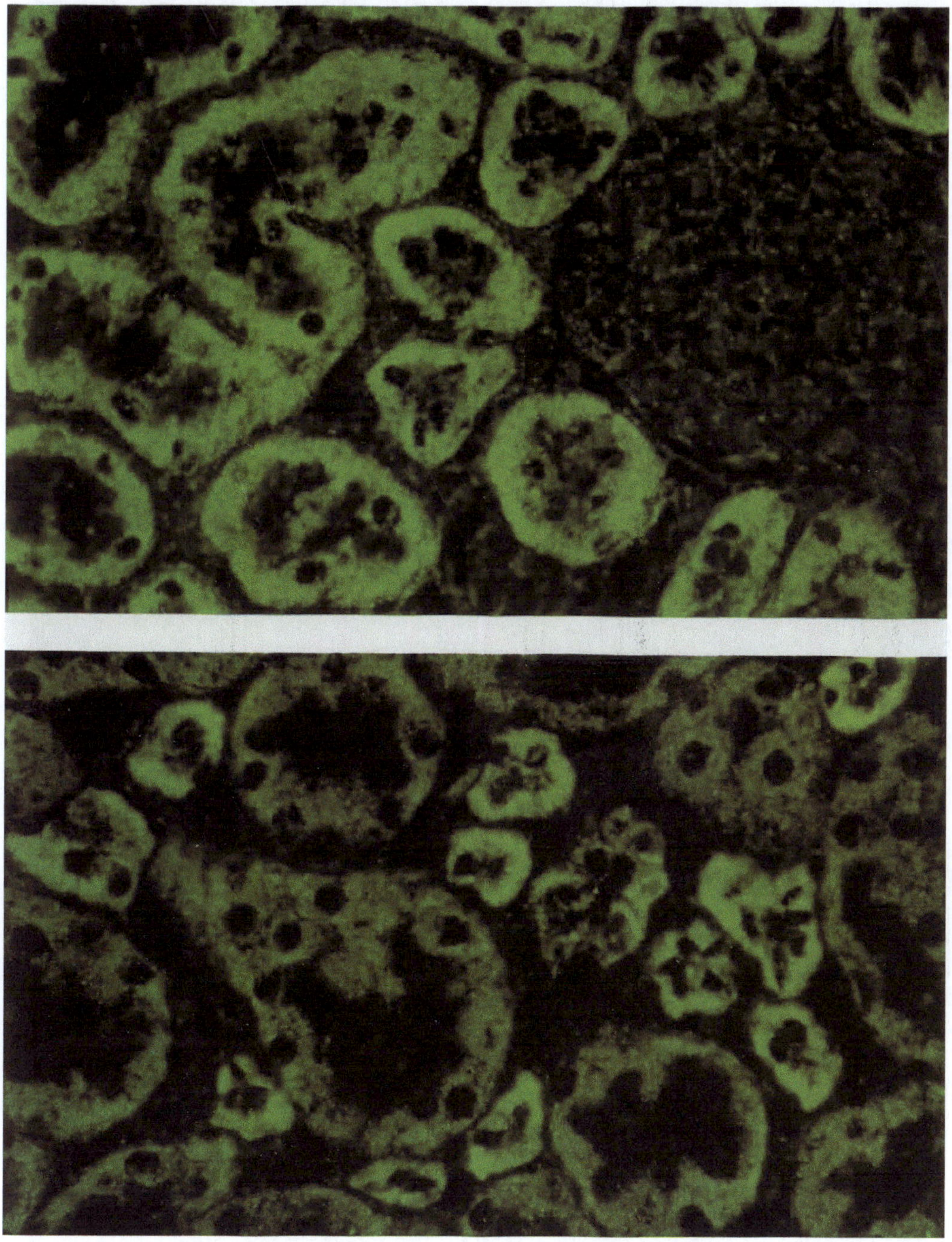

a

b

Figure 75. Antibodies directed against mitochondria type 2; antigen, block containing rat liver, stomach, kidney and heart – (a) *superficial renal cortex*: cytoplasmic fluorescence of all tubules especially the distal ones; the glomeruli show finely granular fluorescence; (b) *deep cortex*: the large tubules (P3 region) react weakly, the small distal tubules more strongly; (c) *stomach*: intense finely granular fluorescence affecting the parietal cells, with the remaining cells and muscle reacting more weakly; (d) *heart*: diffuse, intense, finely granular fluorescence depending on the location of the mitochondria in the muscle fibres (objective ×40, ×500).

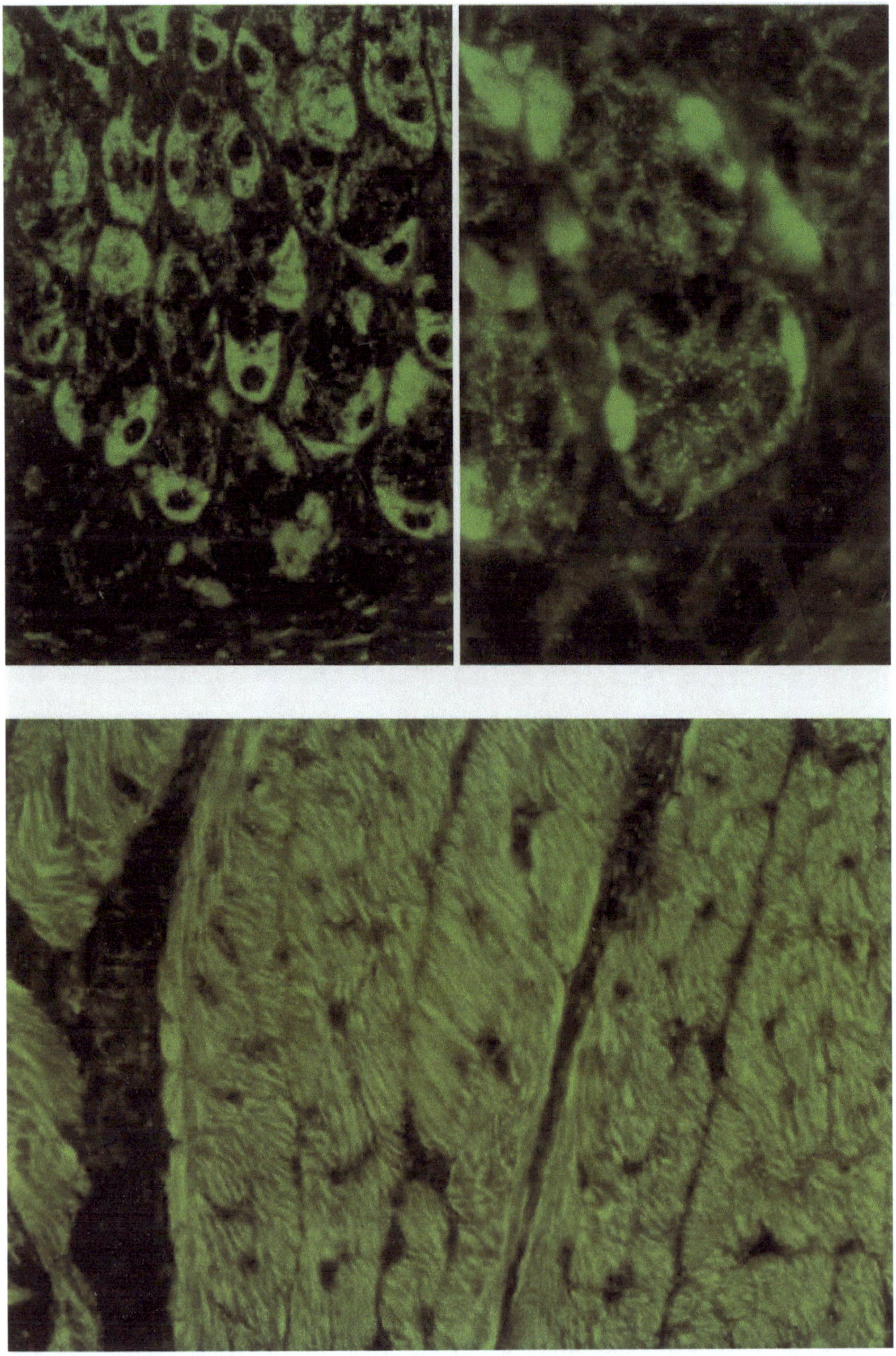

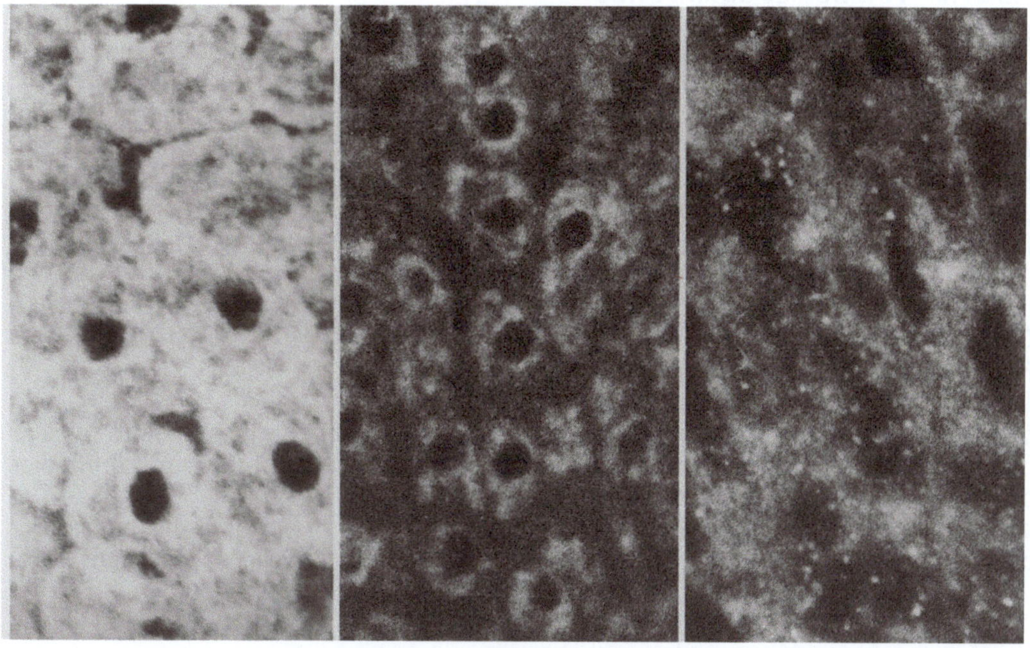

a b c

Figure 76. Antibodies against mitochondria compared with those against microsomes (ER) and ribosomes; antigen, as for Figure 70: *liver* – (a) diffuse cytoplasmic fluorescence for microsomal antibodies (+++); (b) granular fluorescence showing peri-nuclear emphasis for ribosomal antibodies (++); (c) irregular granular fluorescence for mitochondrial antibodies (+); (objective ×40, ×800).

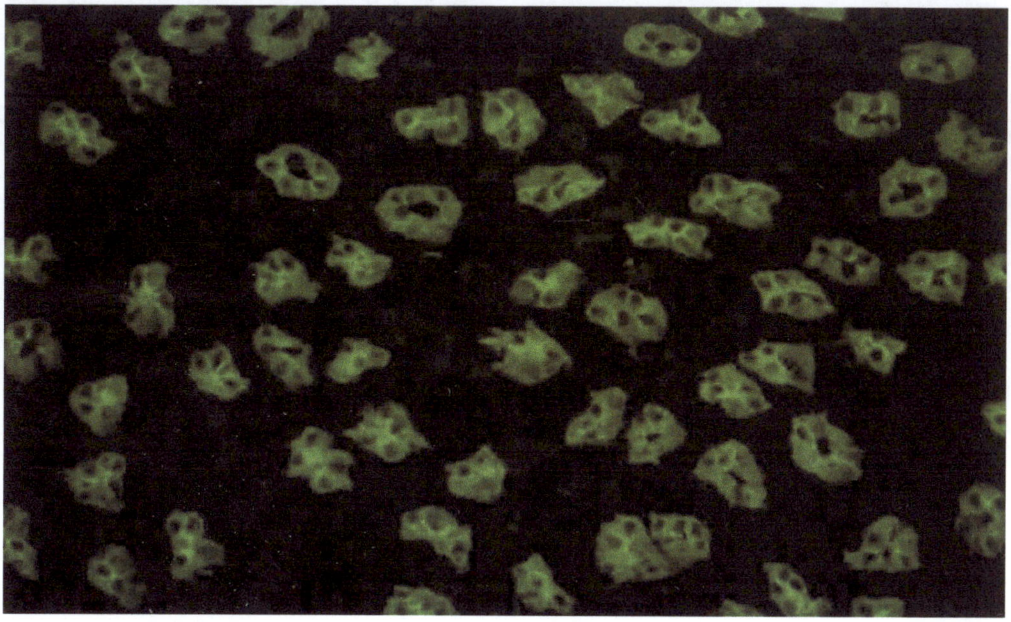

Figure 77. Antibodies against mitochondria type 8; antigen, as for Figure 70 – fluorescence exclusively affecting the distal tubules of the renal medulla and parietal cells, as well as heart and liver; the proximal and remaining distal tubules are negative (objective ×16, ×200).

7.3.2.19 Antibodies against ribosomes

Synonyms/abbreviations: anti-ribosomal antibodies (ARA).
Target antigen(s): ribosomes.
Properties of the antibodies: neither organ, nor tissue or species-specific, typical for the disease, predominantly IgG.
Antigen substrate: organ block containing liver, stomach and kidney.
Conjugate: anti-human IgG.
Pattern of fluorescence: cytoplasmic fluorescence of the chief cells of the stomach (Fig. 78), of hepatocytes with perinuclear emphasis, and a similar pattern of fluorescence affecting proximal and distal renal tubules and the glomeruli, is characteristic.
Tip: confusion with mitochondrial and microsomal antibodies, and antibodies against liver ribosomes, should be avoided (Table 25).
Clinical significance: ribosomal antibodies are evidence of active, untreated SLE or chronic aggressive hepatitis (CAH).
Literature: Sturgill & Preble [1967]; Storch [1973]; Homberg et al. [1974].

7.3.3 Antibodies against cell nuclei

Synonyms/abbreviations: anti-nuclear factors (ANF), anti-nuclear antibodies (ANA).
Target antigen(s): numerous nucleoproteins and derivatives, especially enzymes,

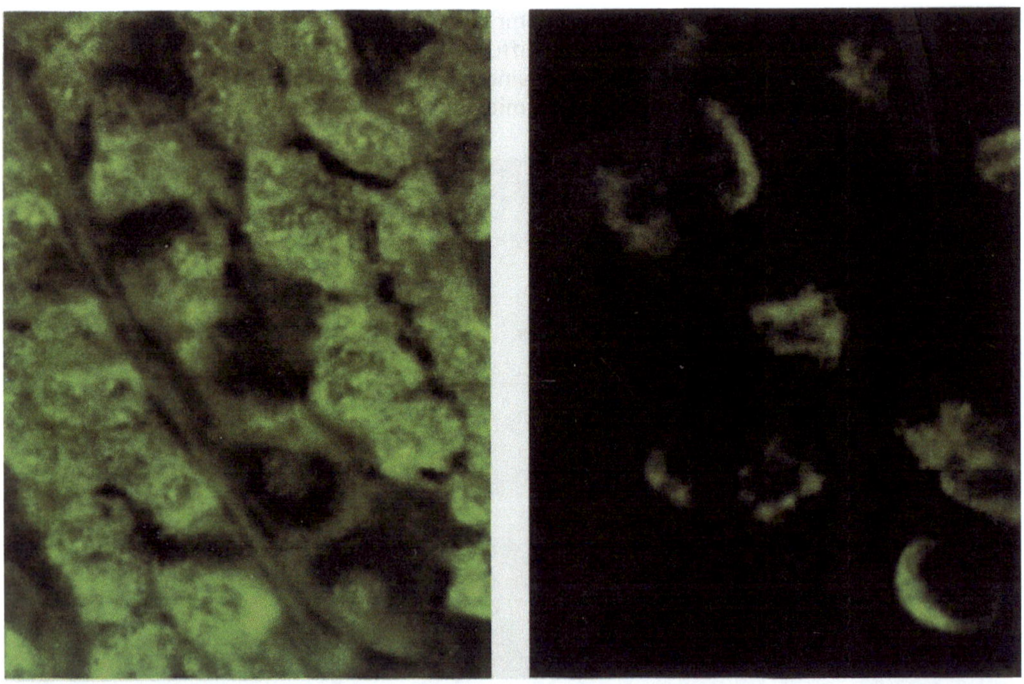

a

Figure 78. Antibodies directed against ribosomes compared with those against parietal cells; antigen, as for Figure 68 – *stomach*: (a) fluorescence mainly of the chief cells is evidence of antibodies against ribosomes, with the parietal cells remaining spared, in general, with only their nuclei reacting as a consequence of the simultaneous presence of ANF; (b) fluorescence exclusively affecting parietal cells in the presence of antibodies directed against parietal cells (objective ×40, ×500; Prof. Seelig, Karlsruhe).

native and denatured DNA, histones, RNA and ribonucleoproteins, in the nucleoplasm and nucleolus; soluble nuclear receptors, proteins associated with mitosis, etc, (Tab. 26).

Properties of the antibodies: all immunoglobulin (Ig) classes, IgG especially clinically relevant, to some extent complement fixing; neither organ, species nor disease-specific.

Antigen substrate: the screening test described in the 1st edition with undiluted serum is hardly used anymore because too many false positives were obtained. Apart from this, a cultured cell line of a human laryngeal carcinoma, HEp–2, is being increasingly used in place of rat liver, which was formerly customary, together with rat kidney and stomach. The cells have a large nucleus so that numerous different patterns can be distinguished, including those during stages of mitosis. In order not to obtain too many positive results, which are mostly meaningless clinically, initial serial dilutions of 1:40 for cryotome sections and 1:80 for HEp–2 cells should be used.

Cultivation of and carrying out immunofluorescence with HEp–2 cells (based on Humbel [1993])

Apparatus and reagents

1. HEp–2 epithelial cells (ATCC CCL–23).
2. Dulbecco's medium (Gibco BRL 042–2501).
3. Penicillin-streptomycin solution (Gibco BRL 061–05075); this contains 500 U/ml penicillin and 5000 µg/ml streptomycin and can be obtained in a lyophilised state.
4. Incubator.

Preparation of the cell culture medium

The following sterile solutions are mixed together:

400 ml distilled water;

50 ml Dulbecco's MEM 10X (Gibco BRL 042–2501);

24 ml 7.5% sodium bicarbonate (Gibco BRL 043–05080);

5 ml 200 mM L-glutamine (Gibco BRL 043–05030);

5 ml 100 mM sodium pyruvate (Gibco BRL 043–01360);

0.2 ml penicillin-streptomycin solution (Gibco BRL 061–05075);

50 ml foetal (neonatal) calf serum.

Cultivation of HEp–2 cells

- ca. 10^5 HEp–2 cells in 5 ml DMEM plus 10% calf serum, placed on slides;
- cell growth at 37°C in 5% CO_2 for 40–48 hours;
- fixation of the cells when the cell culture on the slide shows 80–90% confluence.

Fixation of the cells

Fixation can influence as well as change the antigenicity and pattern of fluorescence obtained in different ways. Thus, different commercial equipment frequently gives completely different results. This is important, for example, for antibodies against SS-A/Ro, which do not allow, most of the time, cytoplasmic staining to be seen using methanol-acetone fixed cells. On the other hand, using formaldehyde fixation one can almost always recognise weak fluorescence affecting the cytoplasm close to the nuclear fluorescence.

Methanol-acetone fixation

- wash three times in cold PBS;
- fix for five minutes in methanol at –20°C;
- immerse briefly twice in fresh (each time) acetone;
- air-dry for 15 minutes;
- store at –20°C in air-tight containers – this is possible for many months;
- before use, thaw quickly using a cold-air blower, avoid drying out at all costs.

Paraformaldehyde-Triton fixation

- wash three times with cold PBS;
- fix for 15 minutes with paraformaldehyde (3% in PBS);
- wash three times with cold PBS;

- immerse for five minutes in Triton X-100 (0.2% in PBS);
- wash three times with cold PBS;
- storage should be avoided.

Preparation of serum

- take a sample of blood without additives;
- centrifuge for 10 min at 2500 g_{av};
- pipette off the serum without disturbing the sedimented cells;
- if the serum is not to be used immediately, or within a week (storage at +4°C is satisfactory), it should be kept at at least –20°C, or better at –70°C. The addition of sodium azide, 0.1–0.3 mg/ml, is recommended;
- dilution of the serum should be done with "PBS-T". This PBS solution contains 0.5 g/l Tween 20 and 0.1 g/l sodium azide. In this way a clean 'picture' will be achieved and bacterial contamination avoided. Background fluorescence can be reduced even further by the addition of serum albumin (20 g/l) to the diluting buffer. PBS: 13 mM Na_2HPO_4; 3 mM KH_2PO_4; 140 mM NaCl; pH 7.4.

Immunofluorescence is carried out as described in Section 4.2.2.2.

Conjugate: anti-human IgG; the working dilution of the conjugate should contain 50 µg immunoglobulin protein with an F/P ratio of three.

Pattern of fluorescence: the former customary sub-division of cryotome sections into four basic categories – homogeneous, peripheral, speckled and nucleolar – has been increasingly abandoned with the introduction of HEp–2 cells because further differentiation is possible; see Table 26.

The characteristic pictures given by cells in interphase and various stages of mitosis need to be described separately (Figs 79 & 80). Because different antibodies frequently coexist, the use of a scheme is to be recommended in order that nothing is forgotten. The following can be included in such a scheme: interphase cells – nuclear membrane, nucleoplasm, nucleolus and cytoplasm – as well as cells in mitosis. The reaction with the chromosomes is especially interesting for those cells undergoing mitosis but the pattern in the respective stages – pro-, meta-, ana- and telophase, as well as cytokinesis – can also be useful for recognising rare or new patterns.

Tip: in order to foster standardisation, a reference conjugate and standard sera should be used (at present available with homogeneous, speckled, centromere-kinetochore and nucleolar patterns of reaction). Furthermore, sera that are also available with specificity against native DNA, Sm, U1-RNP, SS-A/Ro, SS-B/La, Scl–70 and Jo–1, can be useful. These are obtainable from the ANA Reference Laboratory, Immunology Branch, 1–1202 A25, Centre for Disease Control, Atlanta, Georgia 30333, USA, or from the Central Laboratory, Blood Transfusion Service, The Netherlands Red Cross, Plasmalaan 125, 1066 CX Amsterdam, The Netherlands; see also Feltkamp [1993].

Occurrence: ANF occur in numerous diseases and also in healthy patients, especially in the old. One can expect to see positive results in:

active SLE	100%
inactive SLE	90%
drug-induced SLE	100%
discoid lupus erythematosus	30%
Sharp's syndrome (mixed connective tissue disease, MCTD)	100%
systemic scleroderma	80%
Sjögren's disease	50%
Felty's syndrome	40%
rheumatoid arthritis	30%
myasthenia gravis	30%
panarteritis nodosa (PAN)	20%
chronic aggressive hepatitis (CAH)	15%
other autoimmune diseases	10%
healthy patients (20–60 yrs. old)	5%
healthy patients (60–80 yrs. old)	10%

Clinical significance: ANF are generally evidence of an immunological disturbance. Lupus erythematosus has still the greatest relevance in practice. The repeated absence of ANF exclude this diagnosis to a large degree. Both the fluores-

Table 26. Sub-division (classification) of antibodies directed against nuclear material based on the predominant fluorescence picture (according to Humbel et al. [1993] with modifications).

1. **Nuclear membrane fluorescence**
   a. fine, ring-shaped; antigen, lamin.
   b. homogeneous; antigen, lamin.
   c. punctate; antigen, nucleoporin.
   d. perimitotic; antigen, perichromin.

2. **Homogeneous nuclear fluorescence**
   a. diffuse homogeneous; antigen, DNA and histone.
   b. peripheral emphasis; antigen, DNA.

3. **Speckled nuclear fluorescence**
   a. coarse grained; antigen, nuclear matrix, hn-RNP.
   b. medium-sized grains; antigen, Sm, U1-RNP.
   c. fine grains; antigen, S1, Mi and others.
   d. dense grains and nucleolar; antigen, SS-B/La.
   e. fine isolated grains; antigen, SS-A/Ro.
   f. poly- or pleomorphic; antigen, PCNA/Cyclin.

4. **Punctate nuclear fluorescence**
   a. punctate in interphase and mitotic cells; antigen, centromeres and kinetochores.
   b. isolated in interphase cells ("nuclear dots"); antigen, coilin and Sp 100.

5. **Nucleolar fluorescence**
   a. homogeneous; antigen, nucleolin and PM/Scl.
   b. coarse granules; antigen, fibrillarin.
   c. finely stippled; antigen, topoisomerase I and B23.
   d. spotted; antigen, NOR and RNA polymerase 1.
   e. perinuclear; antigen, chromatin.

6. **Mitosis-associated fluorescence**
   a. fluorescence exclusively in metaphase; antigen, NSP.
   b. affecting the mitotic spindle – antigen tubulin:
      i) fibres and spindle;
      ii) in the region of the centrosome ("MSA-1"); antigen, NuMA;
      iii) pole of the spindle; antigen, centriole;
      iv) pericentriolar; antigen, centrosome.
   c. intermediate (mid) body ("MSA-2"); antigen, amongst others the 330 kDa protein.
   d. kinetochore especially in prophase and metaphase; ("MSA-3"), antigen, CENP-F.
   e. homogeneously chromosomal; antigen, "cell division antigen".

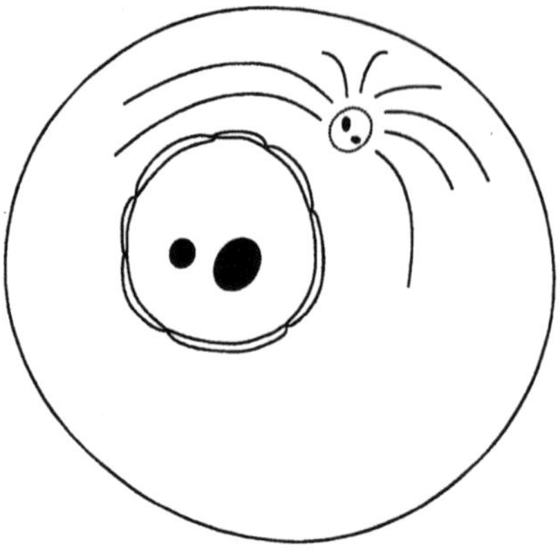

1. Interphase: centrosome with the centriole pair, with microtubules branching off; nucleus with two nucleoli, surrounded by the nuclear membrane (inner and outer layers with nuclear pores).

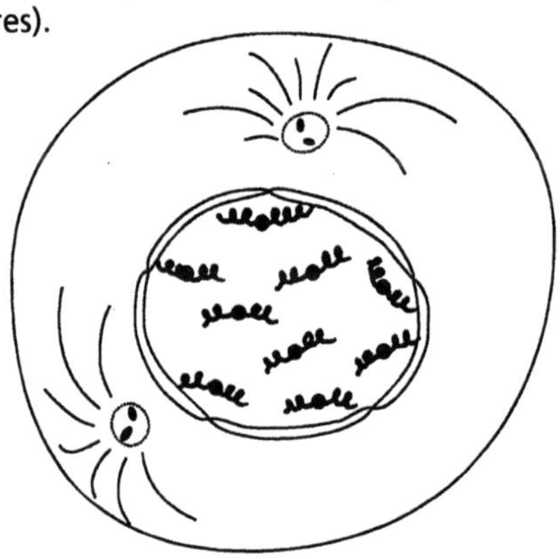

2. Prophase: formation of the mitotic spindles; chromosomes aggregate.

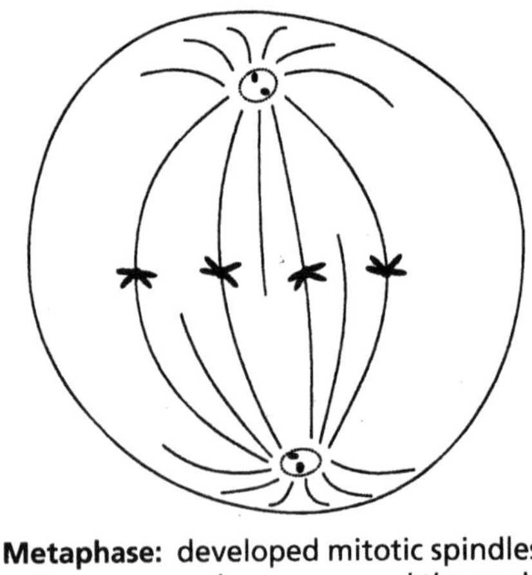

3. Metaphase: developed mitotic spindles; the chromosomes have arranged themselves along the metaphase plate.

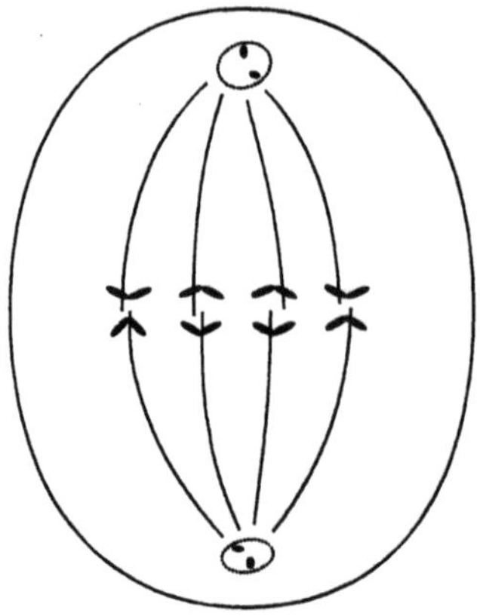

4. Early anaphase: chromosomes travel towards the spindle poles.

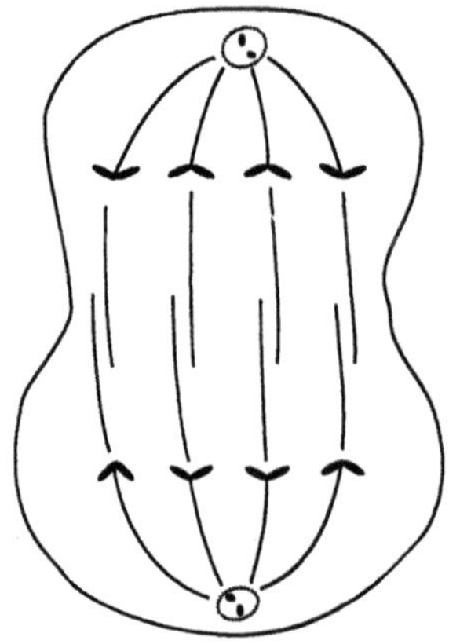

5. Late anaphase: the mid-body and cleavage furrow are clearly visible.

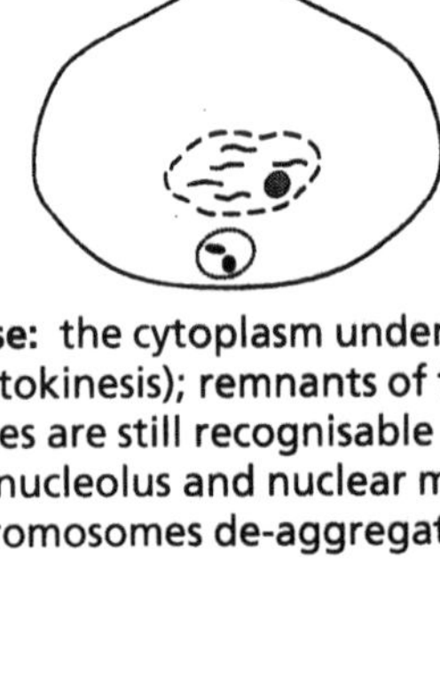

6. Telophase: the cytoplasm undergoes division (cytokinesis); remnants of the microtubules are still recognisable in the mid-body; nucleolus and nuclear membrane reform; chromosomes de-aggregate.

Figure 79. Scheme showing the stages of mitosis.

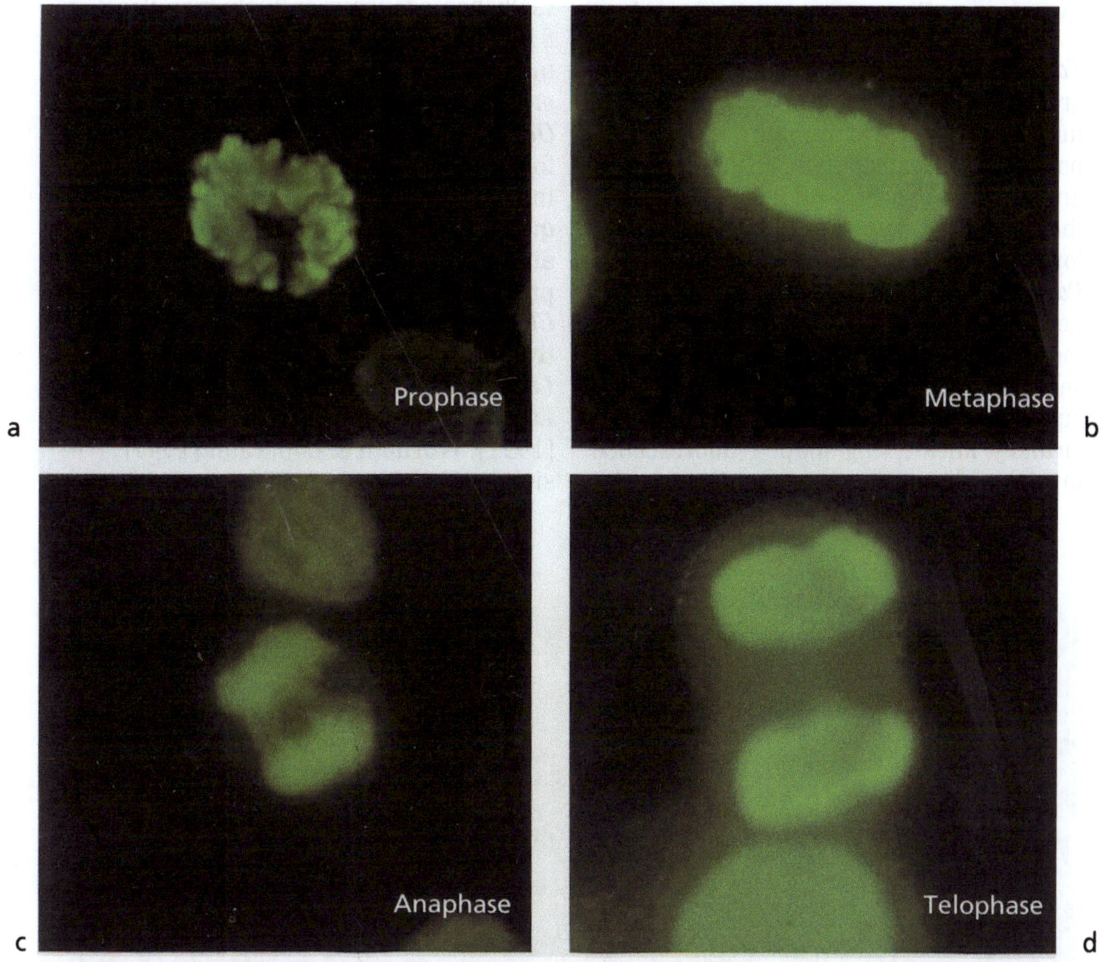

Figure 80. Appearance of the chromosomes in the four stages of mitosis. Incubated with antibody against nuclear material, showing a homogeneous fluorescence picture and reaction of the chromosomes in: (a) prophase; (b) metaphase; (c) anaphase; and (d) telophase (objective ×100, ×1250).

cence picture and the titre have significance. Antibodies against native double-stranded DNA, best detected using *Crithidia*, are to a large extent pathognomic for SLE (see below for the procedure). The association between the remaining fluorescence pictures, causative antigens and the actual illness, is only occasionally clear. An association has been described in individual cases but mostly requires confirmation using other methods, in particular ELISA or blotting procedures.

Literature: because the available literature is extremely large and growing constantly, it is appropriate to mention only a few reviews and monographs, for example: Tan [1982, 1989, 1992]; Seelig [1983]; Nguyen & Markovits [1991]; Conrad et al. [1993]; Fritzler et al. [1984]; Humbel [1993a,b]; Herkner et al. [1994].

7.3.3.1 Antibodies showing nuclear membrane fluorescence

Target antigen(s): laminins, especially groups of nuclear intermediary filaments, nuclear pores, perichromatin and nucleoporin; for an explanation of the structure and function of cell components in general the specialist literature should be consulted, for a review see Alberts et al. [1994].

Pattern of fluorescence: in contrast to a peripheral pattern it is simply the nuclear membrane which fluoresces. The pattern seen depends on the focal plane. In an equatorial plane one can distinguish a fine, virtually linear, ring-shaped fluorescence involving the nuclear margin. Outside the equatorial plane, the nucleus appears virtually homogeneous or finely granular (Fig. 81). One of the main features of perichromatin fluorescence is an additional peripheral staining during metaphase.

Occurrence: McKeon et al. [1983, 1984] found these antibodies for the first time in scleroderma. They occur more frequently, however, in patients with SLE and autoimmune hepatitis, as well as in primary biliary cirrhosis.

Clinical significance: not clear, a relatively rare picture.

Literature: McKeon et al. [1983, 1984]; Lassoued et al. [1988]; Senecal et al. [1991]; Worman & Courvalin [1991]; Wesierska-Gadek et al. [1996].

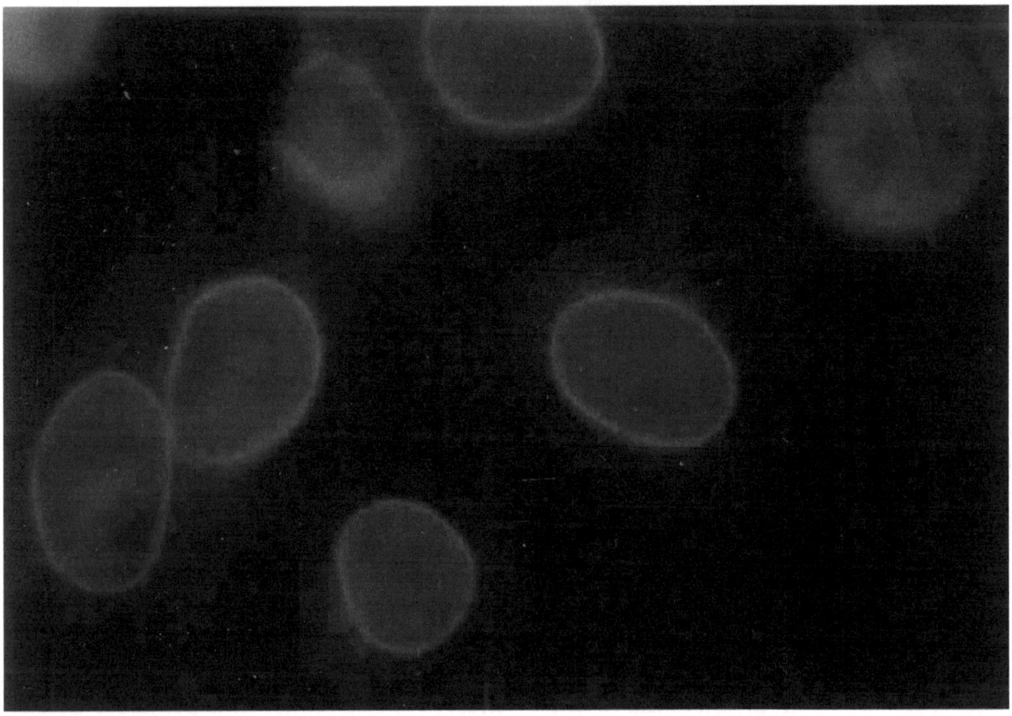

Figure 81. Antibodies directed against nuclei (nuclear membrane pattern).
Nuclear membrane: approximately homogeneous linear fluorescence which, on focusing in the equatorial plane, appears as a ring.
Nucleoplasm: generally negative; the weak staining is caused by the plane of projection.
Nucleoli: negative.
Stages of mitosis: (upper right, in this case pro-metaphase) diffuse, extremely fine, granular fluorescence of the cytoplasm; chromosomes negative.
Cytoplasm: negative.
(objective ×100, ×1250).

7.3.3.2 Antibodies showing homogeneous nuclear fluorescence

Target antigen(s): histones, DNA and other partially unknown antigens.
Pattern of fluorescence: numerous different patterns can occur; thus the nucleoli can be spared, or involved in a recognisable way, or even not be discernible. The nucleus can emit a diffuse homogeneous fluorescence or show a weak granular pattern, which can give rise to difficulties in distinguishing these from other patterns, especially the speckled type (Fig. 82). The reaction of the chromosomes during the stages of mitosis, is important. If this is positive and produces a detectable marked staining of the nuclear periphery, then the presence of double-stranded DNA antibodies should be suspected. This should be confirmed using, for example, the Farr technique or *Crithidia*. Specialised immunofluorescence techniques can be used to test whether histone antibodies are present (see Burlingame & Rubin [1994]).

Detection of antibodies against double-stranded desoxy-ribonucleic acid (ds-DNA) – based on Aarden et al. [1975]
Antigen substrate: drop preparations of *Crithidia luciliae*, a blood flagellate non-pathogenic for man, which has a large kinetoplast reacting almost exclusively with double-stranded DNA antibodies. *Crithidia luciliae* can be obtained from the American Type Culture Collection (ATCC), 12301 Parklawn Drive, Rockville,

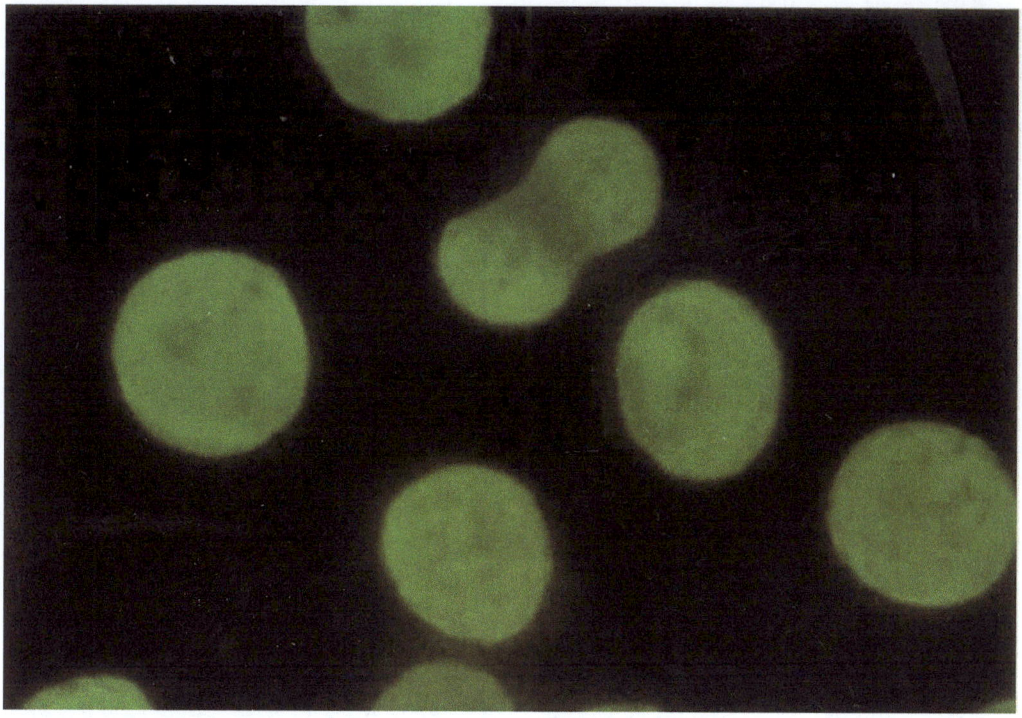

Figure 82. Antibodies directed against nuclei (homogeneous pattern).
Nuclear membrane: not distinguishable.
Nucleoplasm: diffuse, partly homogeneous, partly extremely fine, granular fluorescence.
Nucleoli: negative, standing out as dark patches.
Stages of mitosis: positive, similar picture to that seen for cells in interphase, chromosomes mostly, however, more strongly positive except in telophase. Intermediary bodies stand out as dark areas. Compare with Figure 79. *Cytoplasm*: negative (objective ×100, ×1250).

Maryland, USA. These organisms can be kept at –70°C and grown at 20–25°C (Barthelmes et al. [1983]) A variety of commercial test equipment is available.

Procedure:

- a crithidial suspension, washed three times with PBS, pH 7.4 (see section 2.4.2), and then centrifuged at 3000 g_{av} for ten minutes;
- resuspension in distilled water at a density of 2×10^7 organisms/ml;
- 10 µl aliquots placed on a microscope slide and air-dried using a fan;
- ten minutes fixation in 96% ethanol at room temperature;
- 30 minutes incubation with the patient's serum;
- wash for 30 minutes;
- incubate for 30 minutes in the presence of FITC-labelled anti-IgG (A, M, D, E or polyvalent).
- wash for 30 minutes;
- embed.

Pattern of fluorescence: in the presence of anti-ds-DNA antibodies, the kinetoplast fluoresces homogeneously, whilst the neighbouring nucleus may show a weak ring-shaped fluorescence (Fig. 83). The advantages over the Farr technique are:

1. it is easily carried out in an immunofluorescence laboratory;
2. the immunoglobulin class of the antibody can be determined;
3. there is no interference with C1q-binding to DNA during complement fixation, or from the anti-complement activity of the serum;
4. there is no interference from antibodies against single-stranded DNA.

Tip: conversion of the titre into IU/ml is possible using Reference Serum Wo/80.

Clinical significance: antibodies against native ds-DNA occur really only in 40–50% of cases of SLE; high affinity antibodies, which can be detected using the *Crithidia* test, are, however, pathognomic

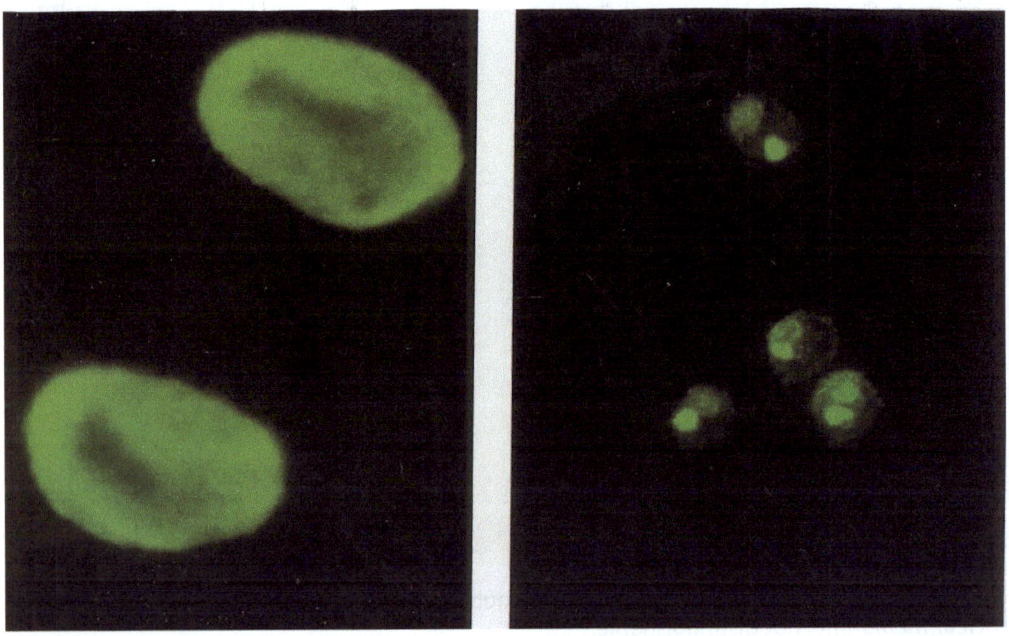

a

Figure 83. Antibodies against nuclei (anti-*ds*-DNA pattern): (a) in HEp–2 cells a similar pattern to Figure 80 with, however, peripheral emphasis of the homogeneous fluorescence; (b) in *Crithidia lucilae*, an intense homogeneous staining of the kinetoplast with the neighbouring nucleus showing a weaker, finely granular fluorescence, with peripheral emphasis in parts (objective ×100, ×1250).

for this disease picture. Although detection using *Crithidia* is less sensitive than the Farr test and above all ELISA, it does almost always show the highest comparative specificity and diagnostic efficiency. This has been only recently confirmed (Wigand et al. [1995]). The antibodies can precede the clinical diagnosis of SLE by up to a year!

Literature: Stingl [1975, 1976]; Smeenk [1993, 1994]; Isenberg et al. [1994]

7.3.3.3 Antibodies showing speckled nuclear fluorescence

Target antigen(s): the nuclear matrix, hn-RNP, U1-RNP, the antigens Sm, S1, Mi, SS-A/La, SS-B/Ro, PCNA/cyclin and others have to be taken into account.

Pattern of fluorescence: the chromosomes during the stages of mitosis are characteristically negative in all patterns; coarse speckling or graininess indicates primarily the nuclear matrix (Fig. 84) and heterogeneous nuclear-(hn)-RNP (Fritzler et al. [1984]); medium-sized granules, which always spare the nucleoli, are typical of U1-RNP antibodies (Fig. 85) – for a review see Tan [1982, 1989]. Assessment of fine granules is particularly difficult, especially as these can also be found in the cytoplasm, depending upon the method of fixation, particularly for SS-A/Ro antibodies (Fig. 86). Finally, the so-called heterogeneous pleo- or polymorphic patterns are easy to recognise (Fig. 87); the

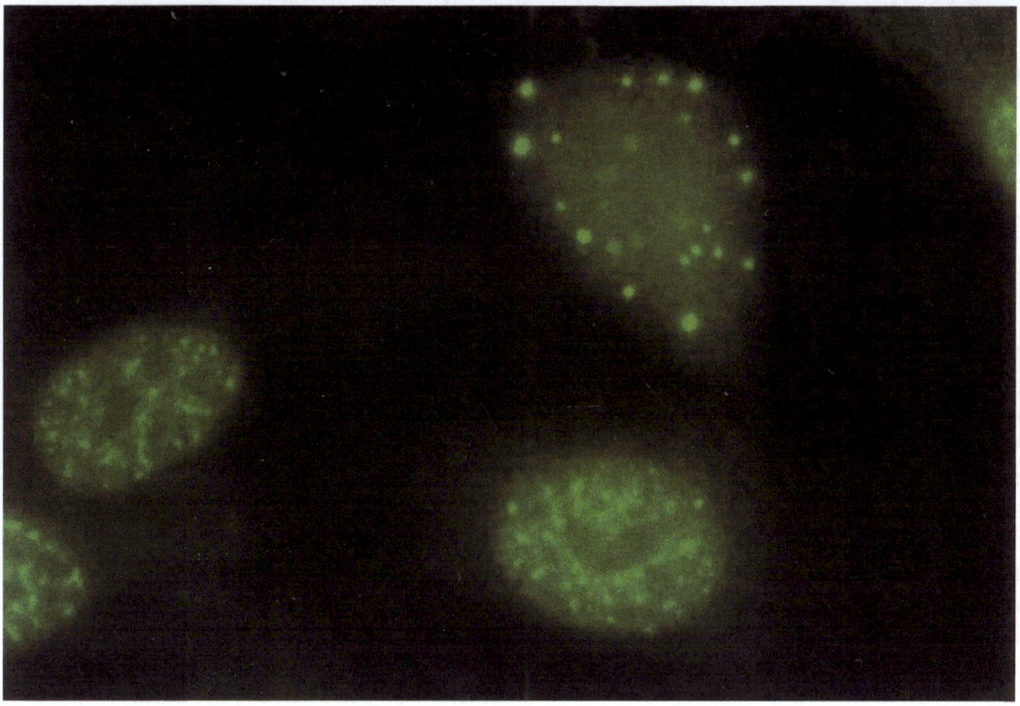

Figure 84. Antibodies against nuclei (matrix pattern).
Nuclear membrane: negative.
Nucleoplasm: isolated, frequently coarse grains or spots of different sizes.
Nucleoli: negative.
Stages of mitosis: various patterns, partly similar picture to that seen in interphase, with the chromosomes negative. *Cytoplasm*: outlines distinguishable (objective ×100, ×1250).

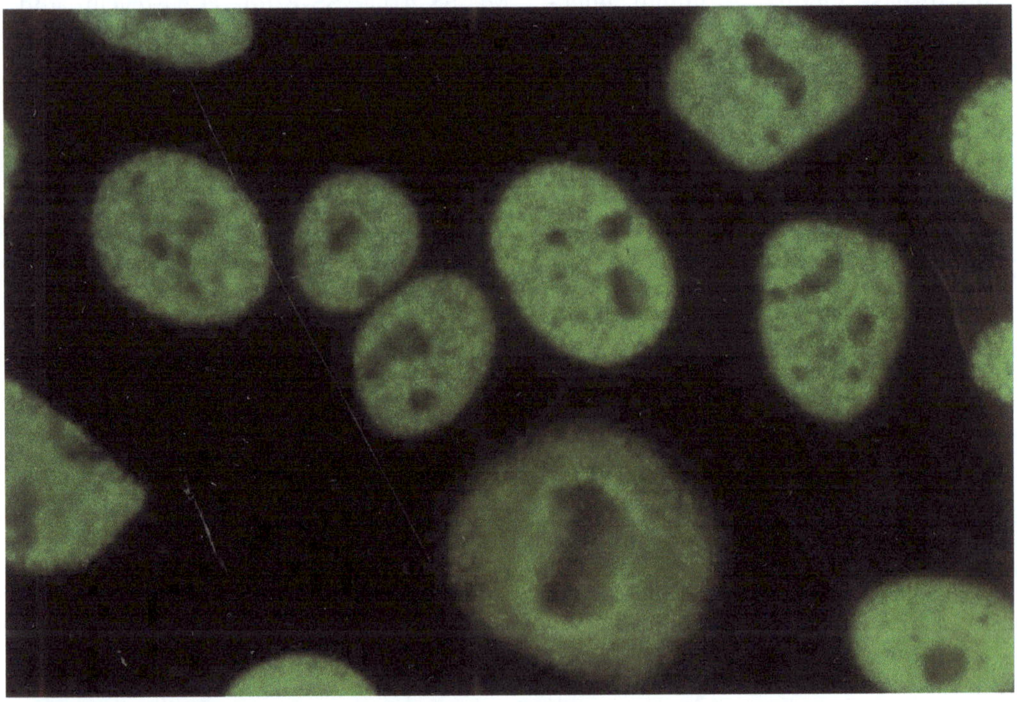

Figure 85. Antibodies against nuclei (RNP pattern).
Nuclear membrane: negative.
Nucleoplasm: very densely packed, mostly medium-sized grains.
Nucleoli: negative, stand out clearly as dark patches.
Stages of mitosis: (bottom centre, in this case metaphase) – cytoplasm shows fading – caused mainly by a loss of definition in regions with different depth of focus – and a similar pattern with emphasis of the region round the chromosomes, which are negative.
Cytoplasm: negative (objective ×100, ×1250).

considerable variability in staining is characteristic. Completely negative cells lie alongside those that are strongly positive, some nucleoli and nuclear membranes may also be stained. Only those cells which are in the synthesis-phase for DNA (the S-phase) react positively. The antibody is directed against PCNA (proliferating cell nuclear antigen) – Miyachi et al. [1978]; this consists of a 36 kDa protein, a helper protein for DNA-polymerase.

Occurrence and clinical significance: best known is the relationship between U1-RNP antibodies and Sharp's syndrome (MCTD = mixed connective tissue disorder, Sharp et al. [1972]), with, by definition, a frequency of 100%. SS-A/Ro and SS-B/La antibodies count as typical markers for Sjögren's syndrome, with a frequency of approx. 70–80%, or 60–70% respectively. These antibodies are also found in other diseases, for example, in SLE SS-A/Ro and SS-B/La are present in 35% and 15% of all cases. Sm antibodies only really occur in SLE, in around 30% of cases, being regarded, however, as largely specific for this disease. This high specificity for SLE also includes PCNA antibodies; the antibody is at any rate very rare (3%).

Literature: for reviews see Tan [1989 1992]; Sm antibodies – Hoch [1994]; SS-A/Ro – Chan & Buyon [1994]; SS-B/La – Pruijn [1994]; PCNA – Muro & Tan [1994].

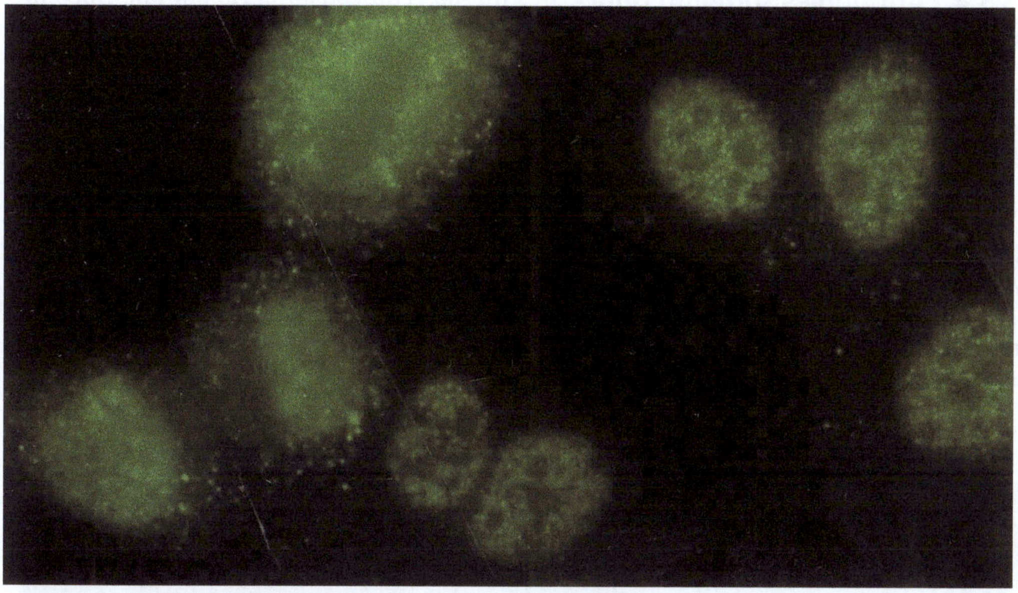

Figure 86. Antibodies against nuclei (SS-A/Ro pattern).
Nuclear membrane: negative; *Nucleoplasm*: mostly fine, isolated grains.
Nucleoli: almost always negative; *Stages of mitosis*: a similar picture for the cytoplasm as seen in interphase, with the chromosomes predominantly negative; *Cytoplasm*: analogous to the nucleoplasm, however, mostly weaker and less strongly pronounced (objective ×100, ×1250).

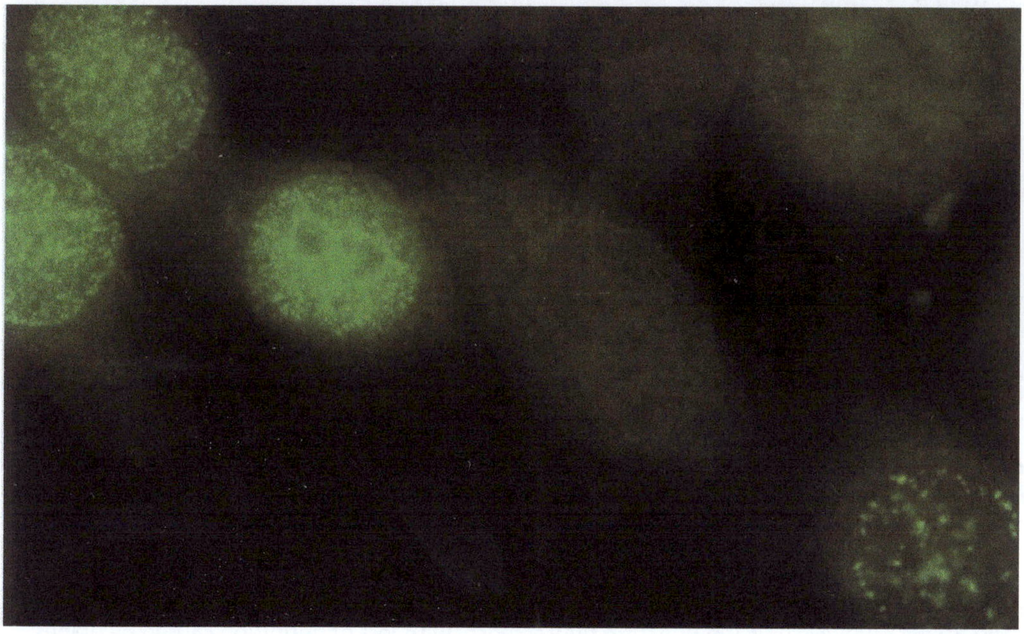

Figure 87. Antibodies against nuclei (PCNA pattern).
Nuclear membrane: occasionally finely granular.
Nucleoplasm: a range of patterns from absolutely negative to intensely positive nuclei is typical, mostly with fine or medium-sized granules of varying density; *Nucleoli*: mostly negative, individually strongly positive (not visible in the picture); *Stages of mitosis*: (upper right, in this case

7.3.3.4 Antibodies showing punctate nuclear fluorescence

This pattern has been distinguished from the speckled form because it produces a very characteristic picture.

7.3.3.4.1 Antibodies against centromeres (kinetochores)

Synonyms/abbreviations: kinetochore antibodies, anti-centromere antibodies.
Definition: one description describes the insertion points of the spindle fibres onto the chromosomes (DNA) as the centromeres. The centromeric spindle fibre insertion point, as part of the chromosome, is described as the kinetochore (for reviews see Alberts et al. [1994]; Brinkley et al. [1992]).
Target antigen(s): primarily the centromeric proteins, CENP-A, CENP-B and CENP-C – for a review see Verheijen [1994].
Pattern of fluorescence: punctate fluorescence (approximately 46) of the cell nuclei at interphase and mitosis. Each fluorescent point marks the position of a kinetochore, or its precursor. The number of

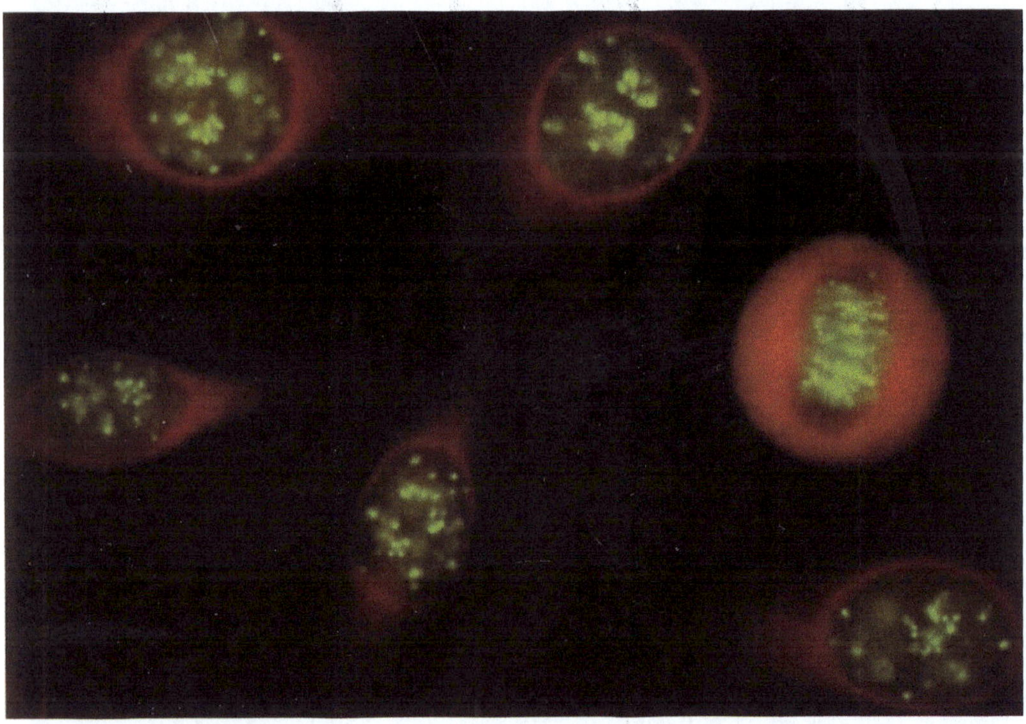

Figure 88. Antibodies against nuclei (centromere or kinetochore pattern). An image is shown recorded with relatively broad-band excitation (450–490 nm) and without blocking of the emitted red light (LP 520), so that one can achieve better orientation for what is in practice an important pattern. The orange-red appearance of the Evans Blue counterstain affects particularly cells in mitosis and also the nuclear membrane and cytoplasm.
Nuclear membrane: negative.
Nucleoplasm: relatively uniform punctate fluorescence, with loss of definition caused by planes with different depth of focus.
Nucleoli: negative.
Stages of mitosis: (in this case metaphase) – cytoplasm generally negative but showing strongly positive chromosomes with a compressed appearance to the kinetochores, otherwise analogous to that seen in interphase. *Cytoplasm*: negative (objective ×100, ×1250).

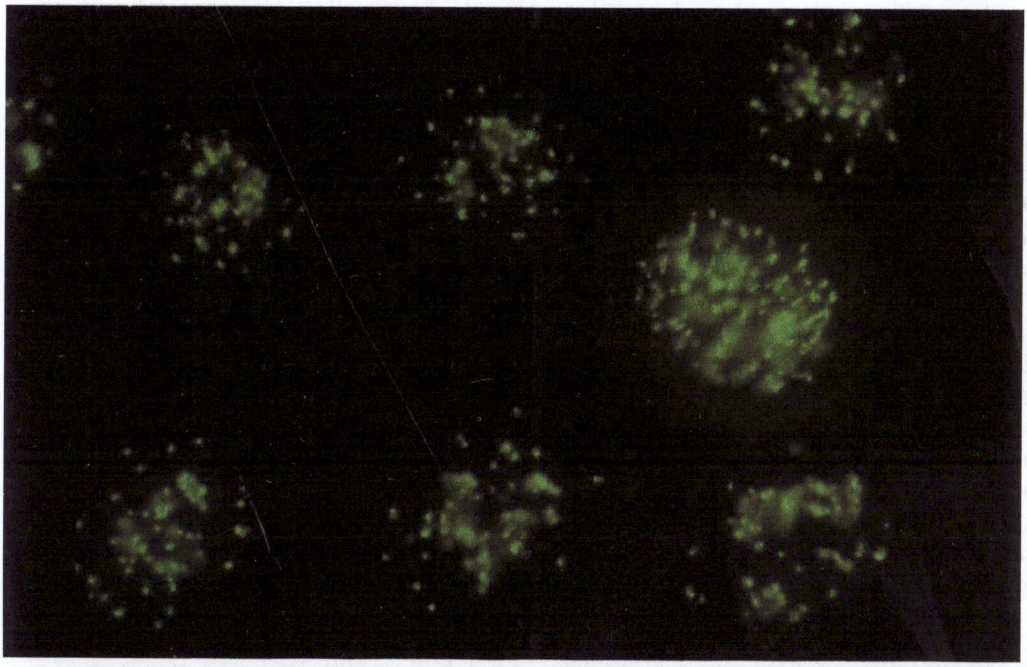

Figure 89. Antibodies against nuclei (centromere or kinetochore pattern). Customary appearance with narrow-band excitation and emission; for a description of the pattern see Figure 88 – note the similarity in the scarcely recognisable example of prophase (centre right) – objective ×100, ×1250.

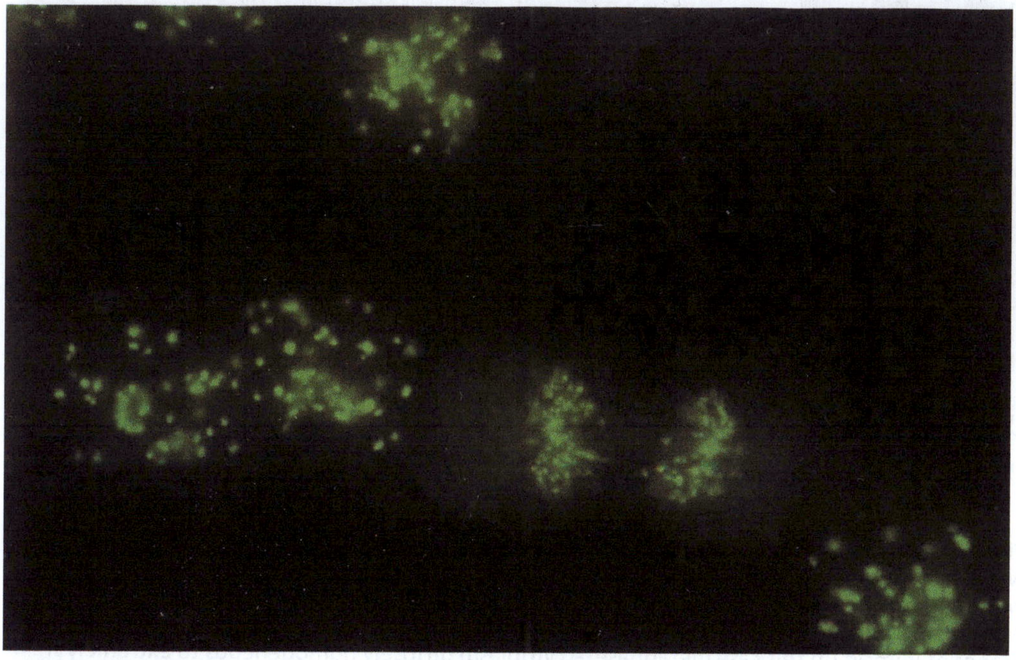

Figure 90. Antibodies against nuclei (centromere and kinetochore pattern). Appearance of mitosis in telophase – compare with Figures 88 and 89 (objective ×100, ×1250).

points in each cell corresponds to the number of chromosomes. In order to identify these antibodies, the patterns for cell nuclei at interphase and mitosis are compared. The same kind of pattern, with a compact appearance during the stages of mitosis, is characteristic (Figs 88–90).

Occurrence and clinical significance: centromere/kinetochore antibodies indicate primarily a benign variant of scleroderma, the CREST syndrome (calcinosis cutis, Raynaud's phenomenon, dysfunction of the oesophagus, sclerodactyly and teleangiectasia), although it does not prove it; positive findings should be expected in approx. 60% of cases. The antibodies appear to be associated with Raynaud's phenomenon, with a frequency in cases of isolated Raynaud's phenomenon of approx. 30% (Vlachoyiannopoulos et al. [1993]); they are also found in other autoimmune diseases, especially in primary biliary cirrhosis (PBC) with a frequency of up to 30% (Chan et al. [1994]).

Literature: Moroi et al. [1980]; Fritzler et al. [1980]; for a review see Verheijen [1994].

7.3.3.4.2 Antibodies showing punctate fluorescence (so-called nuclear dots pattern)

At least two sub-types can be classified.

7.3.3.4.2.1 Antibodies against coilin

Synonyms/abbreviations: p80 coilin antibodies, antibodies against the coiled or convoluted body, few nuclear dots.

Target antigen(s): coilin.

Pattern of fluorescence: a few dots (0–2–6) in the nucleoplasm of cells in inter-

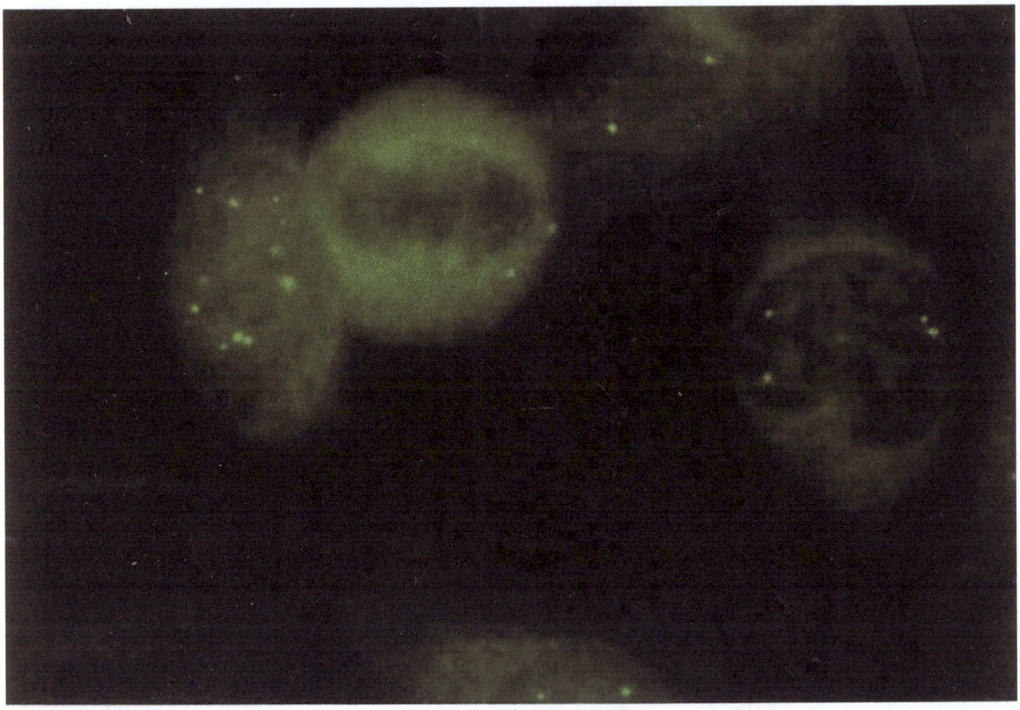

Figure 91. Antibodies against nuclei (nuclear dots and coilin pattern).
Nuclear membrane: negative.
Nucleoplasm: individual or multiple dots of different sizes, irregularly distributed.
Nucleoli: negative.
Stages of mitosis: (in this case metaphase) – cytoplasm diffusely homogeneous to extremely finely granular, weakly positive, individual dots recognisable in the cytoplasm as in interphase, with the chromosomes negative.
Cytoplasm: in this case very weak staining with peri-nuclear emphasis (objective ×100, ×1250).

phase are characteristic. Cytoplasmic staining is almost always visible in mitotic cells; remains of the coiled bodies, mostly smaller than in cells at interphase, are likewise recognisable (Fig. 91). The pattern must be distinguished from others which have also been described as "nuclear dot" patterns, also known as the "pseudo-centromere" pattern, in which 6–10 dots can be observed in the nucleoplasm of cells in interphase but not in mitosis (see §7.3.3.4.2.2).
Occurrence: the antibody is rare and has been observed mostly in primary biliary cirrhosis (PBC) and Sjögren's syndrome.
Clinical significance: unknown.
Literature: Andrade et al. [1991]; Fusconi et al. [1991]; for a review see Carmo-Fonseca et al. [1994].

7.3.3.4.2.2 Multiple nuclear dots pattern
Target antigen: PML, Sp–100.
Pattern of fluorescence: multiple nuclear dots (6–10).
Occurrence: as for coilin-antibodies.
Clinical significance: unknown.
Literature: Bernstein et al. [1984]; Szostecki et al. [1987]; Sternsdorf et al. [1995].

7.3.3.5 Antibodies showing nucleolar fluorescence

Target antigen(s): numerous antigens are known; homogeneous staining is frequently caused by antibodies against nucleolin (Minota et al. [1991]) or against the PM/Scl antigen (Targoff & Reichlin

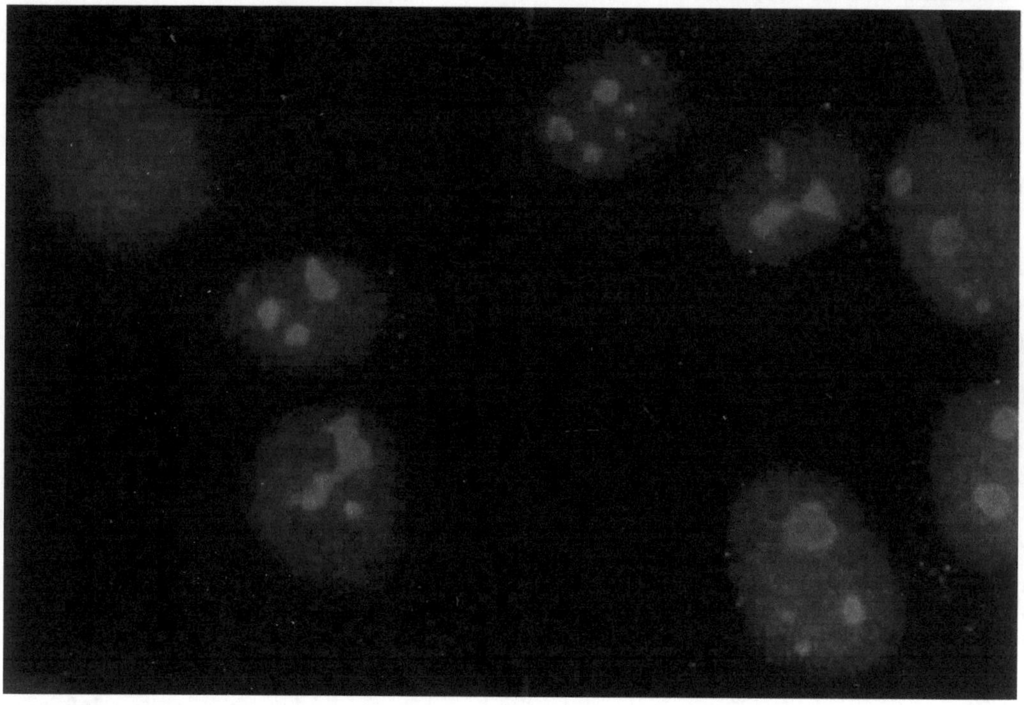

Figure 92. Antibodies against nuclei (Pm/Scl pattern).
Nuclear membrane: negative.
Nucleoplasm: diffusely homogeneous or finely granular, weak staining.
Nucleoli: approximately homogeneous, clearly more intense staining compared to the surroundings.
Stages of mitosis: (upper left, in this case prophase) – cytoplasm shows a similar pattern to that seen for the nucleoplasm of cells in interphase. *Cytoplasm*: negative (objective ×100, ×1250).

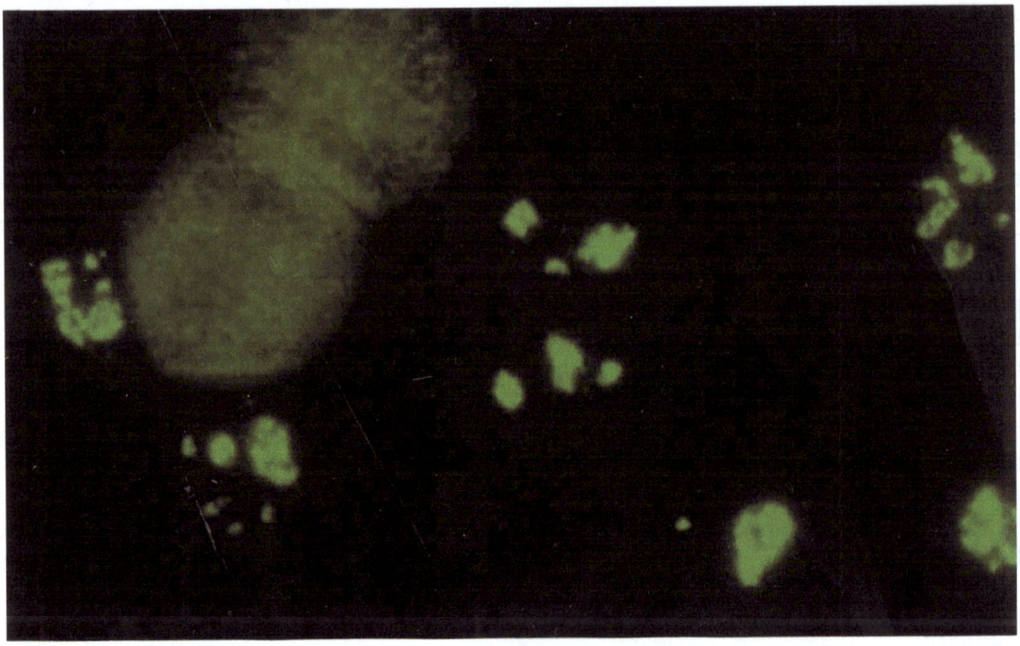

Figure 93a. Antibodies against nuclei (nucleolar pattern, suspicion of fibrillarin). *Nuclear membrane*: negative; *Nucleoplasm*: negative; *Nucleoli*: very intense fluorescence, coarsely granular, fusing together; *Stages of mitosis*: (in this case two cells in pro-metaphase) – cytoplasm finely granular, clearly positive, with negative chromosomes; *Cytoplasm*: negative (objective ×100, ×1250).

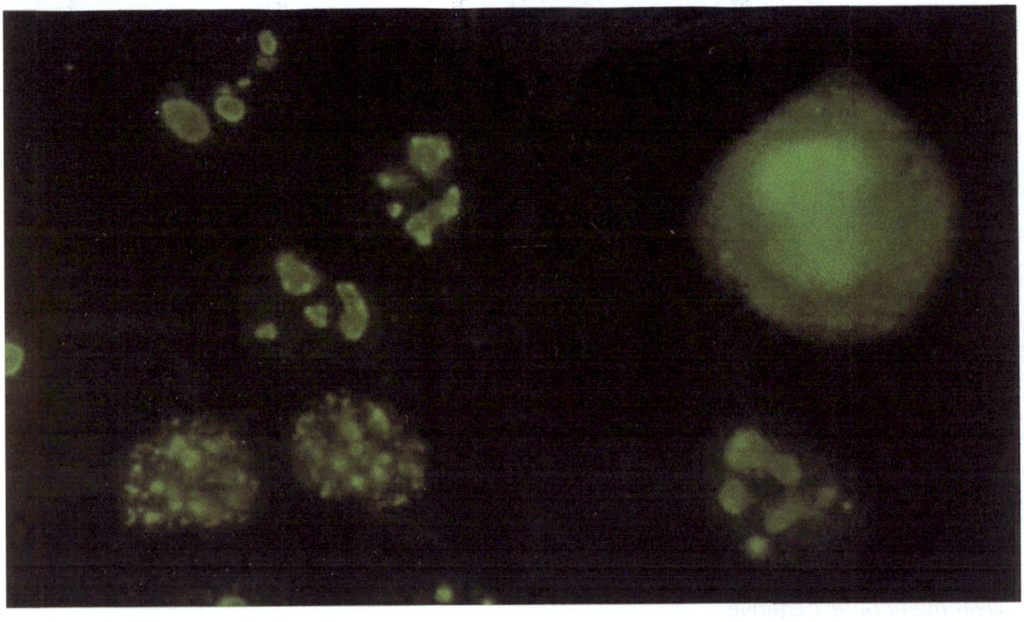

Figure 93b. Antibodies against nuclei (nucleolar pattern, suspicion of topoisomerase I). *Nuclear membrane*: negative; *Nucleoplasm*: weak, finely granular fluorescence, in this case only easily distinguished using a short illumination time; *Nucleoli*: differently sized granules within the nucleoli; *Stages of mitosis*: cytoplasm shows similar but more intense fluorescence than the nucleoplasm, with the chromosomes positive (dense, compact, extremely finely granular); *Cytoplasm*: negative (objective ×100, ×1250).

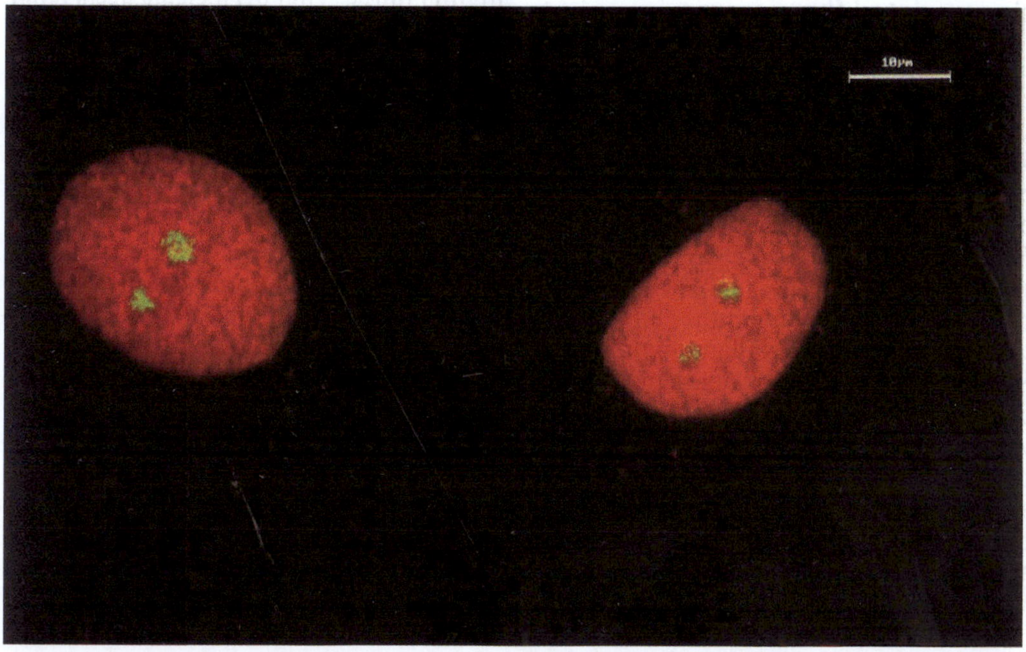

Figure 94. Appearance of nucleolar polymerase I (FITC, green) and of DNA (Texas Red), using animal experiment monoclonal antibodies, PTK cells and a Zeiss LSM4 confocal laser-scanning microscope (objective ×40, ×500; courtesy of Dr. Spring, Heidelberg).

[1985]). A coarsely granular picture is produced by antibodies against fibrillarin (Ochs et al. [1985]) and a finely stippled picture by those against topoisomerase I (Guldner et al. [1986]; Shero et al. [1986]; Maul et al. [1986]). A blotchy pattern occurs in most cases with antibodies against RNA polymerase I (Reimer et al. [1987]). Individual isolated dots stain as a result of antibodies directed against the nucleolar organising region or NOR (Courvalin et al. [1985]). Peri-nucleolar fluorescence may be produced by antibodies against chromatin.

Pattern of fluorescence: these are given in Table 26 and have already been described under the section concerned with target antigens; only in exceptional cases is a clear identification of the target antigen possible from the fluorescence picture alone. The antibody against topoisomerase I typically also stains the nucleoplasm. Differentiation is best achieved by comparing pictures in metaphase (Tan [1992]); Figs 92–94.

Clinical significance: antibodies against nucleoli act as a frequent and characteristic marker for scleroderma and Sjögren's syndrome, as well as for "overlap" phenomena, for example, between scleroderma and polymyositis. The existence of antibodies directed against nucleoli can be taken quite generally to mean that a connective tissue disease, especially of the skin, is active – pulmonary fibrosis should also be considered. Antibodies against topoisomerase I can apparently be induced by exogenous toxins (McHugh et al. [1994]). Long-term investigations lead one to suspect an antigen-induced process (Vasquez-Abad [1995]).

Literature: for reviews on PM/Scl antigen, Bautz & Blüthner [1994]; on fibrillarin, Verheijen [1994c]; and on topoisomerase I, Verheijen [1994a].

7.3.3.6 Antibodies showing mitosis-associated fluorescence

For a better understanding and easier recognition of the typical stages of mitosis, these are explained in Figures 79 and 80.
Target antigen(s): NSP, tubulin, NuMA, centriolar enolase, centrosomes, middle or intermediary (mid) bodies, CENP-F.
Pattern of fluorescence: in contrast to the other patterns, the fluorescence is exclusively, or predominantly, limited to those cells which are undergoing mitosis. Some patterns are characteristic, others difficult to recognise. The fluorescence which is exclusively associated with the microtubules is easy to recognise (tubulin antibodies). The fluorescence produced by antibodies against the nuclear mitotic apparatus protein, NuMA, is likewise relatively easy to recognise as fluorescence in the region of the spindle pole, with radiating fibres. That exclusively associated with the centrioles is characteristic, almost always observed on its own in interphase cells. Fluorescence of the so-called MSA–3 antibody is difficult to recognise reliably, since a different picture can be obtained at practically every stage of mitosis.

7.3.3.6.1 Antibodies against mitotic spindles (microtubules)

Target antigen(s): tubulin.
Pattern of fluorescence: characteristic fluorescence of the microtubules; the poles are generally not visible (Fig. 95).
Occurrence: viral diseases, especially infectious mononucleosis but also in healthy patients.
Clinical significance: unknown.
Literature: Mead et al. [1980].

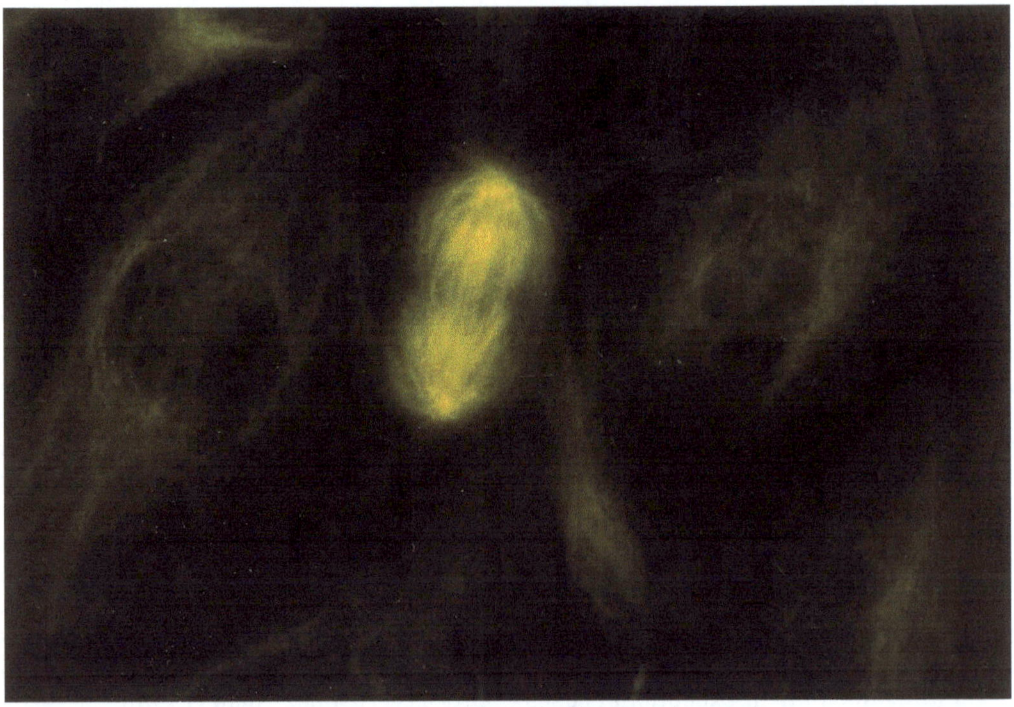

Figure 95a. Antibodies against nuclei (tubulin pattern); *Nuclear membrane*: negative; *Nucleoplasm*: negative, the fibres arise from the cytoplasm; *Nucleoli*: negative; *Stages of mitosis*: (in this case anaphase) – intense fluorescence of the spindle fibres, chromosomes negative; *Cytoplasm*: fine fibres.
(objective ×100, ×1250; courtesy of Prof. Ponstingl, Heidelberg).

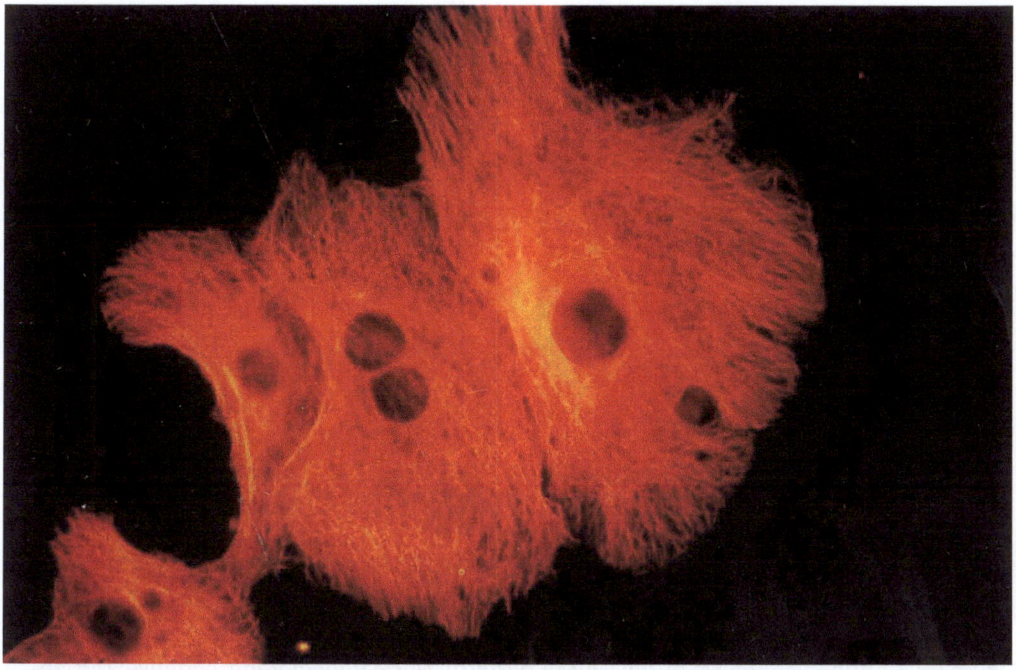

Figure 95b. Appearance of the microtubules in cultivated rat hepatocytes after four days in primary culture; first incubation with animal experiment mouse monoclonal antibodies against α-tubulin; second incubation with TRITC-labelled anti-mouse IgG and IgM (objective ×40, ×412; Dr. Dartsch, Tübingen).

7.3.3.6.2 Antibodies against NuMA

Synonyms/abbreviations: NuMA (nuclear mitotic apparatus), SPN, SP-H, W1, centrophilin (for a review see Compton & Cleveland [1994]), MSA–1 (mitotic spindle apparatus type 1).

Target antigen(s): NuMA protein.

Pattern of fluorescence: fluorescence of the polar regions of mitotic cells, without highlighting the microtubules, in combination with weak fluorescence of the nuclear matrix of cells in interphase (Fig. 96).

Occurrence: a rare antibody, incidence approx. 0.1%; occurs predominantly in connective tissue disorders and autoimmune disease.

Clinical significance: not clear, probably not disease-specific.

Literature: Lydersen & Pettijohn [1980]; Price et al. [1984]; Auer-Grumbach & Archleitner [1994]; for a review see Compton & Cleveland [1994].

7.3.3.6.3 Antibodies against centrioles

Definition: synonymous with the central bodies or the centrosome (refer to this section). A cell organelle that is involved in the formation of the spindle apparatus in mitosis, detectable in interphase as single or double central bodies (diplosomes).

Target antigen(s): centriolar enolase (Rattner et al. [1991]).

Pattern of fluorescence: see Table 26 and Figure 97a.

Occurrence: the antibodies are rare, so that any association with a particular disease is difficult to determine.

Clinical significance: unknown.

Literature: Brenner et al. [1980]; Terreri et al. [1993].

7.3.3.6.4 Antibodies against centrosomes

Definition: the cell centre or major microtubule-organising centre, MTOC; this

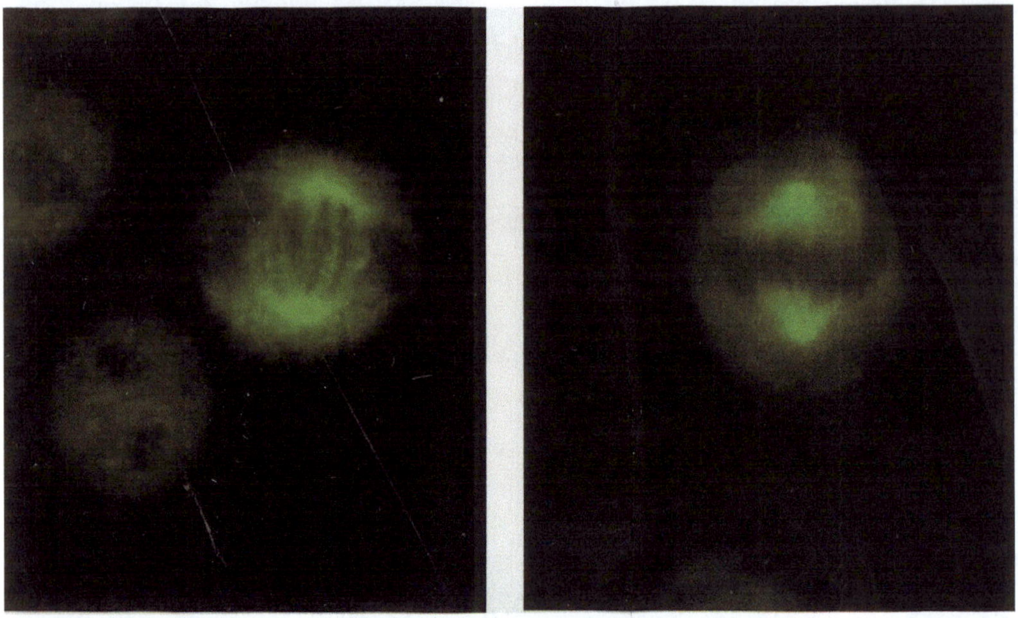

a

Figure 96. Antibodies against nuclei (NuMA pattern); *Nuclear membrane*: negative; *Nucleoplasm*: finely granular, weakly positive; *Nucleoli*: negative; *Stages of mitosis*: (in this case metaphase) – cytoplasm analogous to the nucleoplasm, however, somewhat more intensely positive; chromosomes negative, with intense triangle-shaped fluorescence around the polar region (objective ×100, ×1250).

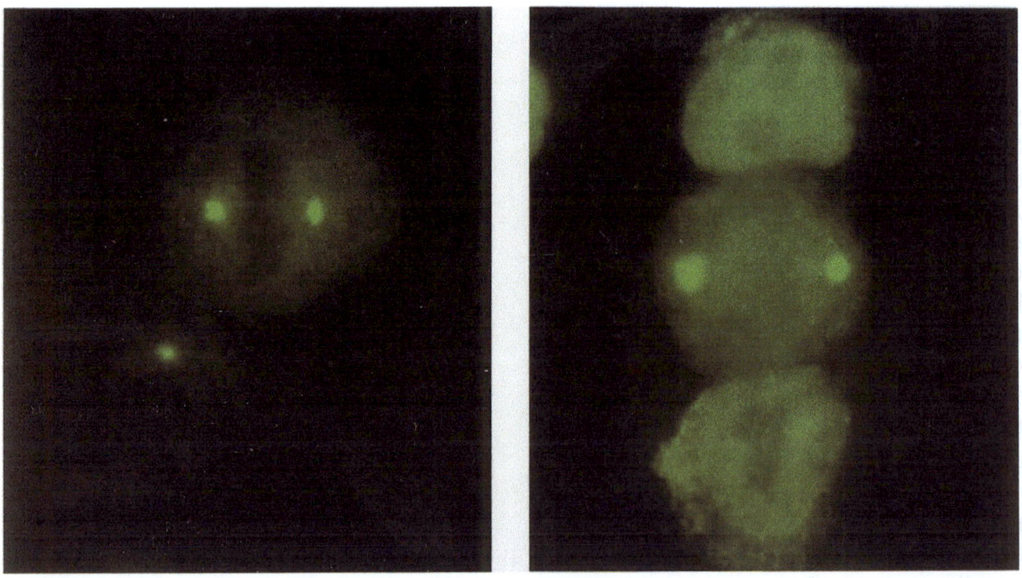

a

Figure 97. Antibodies against nuclei (centriole (a) and centrosome (b) patterns); *Nuclear membrane*: negative; *Nucleoplasm*: negative, or positive and diffuse, finely granular; *Nucleoli*: negative; *Stages of mitosis*: (in this case metaphase) – cytoplasm as for the nucleoplasm, chromosomes negative; the obvious appearance of two spindle poles is characteristic; *Cytoplasm*: variable, fluorescence affecting one pole per cell is typical for centriolar antibodies (objective ×100, ×1250).

lies next to the nucleus and contains mostly a pair of centrioles and so-called peri-centriolar material, the centrosome matrix.
Target antigen(s): unknown.
Pattern of fluorescence: similar to centriolar fluorescence although somewhat more disperse and with mostly no fluorescence affecting cells in interphase (Fig. 97b).
Clinical significance: not known.
Literature: Calarco-Gillaru et al. [1983].

7.3.3.6.5 Antibodies against middle (intermediary) bodies

Synonyms/abbreviations: middle (mid) body antibody, MSA-2 (mitotic spindle apparatus type 2).
Definition: the middle or intermediary bodies appear as a thin bridge between daughter cells especially at the end of cytokinesis; they contain the remains of the microtubules and matrix material.
Target antigen(s): amongst others, a 330 kDa protein (Casanio et al. [1993]).
Pattern of fluorescence: various stage-dependent patterns are observed at the end of telophase and during cytokinesis, sometimes showing only a punctate fluorescence (Fig. 98).
Clinical significance: not clear, occurs rarely but particularly in scleroderma.
Literature: Senécal et al. [1985]; Fritzler et al. [1987].

7.3.3.6.6 Antibodies against the kinetochore, especially in prophase and metaphase

Synonyms/abbreviations: MSA-3 (mitotic spindle apparatus type 3).
Target antigen(s): CENP-F (Rattner et al. [1993]).
Pattern of fluorescence: very different patterns depending on the cell type and stage in the cell cycle; the responsible antigen is visible as a fine granularity in

a 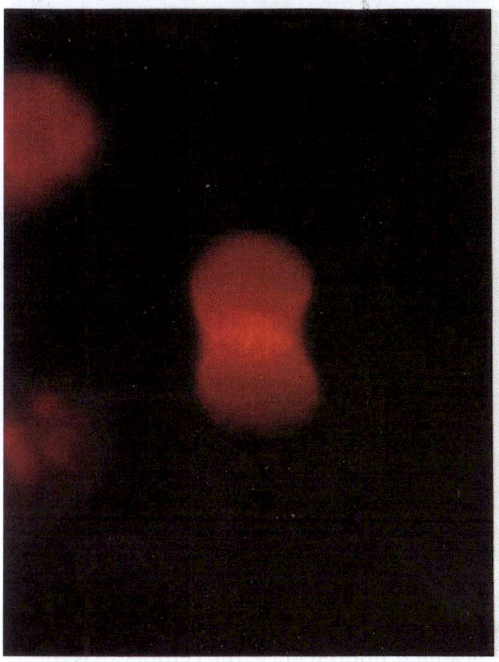 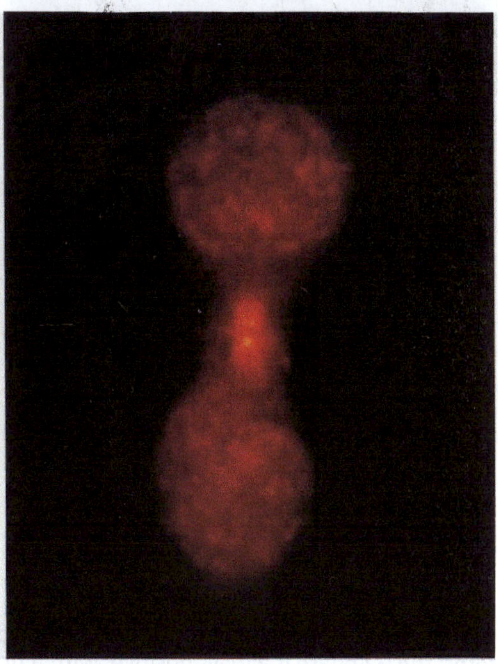 b

Figure 98. Antibodies against nuclei (intermediary body pattern).
Nuclear membrane, nucleoplasm, nucleoli and cytoplasm all negative. Stages of mitosis: various patterns seen – in the present instance, in anaphase (a) with band-like fluorescence and in telophase (b) with punctate fluorescence (objective ×63, ×630; Prof. Paweletz, Heidelberg).

the pre-mitotic G2-phase in HEp–2 cells. It is usually clearly visible in prophase, metaphase and early anaphase (Fig. 99). In contrast to the other mitosis-associated patterns of fluorescence, the spindles and their poles are never stained.
Clinical significance: not clear, occurs rarely but especially in carcinoma.

Literature: Humbel [1986]; Casiano et al. [1995].

7.3.3.6.7 Antibodies against "cell-division antigen"

Synonyms/abbreviations: anti-DCA (dividing cell antigen).

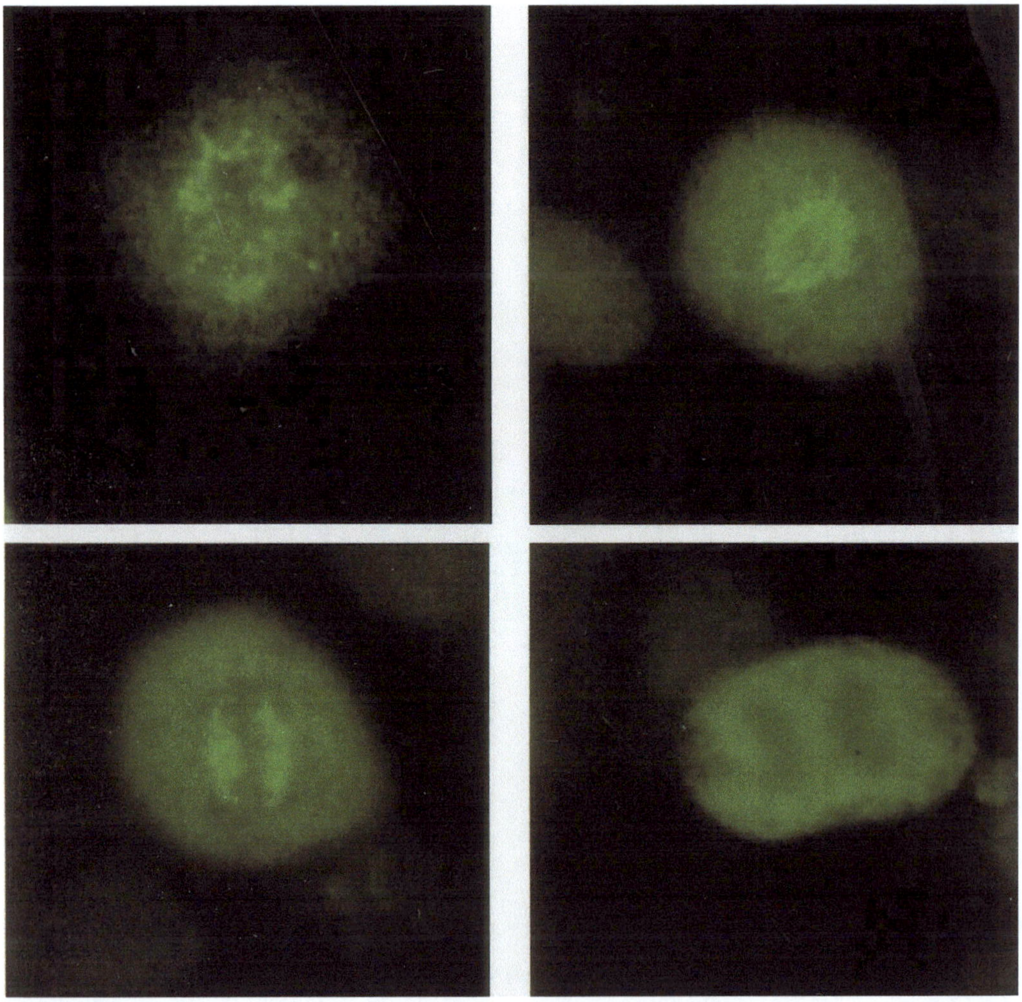

a

c

Figure 99. Antibodies against nuclei (CENP-F pattern): cells in interphase can show various patterns; the G2 phase is always positive, with a finely granular fluorescence; the cytoplasm is frequently weakly positive, mostly finely granular, during the stages of mitosis; the chromosomes show a variety of pictures depending on the phase of mitosis – in prophase and metaphase, (a) and (b), fine granules or dots with central emphasis are particularly easily recognisable (c) at the transition to anaphase – the chromosomes in telophase are negative (d) (objective ×100, ×1250).

Target antigen(s): probably a 60 kDa protein derived from histone (Blaschek et al. [1993]).
Pattern of fluorescence: homogeneous nuclear fluorescence of mitotically active cells; the cell nucleus is negative in interphase.
Occurrence: the antibody has only been found in SLE (in ten out of 183 cases).
Clinical significance: not clear.
Literature: Blaschek et al. [1988].

7.3.4 Antibodies against cytoplasmic antigens in cultured cells

As background information, the following short résumé covers antibodies which can lead to fluorescence of the cytoplasm in cultured cells. Some patterns are so characteristic that confirmation using other methods is not necessary. Although some antibodies, for example, against mitochondria, the endoplasmic reticulum and ribosomes, are detected typically with other antigenic substrates, they are also described here since they can produce a quite different pattern of fluorescence. The ELISA method is in predominant use today for the detection of microfilaments such as actin and myosin, microtubules and intermediary filaments. The antibodies are mentioned below simply for the sake of completeness. The fluorescence patterns observed are detailed in Table 27 and Figures 100–108.

7.3.4.1 Antibodies showing granular cytoplasmic fluorescence

In this group one can include (see Tab. 27): protein (SRP, Reeves et al. [1986]), endoplasmic reticulum, histidinyl tRNA synthetase (Jo–1) and related enzymes like aminoacyl tRNA (Fig. 101, Saito et al. [1989]), ribosomes or especially ribosomal RNP (Fig. 102, Bonfa and Elkon [1994]), lysosomes and peroxisomes (Fig. 103), and the Golgi apparatus (Fig. 104).
Literature: see survey by Kreis and Vale [1993]. This merely emphasises antibodies against the Golgi apparatus.

Table 27. Cytoplasmic fluorescence pattern for HEp–2 cells (based on Humbel [1993], with modifications).

1.	**Granular**
a.	coarse-grained or thread-like; antigen, mitochondria.
b.	fine, isolated granules; antigen, SRP.
c.	perinuclear granules; antigen, endoplasmic reticulum.
d.	diffuse but accentuated fine perinuclear granules; antigen Jo1, PL7, PL12, KJ.
e.	very dense delicate granules; antigen, ribosomes.
f.	irregular granules; antigen, lysosomes and peroxisomes.
g.	compact or speckled juxtanuclear structures; antigen, Golgi apparatus.
2.	**Filamentous**
a.	cable-like; antigen, actin or myosin.
b.	short fibres; antigen, tropomyosin.
c.	radiating fibres; antigen, cytokeratin.
d.	'delicate' fibres; antigen, tubulin.
e.	irregular fibres; antigen, vimentin.
f.	thorn-like fibres; antigen, vinculin.

Antibodies against the Golgi apparatus

Target antigen(s): unknown.

Pattern of fluorescence: juxta-nuclear fluorescence is typical (Fig. 104).

Occurrence: relatively rare; antibodies against the Golgi apparatus have been found primarily in patients with SLE, Sjögren's syndrome, lymphoma, rheumatoid arthritis, autoimmune hepatitis and NonA/NonB hepatitis.

Clinical significance: not clear.

Literature: Rodriguez et al. [1982]; Hong et al. [1992]; Reiner et al. [1994].

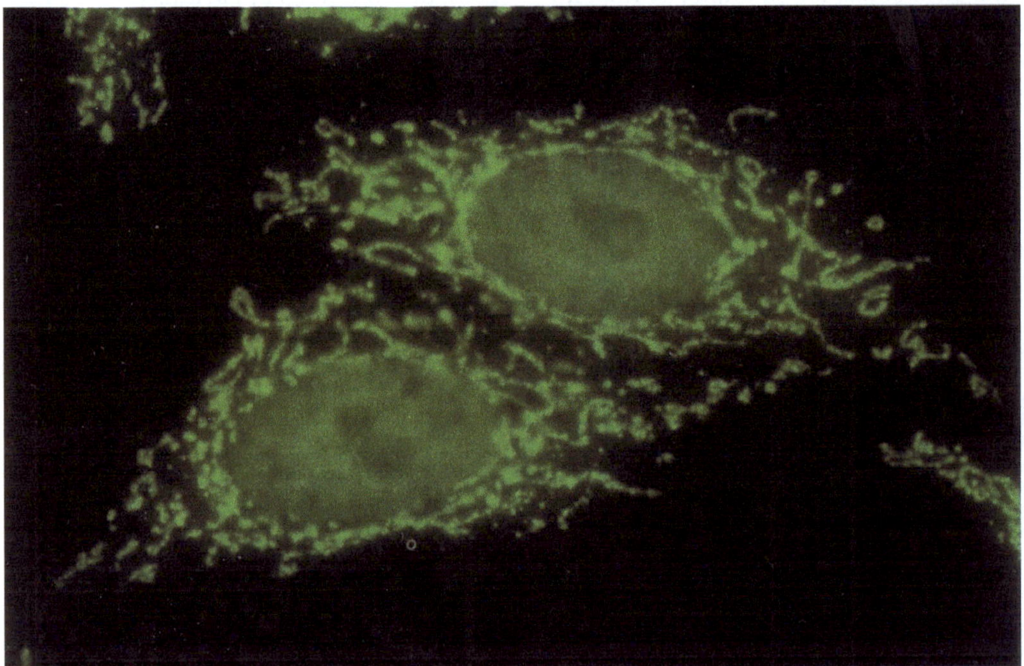

Figure 100. Antibodies against mitochondria; *Nuclear membrane*: frequently weakly positive; *Nucleoplasm*: by itself negative, in this case positive, with extremely fine granularity, caused by the simultaneous presence of ANF; *Nucleoli*: negative; *Stages of mitosis*: various degrees of staining in the cytoplasm, chromosomes negative; *Cytoplasm*: intense, predominantly thread-like and coarse granular fluorescence (objective ×100, ×1250).

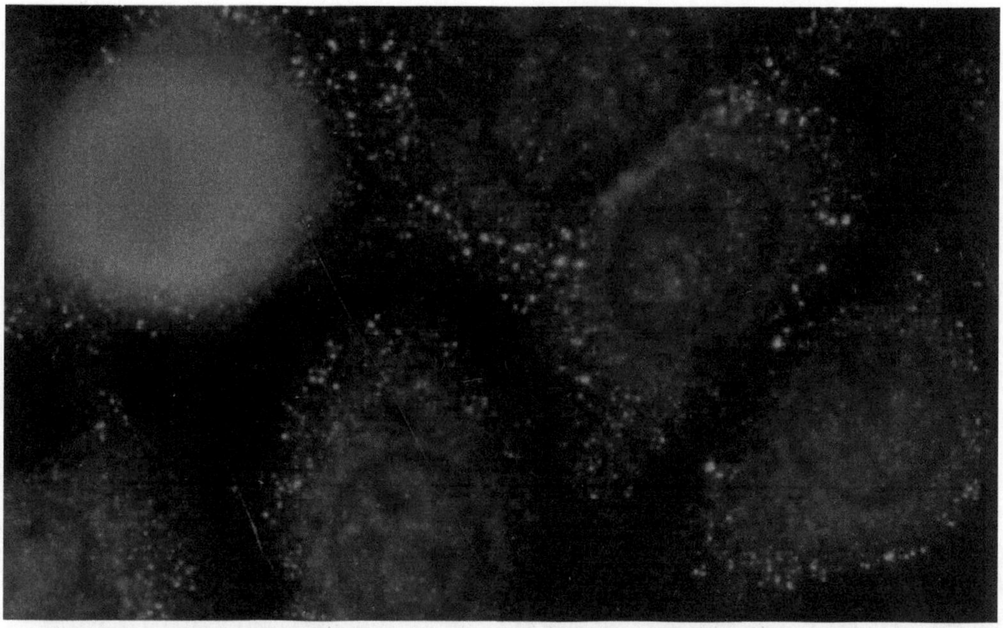

Figure 101. Antibodies against Jo–1; *Nuclear membrane*: negative; *Nucleoplasm*: fine granules, weakly positive; *Nucleoli*: mostly negative.
Stages of mitosis: cytoplasm diffusely positive, chromosomes negative; *Cytoplasm*: mostly fine, isolated granules showing perinuclear emphasis (objective ×100, ×1250).

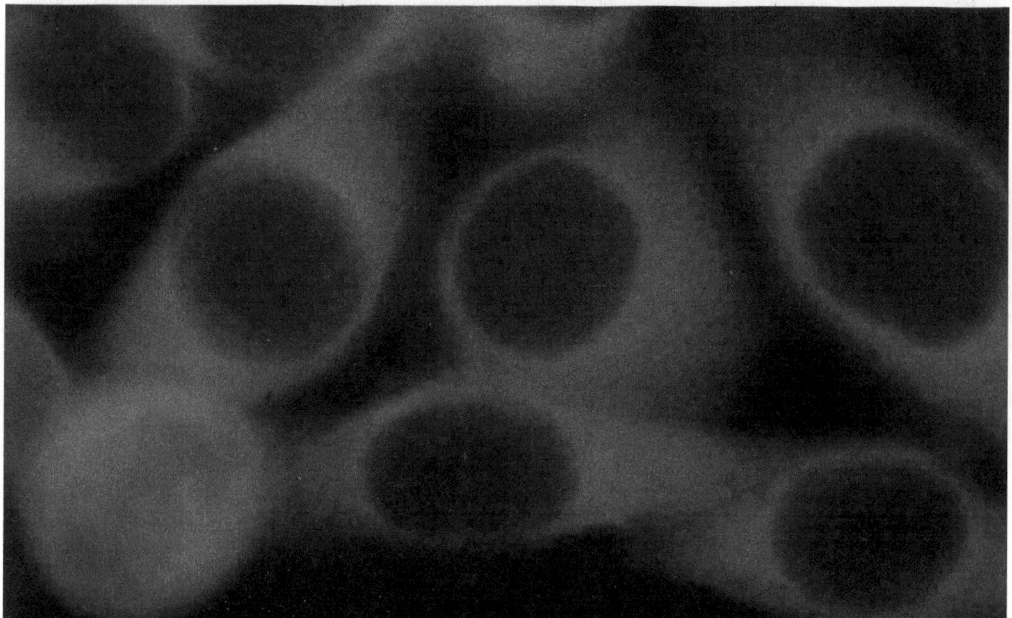

Figure 102. Antibodies against ribosomes; *Nuclear membrane*, nucleoplasm and nucleoli all negative; *Stages of mitosis*: cytoplasm diffuse, extremely fine granular positive; chromosomes negative; *Cytoplasm*: characteristic, somewhat 'washed out', i.e., faded appearance, extremely fine granular, diffuse fluorescence (objective ×100, ×1250).

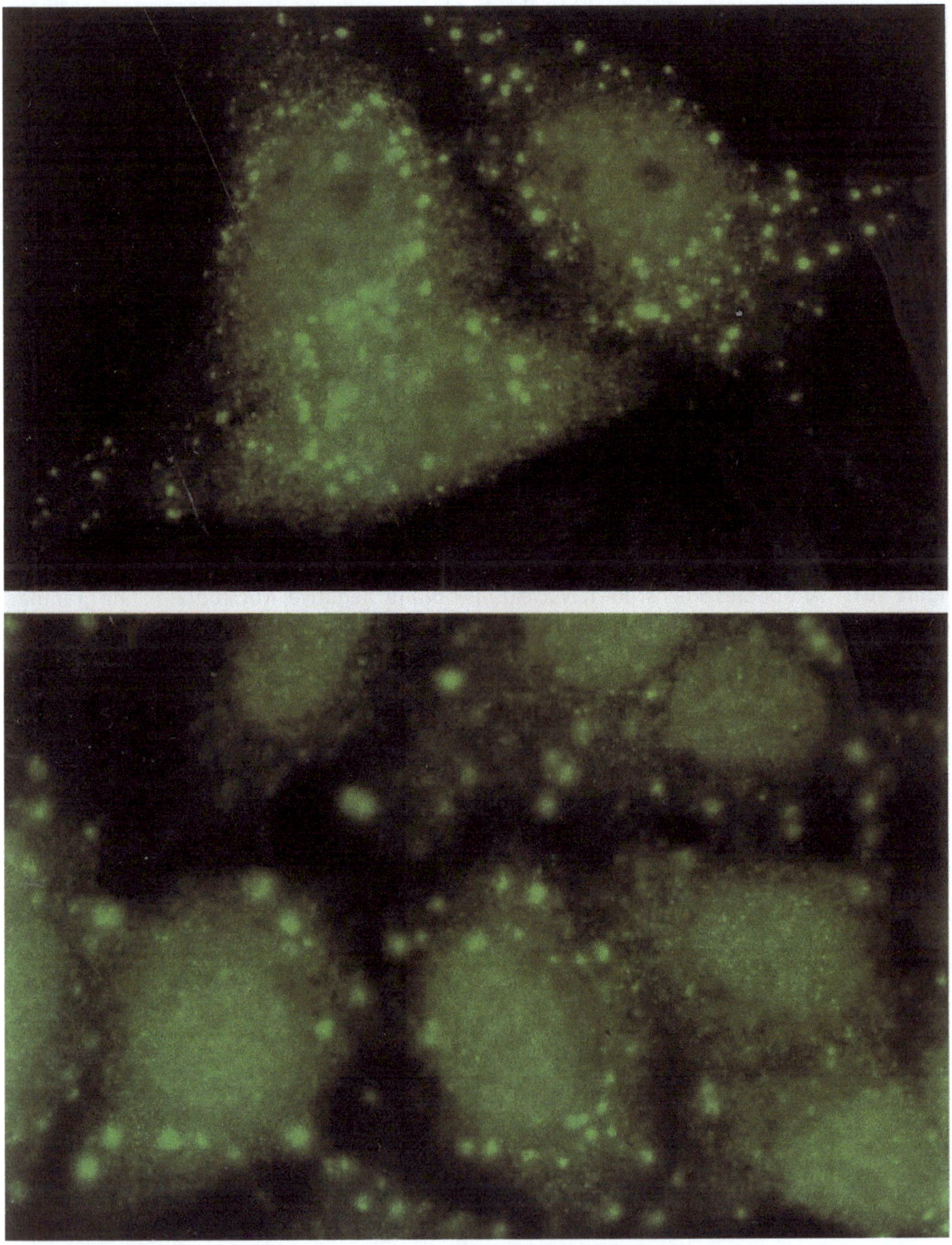

a

b

Figure 103. Antibodies against irregular granules in the cytoplasm (presumptive antibodies against peroxisomes and lysosomes); *Nuclear membrane*: negative; *Nucleoplasm*: various picture caused by simultaneous presence of ANF, in the present case, finely granular fluorescence; *Nucleoli*: negative; *Stages of mitosis*: cytoplasm with various pictures as for the nucleoplasm, chromosomes negative; *Cytoplasm*: irregular and differently sized granules, the smaller of which (a) can indicate peroxisomes and the larger (b) lysosomes (objective ×100, ×1250).

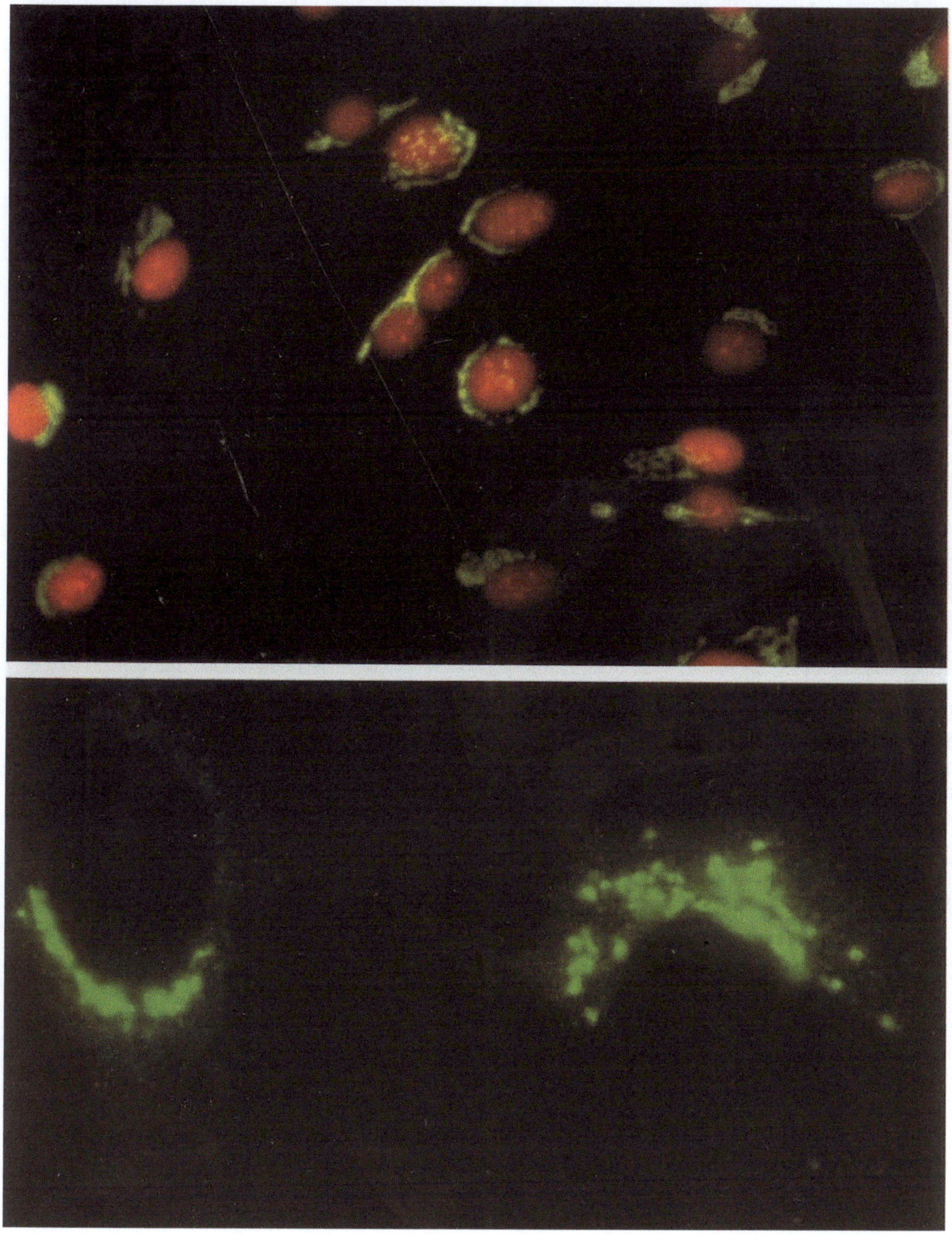

a

b

Figure 104. Antibodies directed against the Golgi apparatus;*Nuclear membrane*, nucleoplasm, nucleoli and chromosomes all negative; *Cytoplasm*: juxtanuclear structures of various forms, partly compact and partly granular; in (a) the nucleus is shown stained red with ethidium bromide to give a better sense of orientation (objective ×40, ×500(a), ×100, ×1250(b)).

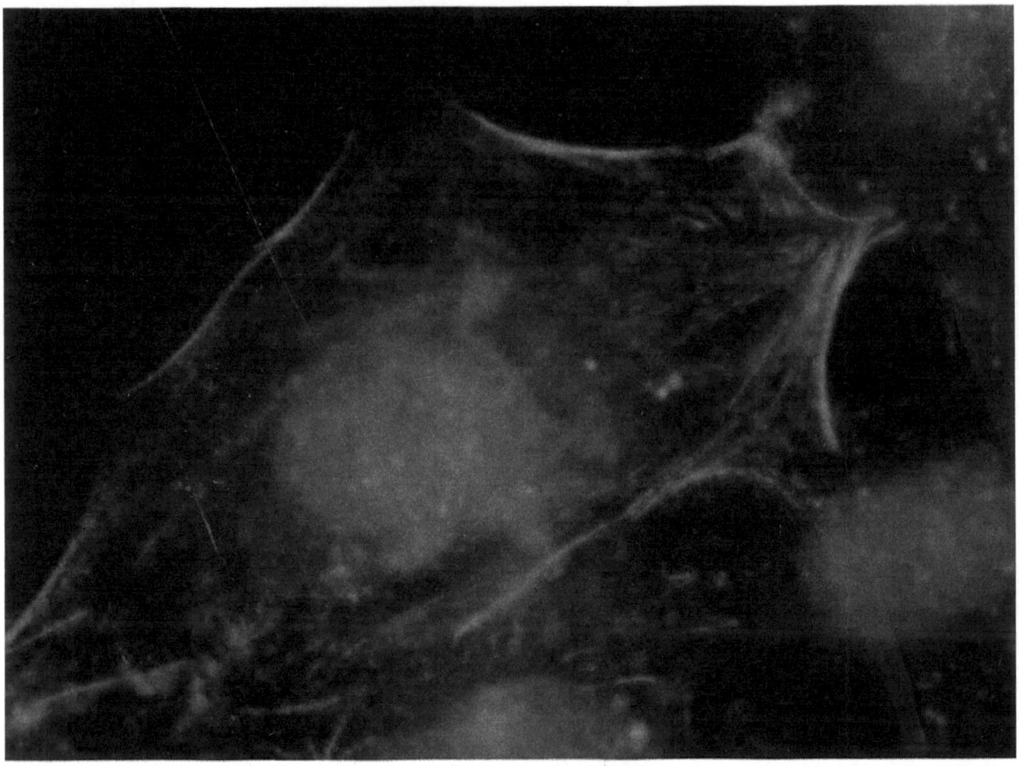

Figure 105. Antibodies against actin; *Nuclear membrane*, nucleoplasm, nucleoli and chromosomes all negative; in the present case the nucleus is stained as a result of the simultaneous presence of ANF; *Cytoplasm*: depending on the fixation procedure many different patterns, in this case a predominant sub-membranous fluorescence as well as a trace of "stress-fibre"-type appearance of the fibres in the cytoplasm (objective ×100, ×1250).

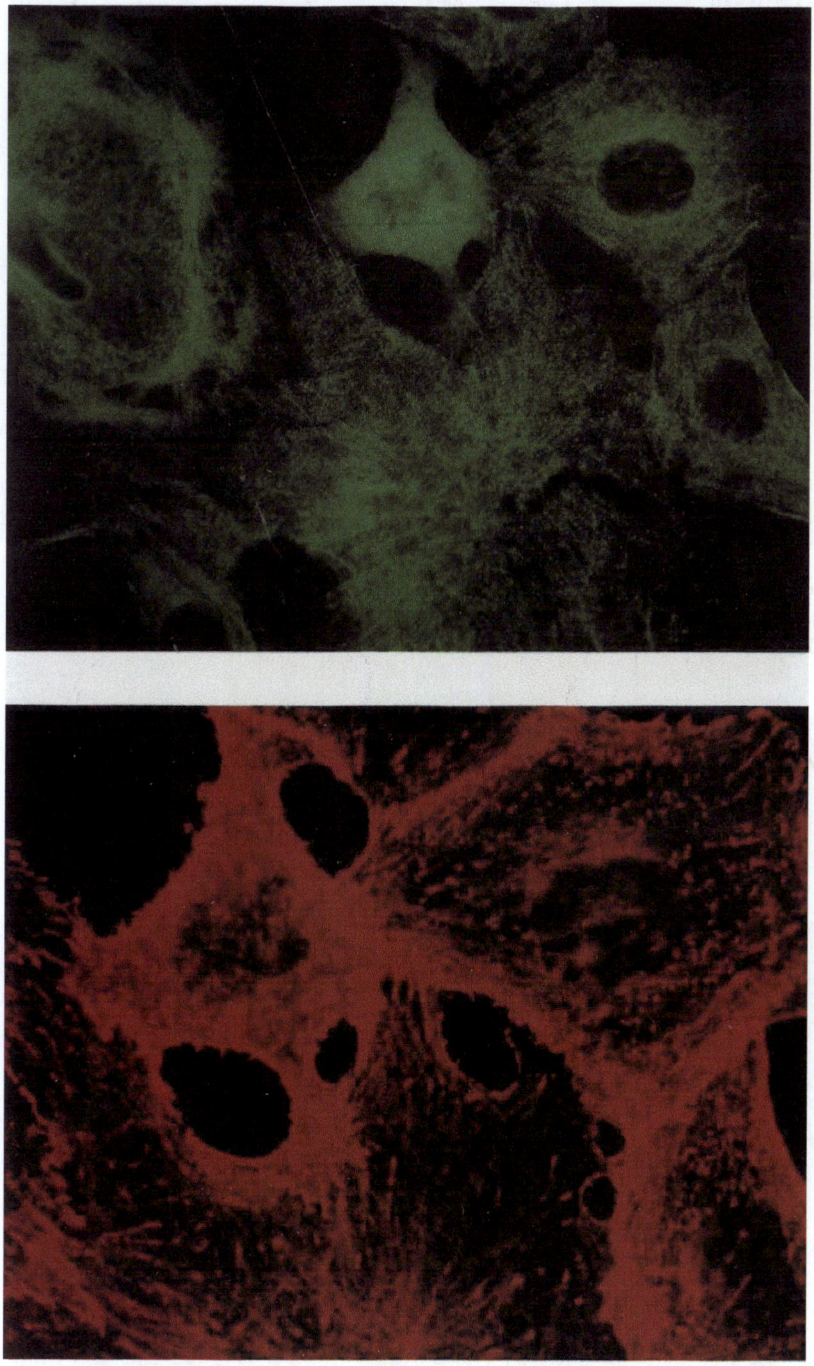

a

b

Figure 106. Antibodies against myosin (a) compared with the appearance of actin; incubation of Vero cells with a mixture of antibodies against cytoplasmic myosin (provided by Prof. Stewart, Darmstadt) and TRITC-phalloidin, which binds to actin – extensive overlapping of the two fluorescence reactions occurs; the cell in mitosis is as such not distinguishable using detection of actin (objective ×63, ×788).

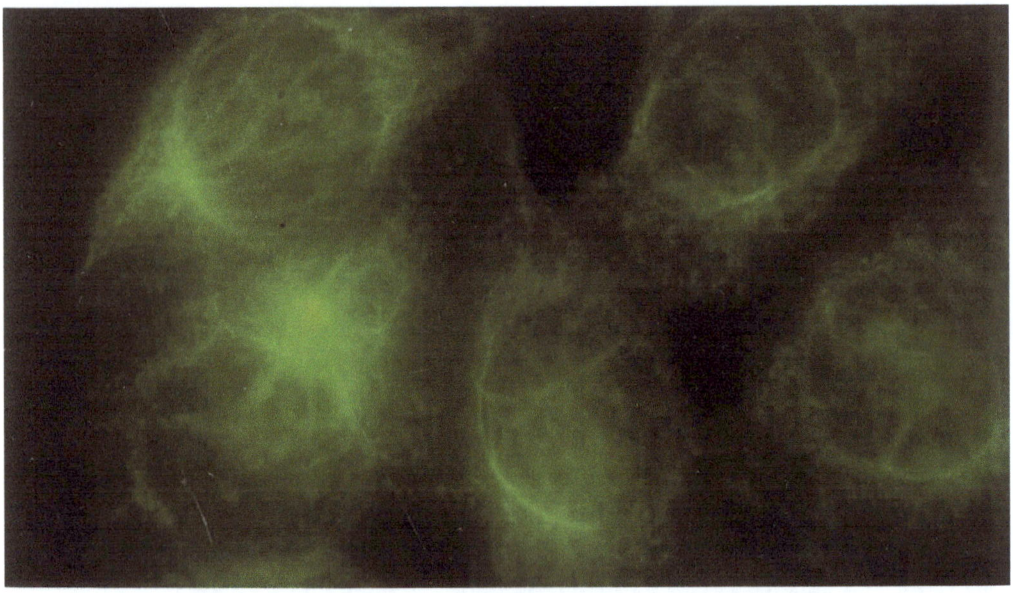

Figure 107a. Antibodies against vimentin. Nuclear membrane, nucleoplasm, nucleoli and chromosomes are all negative. The cytoplasm shows irregular fine fibres, partly with perinuclear emphasis (objective ×100, ×1250).

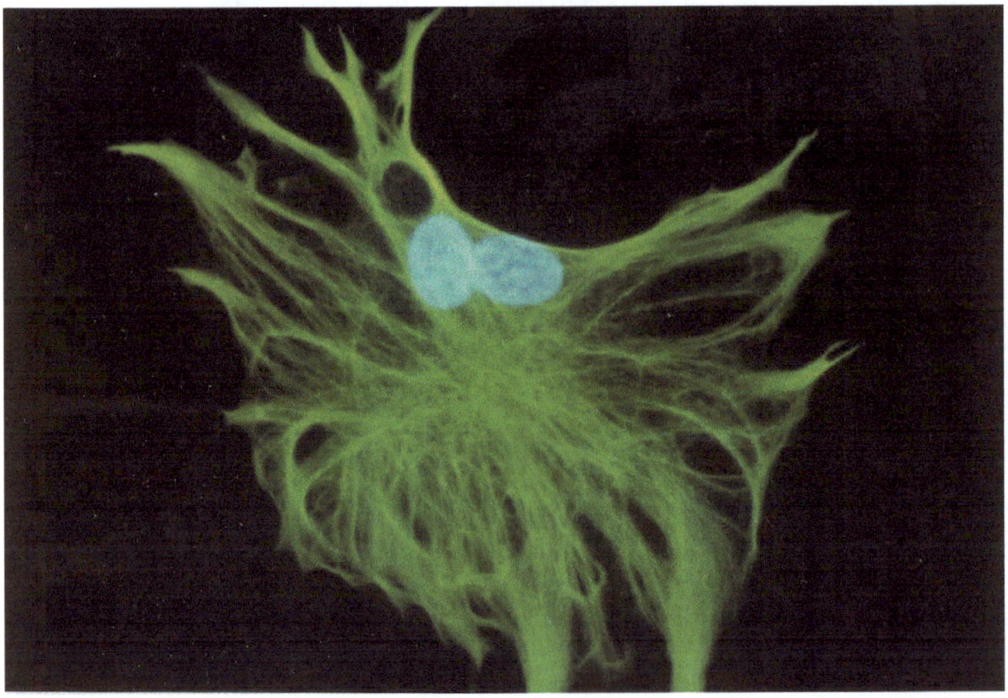

Figure 107b. Appearance of vimentin in a cultured endothelial cell from the great saphenous vein in man, treated with animal experiment mouse monoclonal antibodies followed by FITC-labelled anti-mouse IgG (green); staining of the cell nuclei with DAPI (blue), double illumination with UV-excitation (BP 365/10, TS 400, LP 420) and narrow-band excitation in the blue (objective ×40, ×412; with kind permission of Dartsch [1989]).

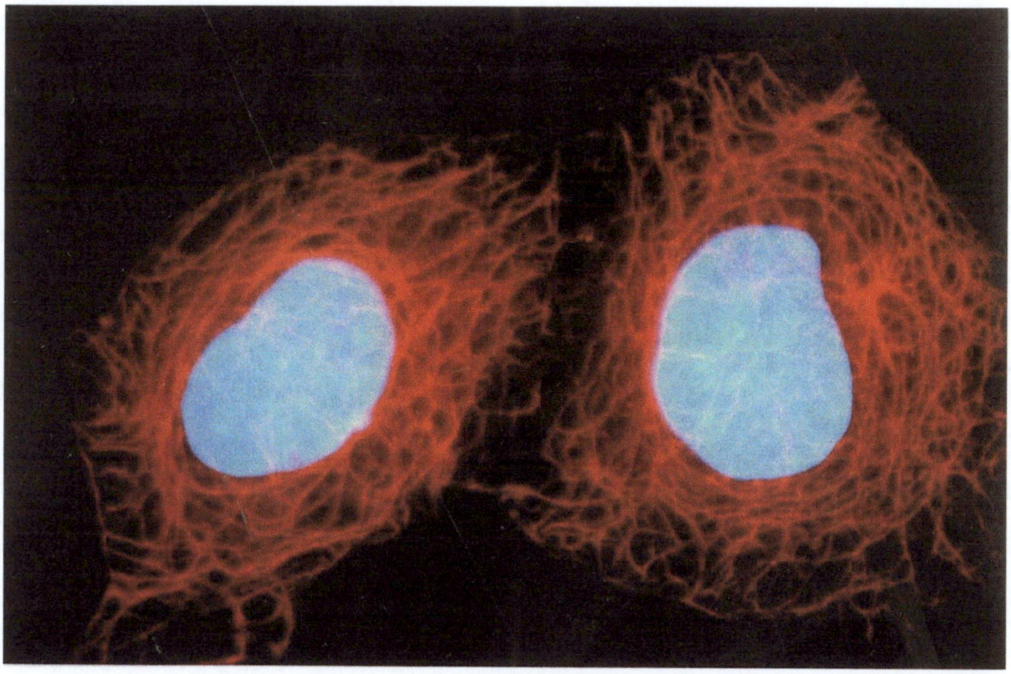

Figure 108. Appearance of cytokeratin 8 in cultured PLC liver cells using animal experiment mouse monoclonal antibodies against cytokeratin 8; the second antibody was Texas Red-labelled anti-mouse antibody (red); cell nuclei stained with DAPI (blue); double illumination with narrow-band green and UV-excitation (objective ×100, ×825; Mrs. Kuhn and Prof. Franke, Heidelberg).

7.3.4.2 Antibodies showing filamentous cytoplasmic fluorescence

In this group one can include (see Tab. 27): antibodies against actin and myosin (Figs 105 and 106), tropomyosin, cytokeratin (Fig. 108), tubulin (Fig. 95), vimentin (Fig. 107), and vinculin.
Literature: see survey by Kreis and Vale [1993],

7.3.5 Antibodies against special components

7.3.5.1 Antibodies against spermatozoa

Synonyms/abbreviations: sperm-antibodies.

Target antigen(s): still insufficiently investigated, at least seven.
Properties of the antibodies: broadly organ-specific and partially species-specific, not disease-specific; predominantly of the IgG type, less so IgM, rarely IgA, to some extent able to fix complement.
Antigen substrate: fresh spermatozoa, fixed for one minute in methanol, use air-drying for the detection of antibodies against the middle portion, trypsin treatment for the detection of antibodies against the head.
Conjugate: anti-human IgG.
Pattern of fluorescence: five different types can be relatively well differentiated, in which fluorescence affecting the following regions can be observed:

1. the acrosome;
2. the equatorial region;
3. the post-acrosomal region;

4. the tail;
5. the nucleus.

Clinical significance: not yet clarified; low titres are unlikely as the cause of infertility resulting from oligospermia or azoospermia, because these are often seen in healthy patients and after vasectomy.

Literature: Cruickshank et al. [1959]; Tung [1975].

7.3.5.2 Antibodies against epidermis

Antibodies which react with the epidermis will be discussed briefly in the form of a summary differing somewhat from the other classification schemes used.

1. Antibodies against the surface of the *stratum corneum*.
2. Antibodies against the *stratum corneum*.
3. Antibodies against layers of the *epidermis*.
4. Antibodies against the upper cell layers of the *epidermis*.
5. Antibodies against the lower cell layers of the *epidermis*.
6. Antibodies against the cell surface (formerly the inter-cellular substance), especially against the *stratum spinosum* – so-called pemphigus-antibodies.
7. Antibodies against the basal cells of the *epidermis*.
8. Antibodies against the basement membrane between the *epidermis* and the *corium* of the squamous epithelium – so-called pemphigoid-antibodies.

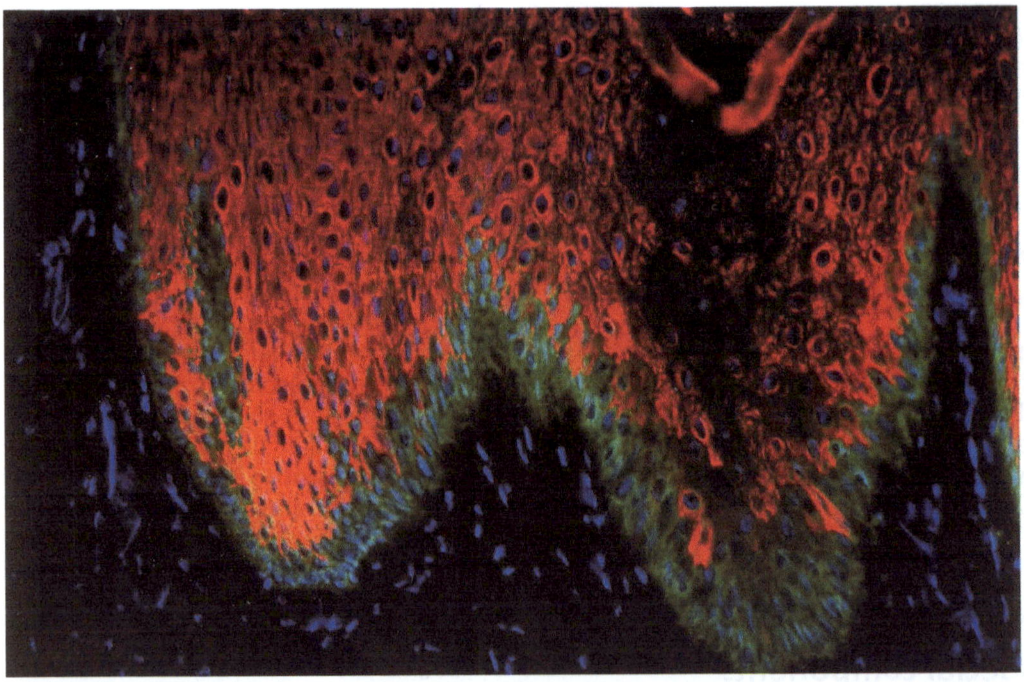

Figure 109. Antigen, human epidermis from the sole of the foot; demonstration of cytokeratin 5, using animal experiment monoclonal antibodies and Cy 2-labelled secondary antibodies (green or yellow-orange when overlapping), in the *stratum basale* with the beginnings of a trace in the *stratum spinosum*; and of cytokeratin 9 (likewise using monoclonal antibodies and Cy 3-labelled secondary antibody) in the whole of the *stratum spinosum* and *granulosum* (red); nuclear staining with DAPI (blue); three-fold illumination with blue, green and UV-excitation (objective ×25, ×206; Dr. Langbein and Prof. Franke, Heidelberg).

9. Antibodies against perinuclear antigens, so-called anti-perinuclear factors.

The classifications and descriptions in the literature on antibodies which react with the *epidermis*, differ considerably. The clinical significance depends on each antigen substrate used. Because of this, completely different points of view can, in part at least, be supported. The significance of the so-called pemphigus-antibody and the pemphigoid-antibody, (6) and (8) respectively, in the differential diagnosis of bullous dermatitis, is certain (see below). Antibodies against the surface of the *stratum corneum*, which react in a linear fashion, can be detected with ease, using primarily rat tongue. Antibodies which react diffusely with the *stratum corneum* count as having no significance because almost every healthy subject has these antibodies (Krogh [1969]). More recently, antibodies which react with the *stratum corneum* in a laminar fashion, have come to be regarded as clinically relevant (see below). Likewise, antibodies which react with the cytoplasm of all the cells of the upper and lower layers of the *epidermis* are considered without significance. These are often described as so-called cytoplasmic keratinocyte antibodies (for reviews see Beutner et al. [1987]; Saurat & Didierjean [1987]); they are directed predominantly against the cytokeratins (review, Moll et al. [1982]; Fig. 109). Antibodies against the basal cells are often classified in the same group as antibodies against "keratinocytes"; they are, however, mentioned separately here since their differentiation, based on the antigen substrate used, may possibly achieve clinical significance.

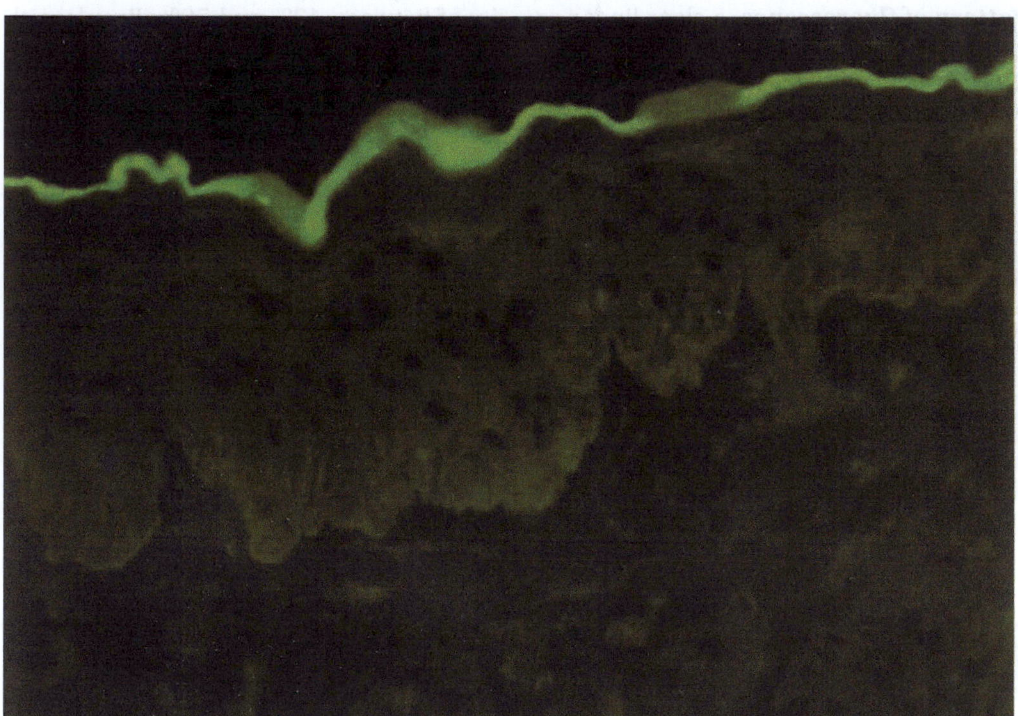

Figure 110. Antibodies against the surface of the *stratum corneum*; antigen, rat tongue – linear fluorescence of the surface of the *stratum corneum* (objective ×16, ×315).

7.3.5.2.1 Antibodies against the surface of the *stratum corneum*

Target antigen(s): unknown.

Properties of the antibodies: IgG, obviously species-specific.

Antigen substrate: rat tongue.

Conjugate: anti-human IgG.

Pattern of fluorescence: linear fluorescence (Fig. 110)

Clinical significance: not clear; they are seen primarily in autoimmune hepatitis but also in healthy patients.

Literature: Storch [1989, 1992].

7.3.5.2.2 Antibodies against the *stratum corneum*

Synonyms/abbreviations: anti-keratin antibody (AKA).

Target antigen(s): filaggrin (Simon et al. [1993]).

Properties of the antibodies: neither organ, nor tissue or species-specific, typical for the disease, predominantly IgG.

Antigen substrate: rat oesophagus (middle third).

Pattern of fluorescence: a clear linear or laminar fluorescence affecting the *stratum corneum* is characteristic; homogeneous fluorescence affecting the *stratum corneum*, fluorescence of the *stratum spinosum* and of the *stratum basale*, as well as surface fluorescence, must be excluded – Figures 111 and 112.

Tip: sections should be washed for five minutes at room temperature using PBS-T buffer (PBS plus 0.5% w/v Triton X-100) before being incubated (Hoet [1993]). The initial serum dilution can be 1:10. The serum should have been centrifuged beforehand for five minutes at 13,000 g_{av}.

Clinical significance: the antibody counts as typical for rheumatoid arthritis. In spite of numerous positive reports, the significance and necessity for this test have been assessed in a variety of ways. Aho et al. [1994] quote the following frequencies: rheumatoid arthritis, mostly rheumatoid factor positive, 40%; early stages of rheumatoid arthritis, 20%; healthy patients, 1%. Using immunoblotting, Gomès-Daudrix et al. [1994] have recently established that these antibodies have a specificity of 99%, with a sensitivity of between 43% and 70%. People with antibodies against the *stratum corneum* have mostly anti-perinuclear factors as well, and *vice versa*.

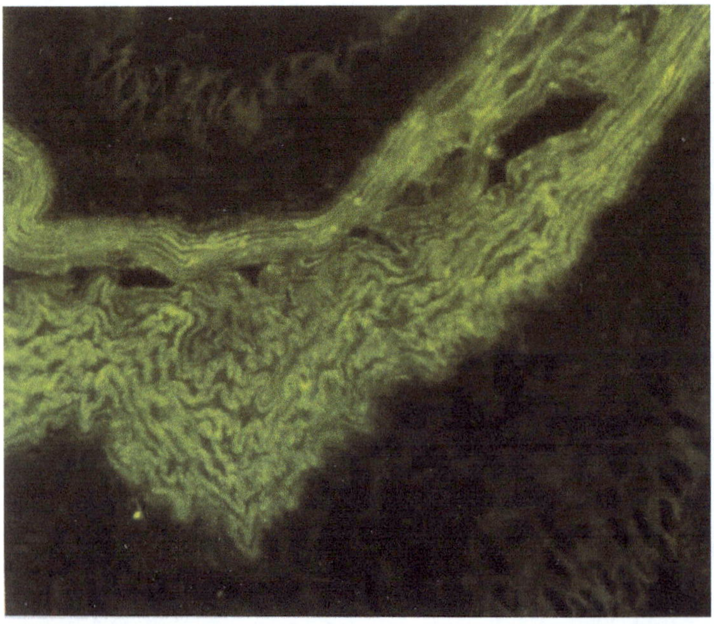

Figure 111. Antibodies against the *stratum corneum*: antigen, rat oesophagus – laminar fluorescence of the *stratum corneum* (objective ×40, ×500; Dr. Vincent, Toulouse).

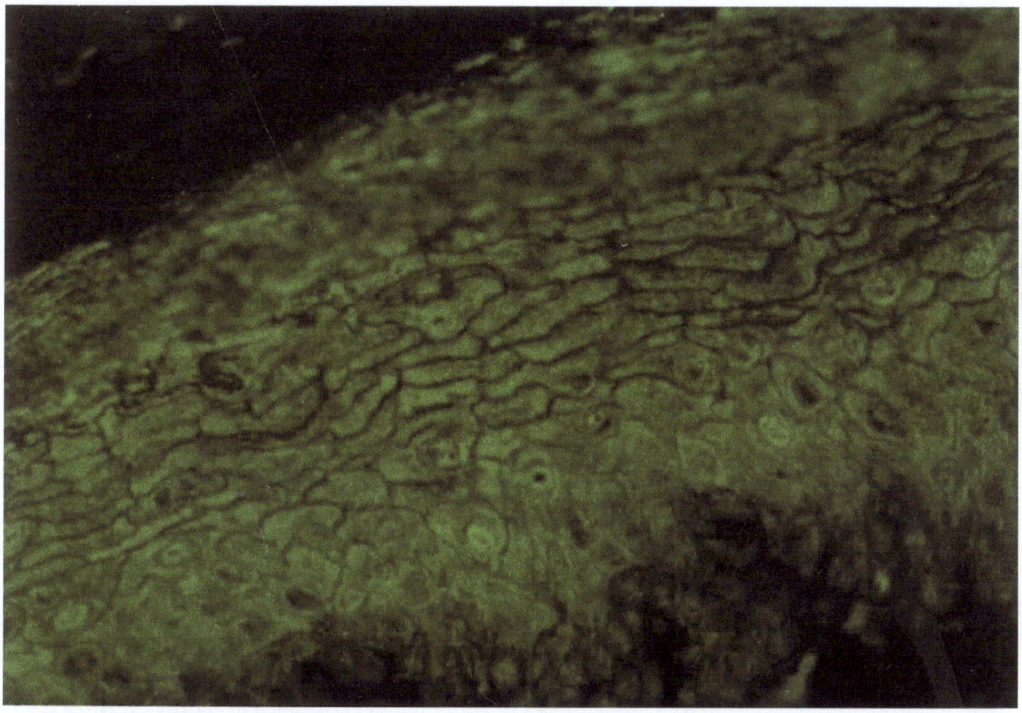

Figure 112. Antibodies against the *stratum corneum, spinosum* and *basale*: antigen, rat oesophagus – diffuse fluorescence of many layers, additionally of the nuclei caused by low titres of ANF; frequently seen pattern without clinical significance (objective ×40, ×500).

Literature: Young et al. [1979]; Vincent et al. [1989]; Kurki et al. [1992]; Aho et al. [1994].

7.3.5.2.3 Antibodies against the cell surface (formerly the inter-cellular substance) of layered squamous epithelia

Synonyms/abbreviations: Pemphigus-antibodies, antibodies against prickle-cell desmosomes.

Target antigen(s): the cell surface of squamous epithelia especially of the *stratum spinosum*, in particular the extracellular domain of desmoglein–1 which belongs to the calcium-dependent cell adhesion molecules of the Cadherin family – glycoproteins of the desmosomes (Emery et al. [1995]; for reviews see Stanley [1993], Furukawa et al. [1994], Amagai [1995]).

Properties of the antibodies: IgG, in particular IgG–4 (for example, Rock et al. [1989]), occasionally IgA and IgM, rarely fix complement, not organ or species-specific but tissue-specific.

Antigen substrate: monkey oesophagus is ideal but rabbit tongue, or the patient's own skin, can also be used. Fix for 30 sec to 1 min in acetone. Rat bladder is recommended for the detection of antibodies in para-neoplastic pemphigus (Anhalt et al. [1990]; Liu et al. [1993]).

Conjugate: anti-human IgG, molar F/P ratio 1.0–1.4; antibody content at the working dilution, approx. 50 µg/ml.

Pattern of fluorescence: fluorescence strictly limited to the inter-cellular region is typical (Fig. 113).

Tip: if blood vessels fluoresce at the same time, one should suspect blood-group antibodies; absorption with anti-A and anti-B, or stomach mucosa, is necessary.

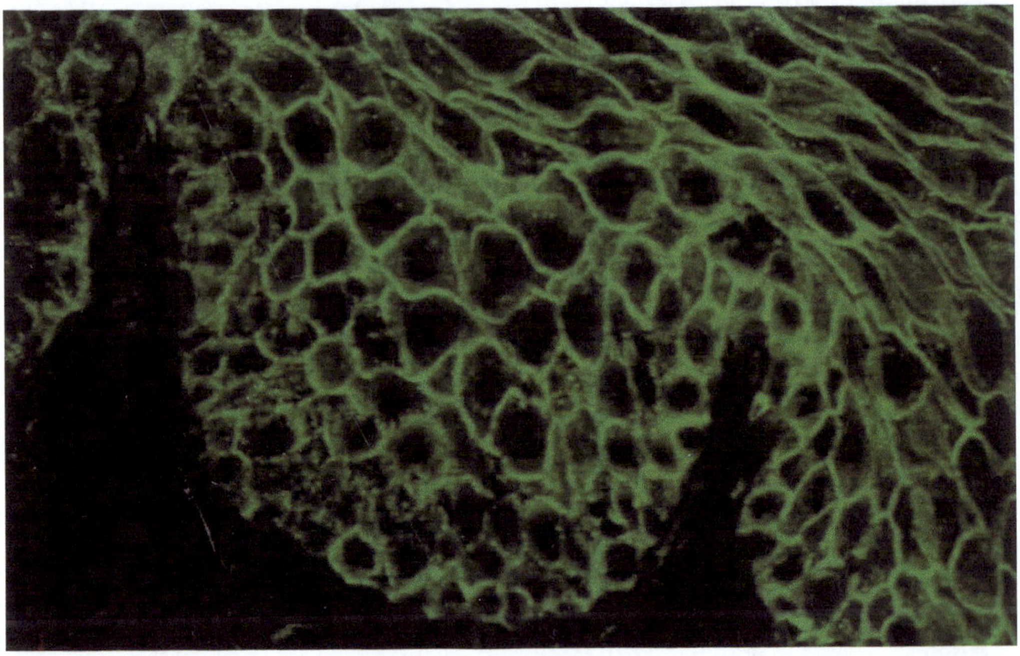

Figure 113a. Antibodies against cell surfaces (desmosomes, formerly inter-cellular substance): antigen, monkey oesophagus – fluorescence along the cell surface especially of the *stratum spinosum* (objective ×40, ×500).

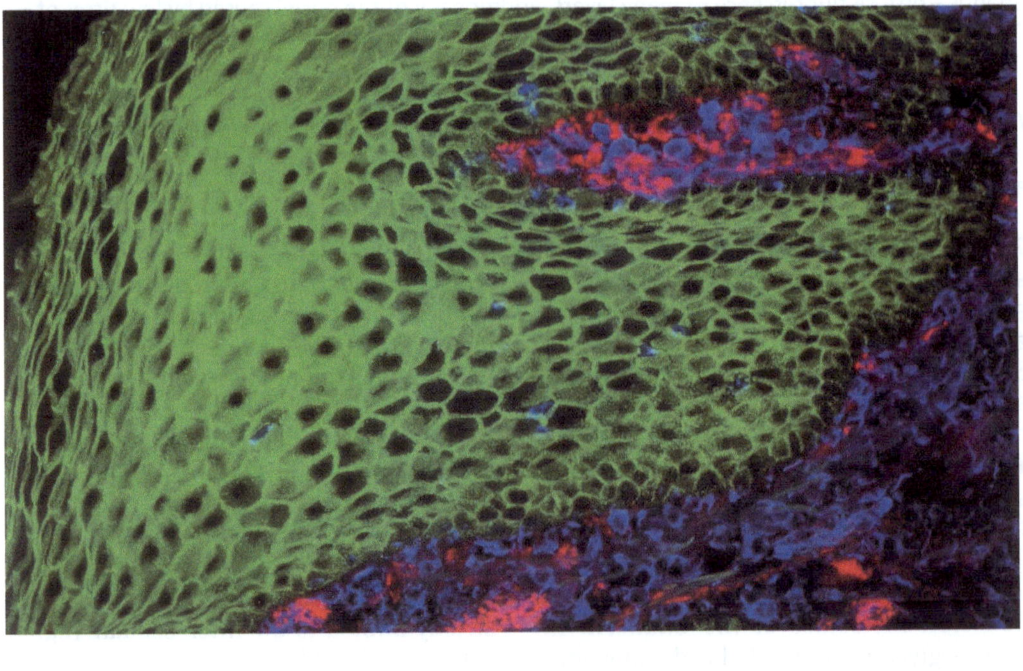

Figure 113b. Demonstration of desmoplakin along the desmosomes (green) using animal experiment Cy 2-labelled antibodies, Factor VIII in the blood vessels (red) using Texas Red-labelled antibodies, and vimentin using AMCA-labelled antibodies (blue): antigen, human tongue – three-fold illumination with blue, green and UV-excitation (objective ×25, ×208; courtesy Mrs. Kuhn and Prof. Franke, Heidelberg).

Clinical significance: Pemphigus-antibodies occur in virtually 100% of cases with *Pemphigus vulgaris*. The detection of antibodies is diagnostic for this disease; titres correlate with activity. So-called pemphigus-antibodies are also found in other dermatoses, for example, bullous pemphigoid, healed pemphigoid, SLE, toxic epidermal necrosis, *Lichen planus*, burns, and drug-induced exanthemata especially in response to penicillin and penicillamine. Other diseases also occasionally react positively, especially autoimmune diseases like myasthenia. Direct immunofluorescence can be of help in difficult cases.

Literature: Beutner & Jordon [1964]; for a review see Beutner et al. [1987]; for para-neoplastic pemphigus, Zillikens & Bröcker [1994]; for a review on the use of direct immunofluorescence for skin biopsies, refer to Gately & Nesbit [1994]).

7.3.5.2.4 Antibodies against the basal cells of the *epidermis*

Target antigen(s): still insufficiently investigated.

Properties of the antibodies: neither organ nor species-specific, tissue-specific, primarily IgG, to some extent fix complement.

Antigen substrate: rat tongue and oesophagus; for the detection of melanocyte antibodies use human skin naevi.

Conjugate: anti-human IgG.

Pattern of fluorescence:

1. Broad homogeneous cytoplasmic fluorescence; "diffuse" = virtually all cells, or "focal" = some basal cells;
2. cytoplasmic fluorescence with emphasis of the periphery (Fig. 114);
3. fluorescence of the cell membrane;
4. perinuclear granular fluorescence;
5. nuclear fluorescence;
6. fluorescence involving melanocytes.

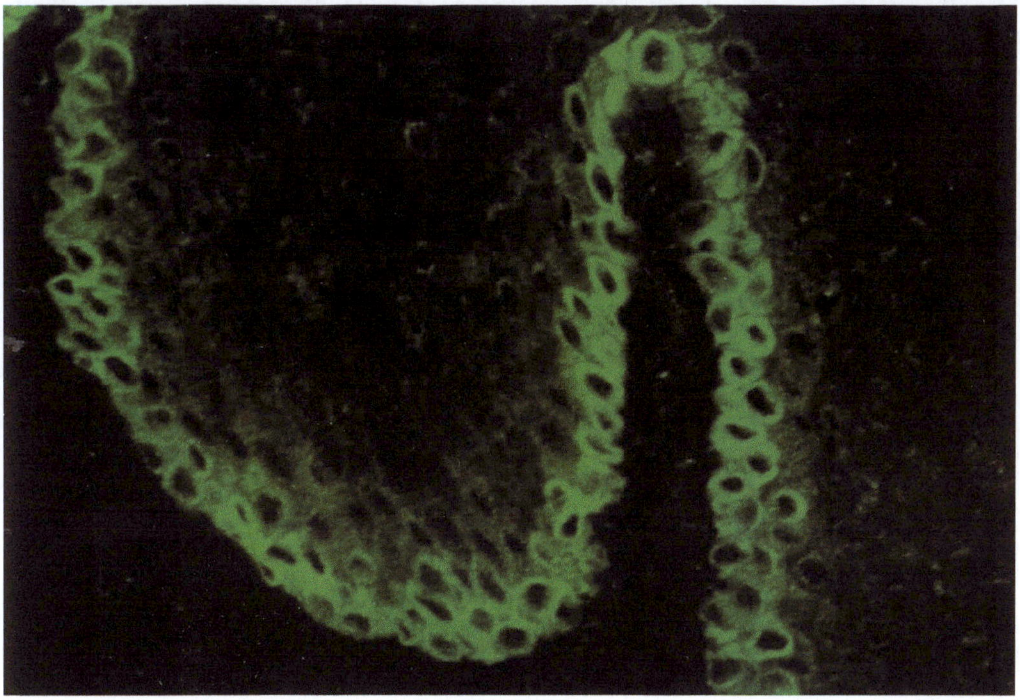

Figure 114. Antibodies against the basal cells of the epidermis: antigen, monkey oesophagus – fluorescence mainly of the periphery of the basal cells (objective ×40, ×500).

In addition to these types, there are numerous combined types involving, for example, the epidermal basement membrane, keratinocytes, basal cells and the hair follicles. Antibodies of type 3 frequently also react with epithelial thymus cells (Lenkei et al. [1985]).

Clinical significance: partly contradictory and not at all clear; type 3 (fluorescence of the cell membrane) is often associated with drug reactions (van Joost [1974]), hepatitis B or D (Zauli et al. [1984]), and with oral *Lichen planus* (Lin et al. [1991]). A combination of types 4 and 5 has been observed in autoimmune hepatitis and NonA/NonB hepatitis (Storch [1989, 1992]). Antibodies against melanocytes occur primarily in vitiligo (Hertz et al. [1977]; Naughton et al. [1983])

Literature: Bystryn [1977]; review, Saurat & Didierjean [1987].

7.3.5.2.5 Antibodies against the basement membrane between the *epidermis* and *corium* of squamous epithelia (pemphigoid-antibodies)

Synonyms/abbreviations: pemphigoid-antibodies.

Target antigen(s): various proteins from the basement membrane of squamous epithelia, in particular a 180 kDa and 230 kDa protein (Kitajima et al. [1994]; review, Stanley [1993]).

Properties of the antibodies: neither organ nor species-specific, not typical of the disease, broadly tissue-specific, predominantly IgG, primarily IgG–4 (e.g., Hadi et al. [1988]), also IgA and IgE (Soh et al. [1993]), fix complement.

Antigen substrate: as for pemphigus-antibodies (but not fixed!!), or human skin that has been cleaved along the *lamina lucida* by means of sodium chloride (the "salt-split" method). Thus small pieces of

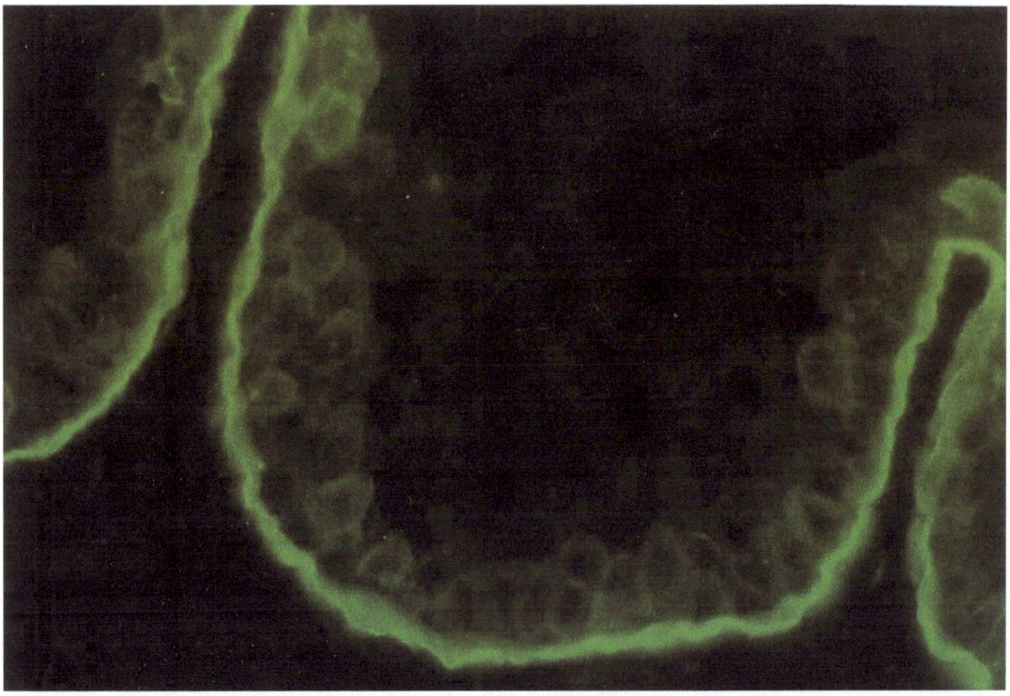

Figure 115. Antibodies against the basement membrane between the epidermis and the corium: antigen, monkey oesophagus – fluorescence of the basement membrane and, in addition, of the basal cells (typical pattern for high titre antibodies – objective ×40, ×500).

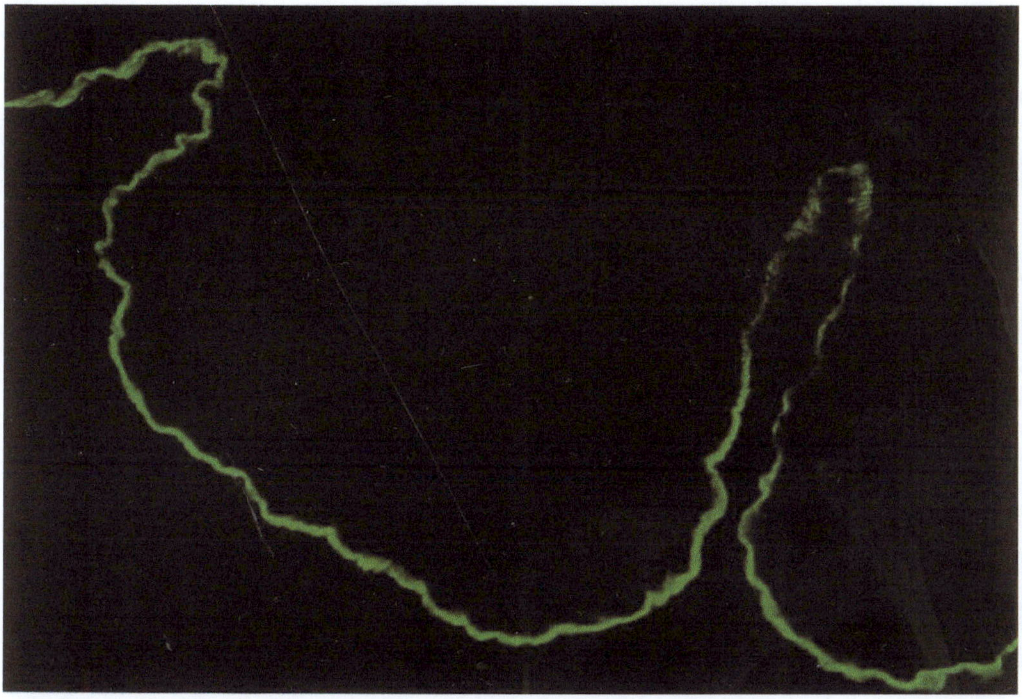

Figure 116. Antibodies against the basement membrane between the epidermis and the corium: antigen, as for Figure 115 – recorded using confocal laser-scanning microscopy, with summation of images from four layers of the basement membrane; note the extreme sharpness of the picture in the absence of background staining (objective ×40, ×500).

normal human skin, unexposed to light, are put into 1 M sodium chloride solution for 72 h at +4°C, constantly shaken, and afterwards the *epidermis* separated from the *dermis* with forceps. Finally, the pieces are frozen in liquid N_2 and stored at at least –70°C until required for cryosectioning (Gammon et al. [1984, 1992]). Addition of protease enzyme inhibitors, e.g., phenyl-methyl-sulphonyl fluoride, is a good idea (Jenkins et al. [1994]).

Conjugate: as for pemphigus-antibodies.

Pattern of fluorescence: linear fluorescence of the epithelial basement membrane; high serum titres in patients with bullous pemphigoid also show fluorescence affecting the basal cells (Figs 115 and 116), in particular of the apical and lateral cell membrane. Use of split skin allows one to distinguish three patterns: a fluorescence exclusively affecting the epidermal portions of the basement membrane; a fluorescence of the epidermal and dermal portions of the basement membrane; and a fluorescence affecting only the dermal portion of the basement membrane (Fig. 117).

Clinical significance: foremost is the significance for the differential diagnosis of bullous dermatoses; antibodies against the basement membrane can be detected in 70% of cases using the standard method, and in virtually 100% of cases using the skin cleavage method. The antibody reacts typically with only the epidermal portion in approx. 90% of cases, with both epidermal and dermal portions involved in 10% of cases, and only in exceptional cases with the dermal portion alone. In *Herpes gestationis* the antibodies react with both portions of the basement membrane, also in healed pemphi-

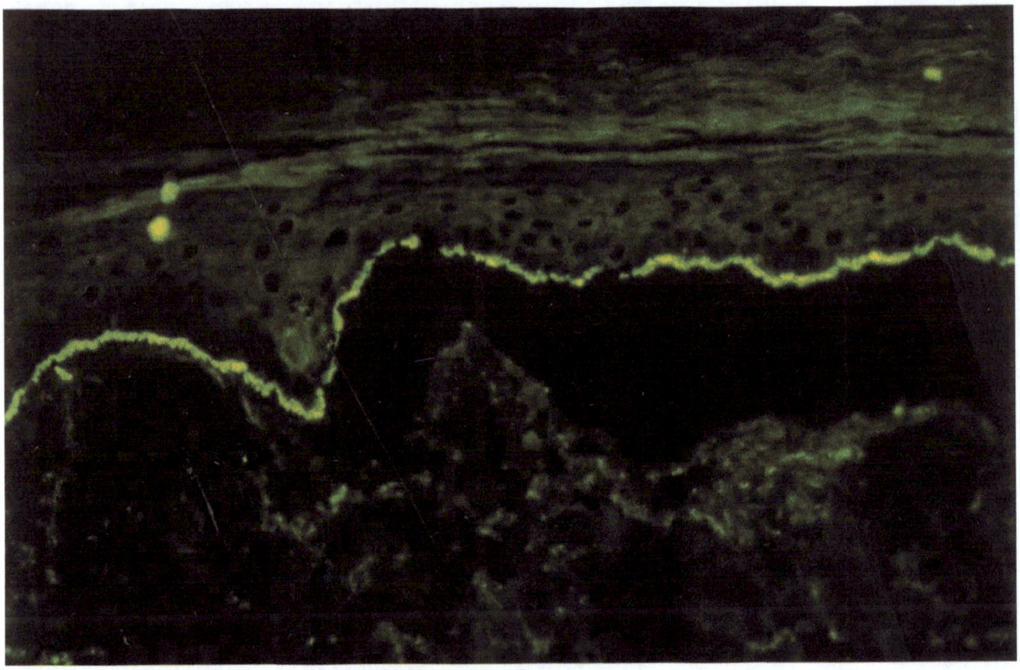

Figure 117. Antibodies against the basement membrane between the epidermis and the corium in bullous pemphigoid: antigen, human 'split' skin (for an explanation of this technique, see text) – fluorescence on the top surface of the artificial cleavage plane or cleft (objective ×25, ×264; Dr. Wever and Prof. Bröcker, Würzburg).

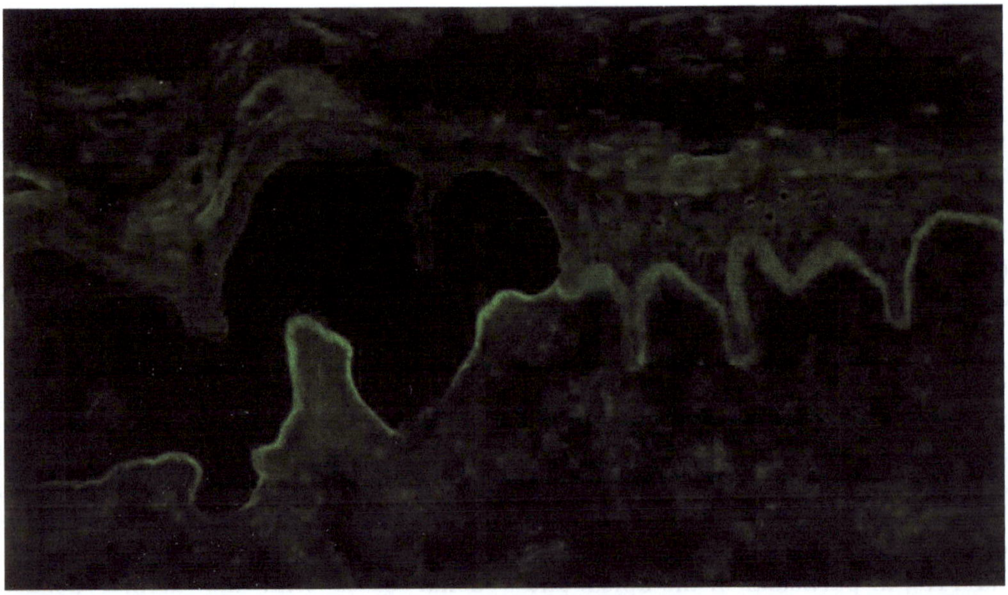

Figure 118. Antibodies against the basement membrane between the epidermis and the corium in acquired bullous epidermolysis: antigen, human 'split' skin as for Figure 117 – fluorescence of the floor of the artificial cleavage plane or cleft (objective ×25, ×264; Dr. Wever and Prof. Bröcker, Würzburg).

goid and linear IgA-dermatosis. In acquired bullous epidermolysis and in bullous SLE, fluorescence exclusively affecting the dermal region is practically always observed (Fig. 118).

Literature: Beutner et al. [1965]; Jordon et al. [1967, 1992]; Stanley et al. [1984]; Parodi & Rebora [1992]; Kormann [1993]; for a review on the use of direct immunofluorescence techniques for skin biopsies, see Gately & Nesbit [1994].

7.3.5.2.6 Antibodies against perinuclear antigens

Synonyms/abbreviations: anti-perinuclear factors (APF).

Target antigen(s): probably filaggrin as for the *stratum corneum* antibodies (Sebbag et al. [1995]; for a review, see Berthelot et al. [1994]).

Properties of the antibodies: neither organ, tissue, or species-specific, typical of the disease, predominantly IgG, IgA and IgM.

Antigen substrate: dried suspension of human buccal mucosa cells; for their preparation, see below.

Conjugate: anti-human IgG.

Production of a preparation containing buccal mucosa cells (based on Hoet [1993])

1. The donor should clean their teeth and then wash out their mouth a number of times.
2. Mucosal cells are taken from the buccal mucosa of the oral cavity using a sterile plastic-foam sponge, which should be not too hard and not too soft!
3. The cells are then suspended in PBS by washing the sponge with 10 ml PBS, pH 7.4.
4. The cells are washed twice with 5 ml PBS followed by centrifugation at 800 g_{av} for five minutes.
5. This is followed by washing with 5 ml PBS-T buffer (0.5% w/v Triton X–100) and centrifugation at 800 g_{av} for five minutes.
6. The cells are then washed again with 5 ml PBS followed by centrifugation as before.
7. The cells are finally suspended in 1–5 ml PBS to give an end-concentration of approx. 10^5 cells/ml.
8. 10 µl of the suspension, containing approx. 10^3 cells, are placed in each well of a slide with multiple wells, and dried using a cold-air fan.
9. The slide, which is now ready for use, can be kept for up to two weeks at −70°C.

Carrying out immunofluorescence (based on Hoet [1993])

1. The serum being investigated is diluted 1:5 with PBS and then centrifuged at 13,000 g_{av} for five minutes.
2. Cover slide with diluted serum and incubate for 90 minutes at room temperature in a humidity chamber.
3. Wash three times with PBS, each time for ten minutes.
4. Incubate with FITC-labelled trivalent anti-human IgG, IgA and IgM, e.g., Dakopatts F200, Glostrup, Denmark, at a dilution of 1:100 in PBS.
5. Wash three times with PBS, each time for ten minutes.
6. Embed in glycerol/PBS (1:1 v/v) solution in the presence of 0.5 mg/ml ethidium bromide to counterstain the nuclei.
7. Assess the results by fluorescence microscopy.

Pattern of fluorescence: a positive result consists of typical perinuclear fluorescence (Fig. 119).

Tip: Triton X–100 treatment should improve the sensitivity by increasing cell transparency and reducing background staining. The APF-activity can be converted into IU/ml using a rheumatoid arthritis reference serum. WHO rheumatoid arthritis reference serum contains by definition 100 IU/ml, and is obtainable through CLB, PO Box 9190, 1000 AD Amsterdam, the Netherlands.

Clinical significance: anti-perinuclear factors count as characteristic markers for rheumatoid arthritis, even in rheumatoid factor-negative cases. Sensitivity and

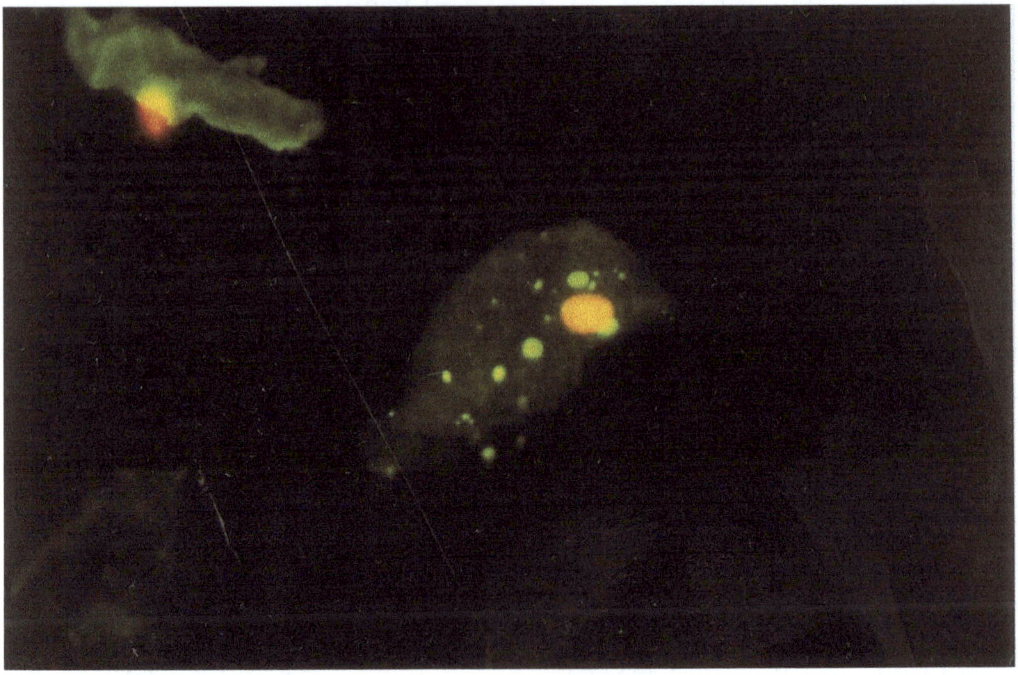

Figure 119. Antibodies against peri-nuclear antigens: antigen, smears of buccal mucosal cells – fluorescence of peri-nuclear granules (objective ×63, ×800; Dr. Geertzen, Amsterdam).

specificity depend on both the serum dilution and the conjugate. An agreed uniform interpretation is not currently available. Sera which contain anti-perinuclear factors also mostly contain antibodies against the *stratum corneum*, and *vice versa*, so that it is possible that these are one and the same antibody.
Literature: Nienhuis & Mandema [1964]; Aho et al. [1994].

7.3.5.3 Antibodies against smooth muscle

Synonyms/abbreviations: smooth muscle antibodies (SMA), anti-smooth muscle antibodies (ASMA).
Target antigen(s): apart from actin (Gabbiani et al. [1973]), numerous different microfilaments, intermediary filaments and microtubules, as well as still unknown protein components, must be considered – for a summary see the 1st edition of this book and Table 28.
Properties of the antibodies: generally neither organ- nor tissue- nor species-specific, not disease-specific, although individually typical antibodies do appear to occur; predominantly IgG, less so IgM, very rarely IgA or monoclonal precursors, occasionally fix complement.
Antigen substrate: a block containing stomach, liver, kidney, if possible skeletal and heart muscle in order to recognise cross-reactions.
Conjugate: anti-human IgG.
Pattern of fluorescence: typical antibodies produce fluorescence of smooth muscle fibres in the *muscularis* of blood vessels, in the *muscularis mucosae* and in the *muscularis* of the stomach, and of the inter-glandular contractile filaments of the gastric mucosa (Fig. 120). In addition and sometimes even exclusively, one

can observe staining of the following structures, for example (Storch [1975]) -

Liver:

1. linear fluorescence of the hepatocyte periphery (hexagonal or polygonal fluorescence);
2. interrupted double-lines around hepatocytes;
3. spotty fluorescence round hepatocytes;
4. net-like fluorescence of hepatocytes;
5. finely granular or homogeneous fluorescence affecting hepatocytes;
6. linear fluorescence of the apical regions of the 'lining' cells of the small bile ducts in the periportal areas;
7. broad, diffuse staining of the wall of the large bile ducts.

Kidney:

1. glomeruli as a diffuse, mesangial or linear pattern of fluorescence;
2. small peritubular vessels and capillaries with a preference for the endothelium;
3. peritubular fibrils;
4. apical portions of individual tubular cells;
5. the brush-border of the proximal tubules;
6. diffuse cytoplasmic fluorescence of the proximal tubules;
7. epithelial cells of the renal pelvis;
8. the basement membrane of the renal pelvis.

Table 28. Antibodies directed against smooth muscle and the cytoskeleton (selected from Abuaf et al. [1991]).

Filament	Antigen	Disease
Microfilaments	Actin	lupoid hepatitis
	Myosin	primary biliary cirrhosis
	Troponin	drug-induced hepatitis
	Tropomyosin	
	Filamin	
	Calmodulin	
	Gelosin	
Intermediary filaments	Vimentin	viral hepatitis A
	Desmin	viral hepatitis B
	Keratin	viral hepatitis D
	Skeletin	chicken-pox (varicella)
		measles (morbilli)
		mumps
	fibrillar glial	multiple sclerosis
	protein	Mycoplasma pneumonia
		acute malaria
Microtubules	Tubulin	alcoholic liver damage
		infectious mononucleosis
		parasites; demyelinating
		disease; autoimmune
		disorders of the thyroid

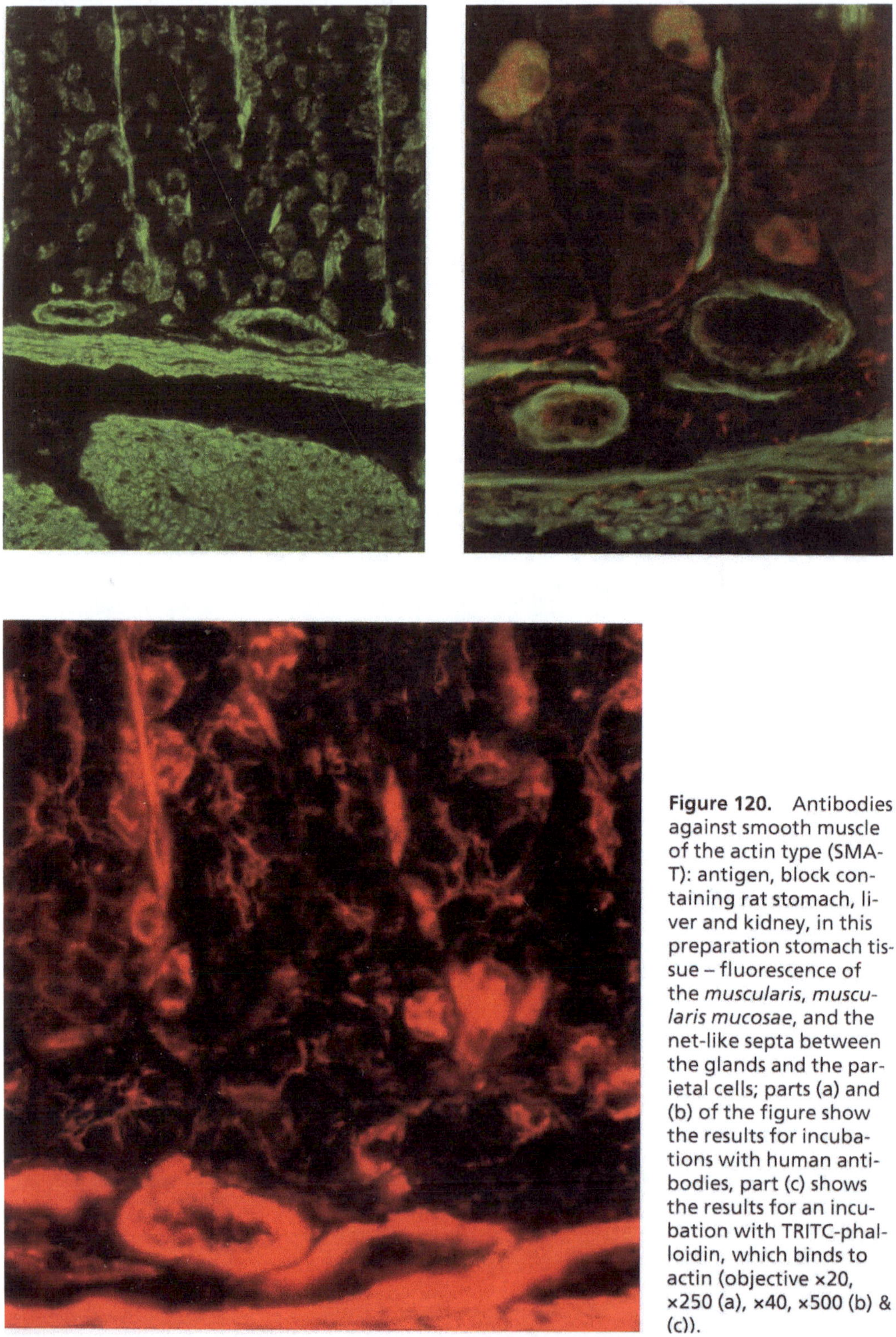

Figure 120. Antibodies against smooth muscle of the actin type (SMA-T): antigen, block containing rat stomach, liver and kidney, in this preparation stomach tissue – fluorescence of the *muscularis*, *muscularis mucosae*, and the net-like septa between the glands and the parietal cells; parts (a) and (b) of the figure show the results for incubations with human antibodies, part (c) shows the results for an incubation with TRITC-phalloidin, which binds to actin (objective ×20, ×250 (a), ×40, ×500 (b) & (c)).

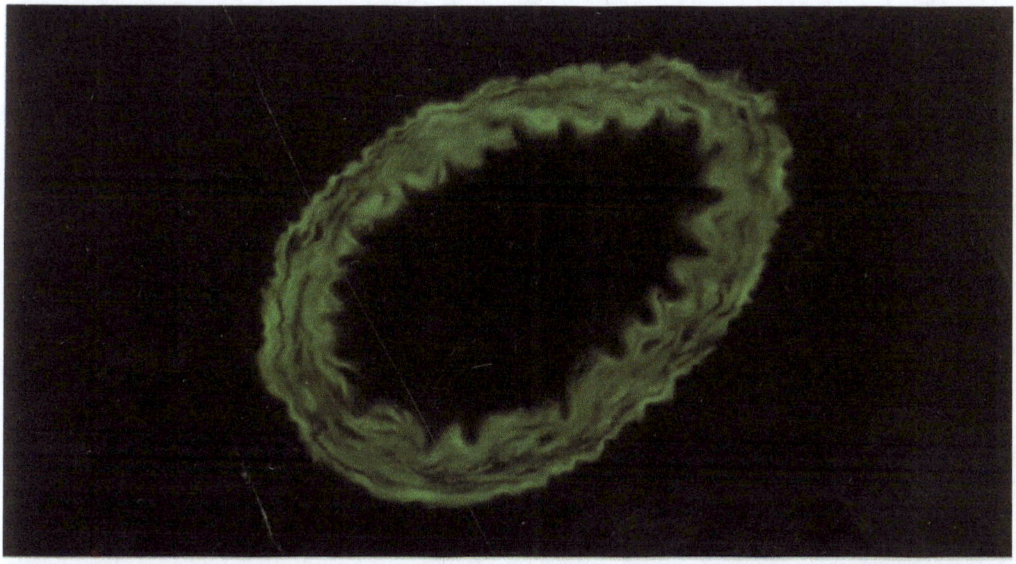

Figure 121. Antibodies against smooth muscle (type SMA-V): antigen, as for Figure 120, in this instance kidney – fluorescence exclusively affecting the media of a small artery; glomeruli and tubules are negative (objective ×40, ×500).

Stomach:

1. net-like fluorescence of the *muscularis*;
2. exclusive fluorescence of the *muscularis mucosae*;
3. apical portions of luminal or basal epithelial cells of the mucosa (Figs 76–80);
4. parietal cells.

Fluorescence has, furthermore, been observed affecting the membrane region of epithelial thyroid cells and in lymphocytes and thymocytes, in numerous other non-muscle cells, cross-striations (particularly the I-band) and especially the intercalated disks in cardiac muscle. Antibody heterogeneity is considerable; even in 1980, 123 different patterns of fluorescence out of 170 positive sera investigated had been reported (Storch [1980]).

The classification of kidney sections into SMA-V (fluorescence only affecting vessel walls), SMA-G (vessel walls and glomeruli) and SMA-T (vessel walls, glomeruli and tubule fibrils), is useful in practice (Bottazzo et al. [1976]; Figs 121–123).

Tip: non-specific staining and cross-reactions, occurring primarily with antibodies against connective tissue and vascular endothelium, must be taken into consideration.

Clinical significance: persistent antibodies against smooth muscle and the cytoskeleton, with a titre of 1:160 or more, are evidence of chronic aggressive hepatitis (CAH), above all if these are of the actin type (type SMA-T: in the kidneys, apart from fluorescence of the blood vessels, react simultaneously with glomeruli and peritubular fibrils, *vide supra*). SMA occur transiently and at low titre in numerous diseases (Tab. 28), i.e., in yellow fever, cytomegaly, wart carriers, infections with *Mycoplasma pneumoniae*, malignant disease, as well as with banal influenza-like infections and in healthy patients. They often appear as part of the integrated physiological reaction to cell destruction, primarily caused by viruses. This is understandable since different

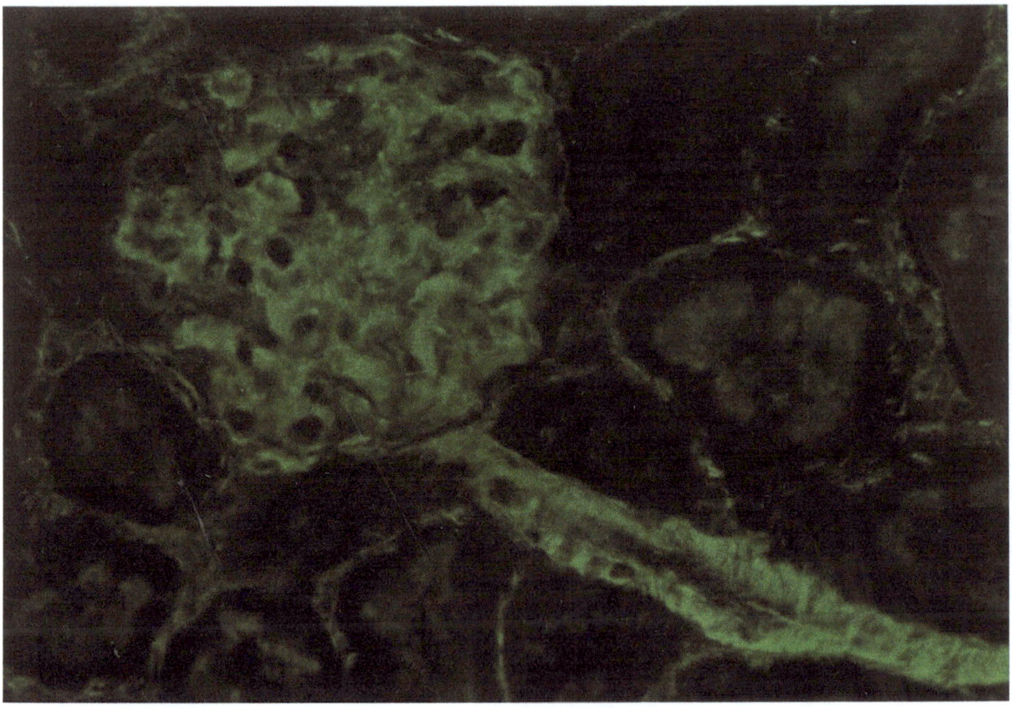

a

Figure 122. Antibodies against smooth muscle of the actin type (SMA-T): antigen, as for Figure 120; (a) and (b) *kidney* – fluorescence of the artery, glomerulus (mesangial), the peritubular fibrils and the brush-border of the tubule cells; (c) *liver* – fluorescence of the periphery of the liver cells and bile canaliculi ('double-contouring'); (a) and (c) incubated with human antibodies, (b) with TRITC-phalloidin (objective ×40, ×500).

forms of actin are components of almost all eukaryotic cells (Garreis & Gibson [1976]).

The combination of high titre antibodies against smooth muscle and against cell nuclei is pathognomic for so-called "lupoid" hepatitis, a special form of CAH. The combination of high titre SMA and AMA indicates another special form of CAH (Berg [1976]), or cryptic cirrhosis. The simultaneous appearance of SMA, of antibodies against vascular endothelium and of those directed against connective tissue, indicates a viral pathogenesis for the disease in question.

Literature: Johnson et al. [1965]; Whittingham et al. [1966]; Storch [1992].

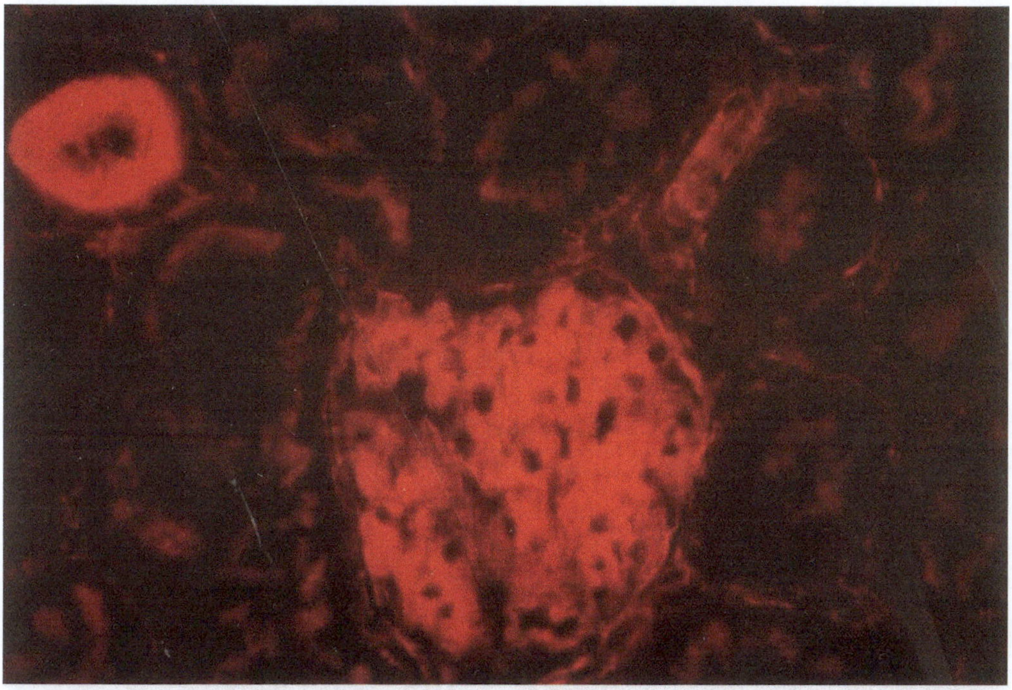

b

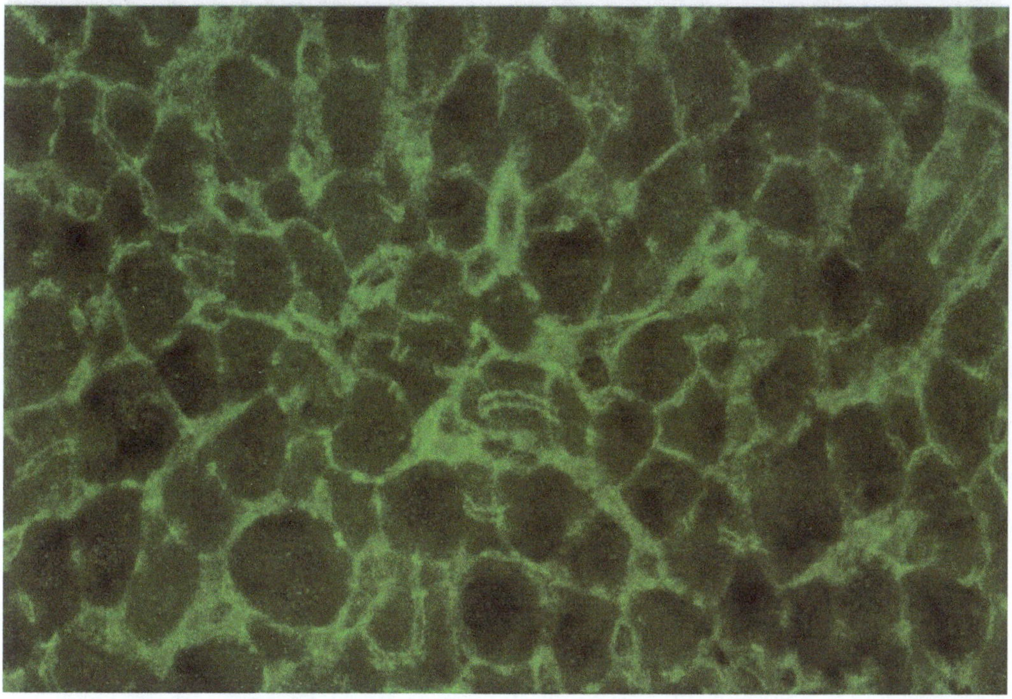

c

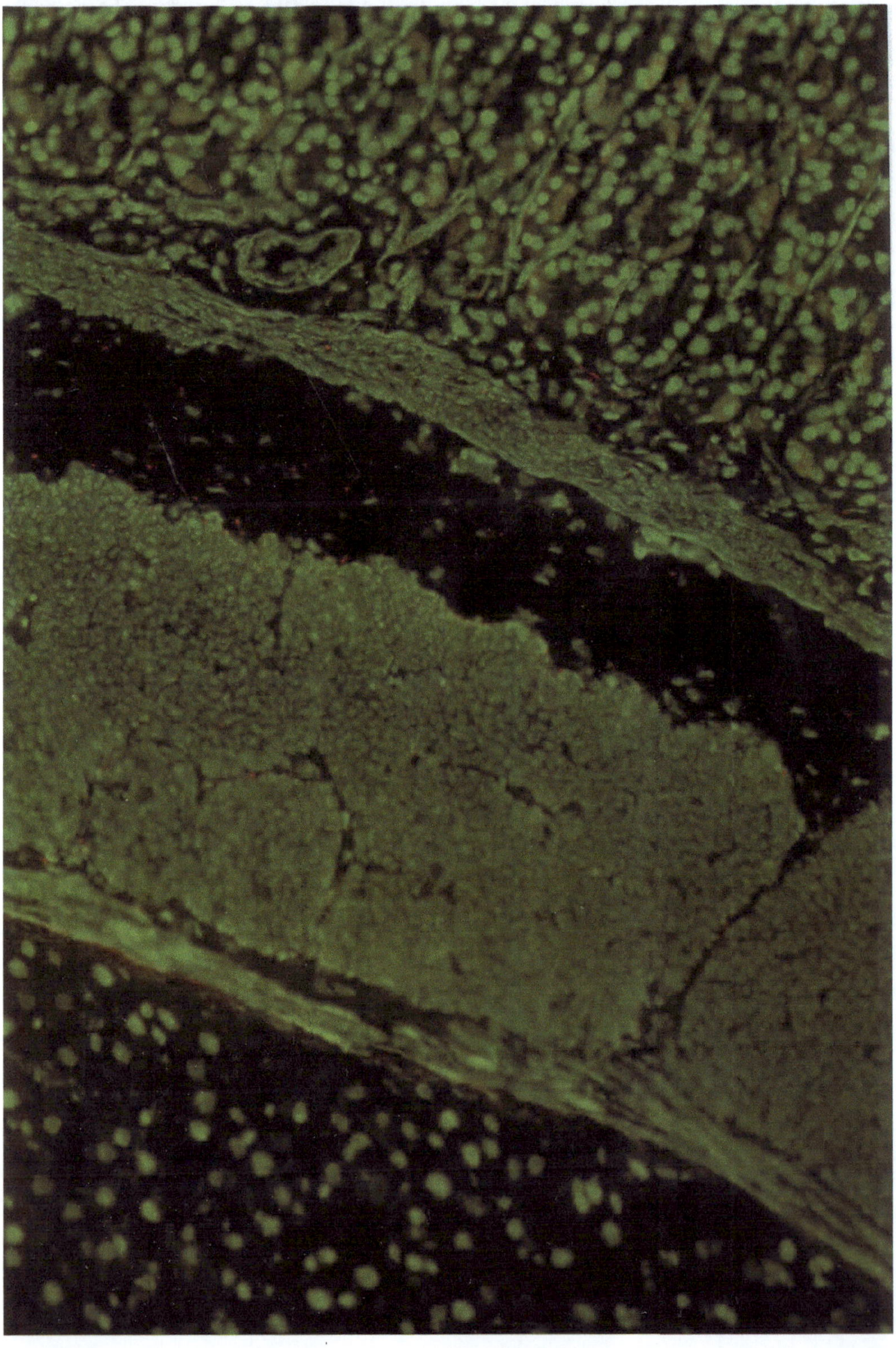

a

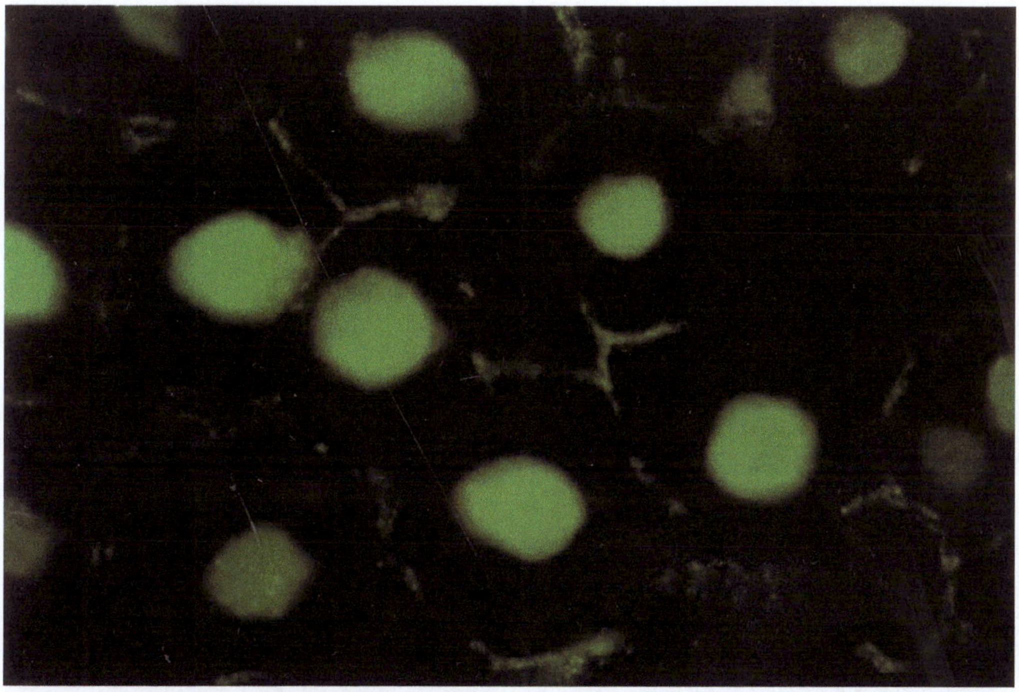

b

Figure 123. Antibodies against smooth muscle of the actin type together with ANF: antigen, as for Figure 120 – (a) the border between stomach tissue (upper) and liver tissue (lower); (b) liver tissue (objective ×20, ×250 (a), ×100, ×1250 (b)).

7.3.5.4 Antibodies against skeletal muscle

Target antigen(s): cytoplasmic components of the cross-striations of skeletal and cardiac muscle as well as of thymus cells, predominantly in the I-bands, also in the A-bands and the Z-striae; furthermore in the sub-sarcolemmal and sarcoplasmic reticulum. Biochemically speaking, myosin, heavy meromyosin, actin and other components especially titin (Aarli et al. [1990]), should be considered as antigens.

Properties of the antibodies: they are generally tissue-specific but not species-specific; they belong predominantly to the IgG type; type SH reacts with skeletal and cardiac muscle, it does not fix complement; type S reacts only with skeletal muscle and fixes complement (Beutner et al. [1962]).

Antigen substrate: highly extended muscle fibres from man, monkey or cow; fixed (30 sec in acetone) or unfixed tissue is suitable. We prefer unfixed bovine muscle in order to reduce non-specific binding to F_c-receptors. If only antibodies against skeletal muscle are to be investigated, the use of isolated myofibrils has advantages. Preparations can be made by extraction with PBS-glycerol (50% v/v), pH 7.2 – Peers et al. [1977].

Conjugate: anti-human IgG; when using human tissue, if possible the $F(ab')_2$ fragment.

Pattern of fluorescence: assessment should be carried out by comparison with the appearance under phase contrast in order to be able to distinguish staining of the I-bands and A-bands (Figs 124 & 125).

a

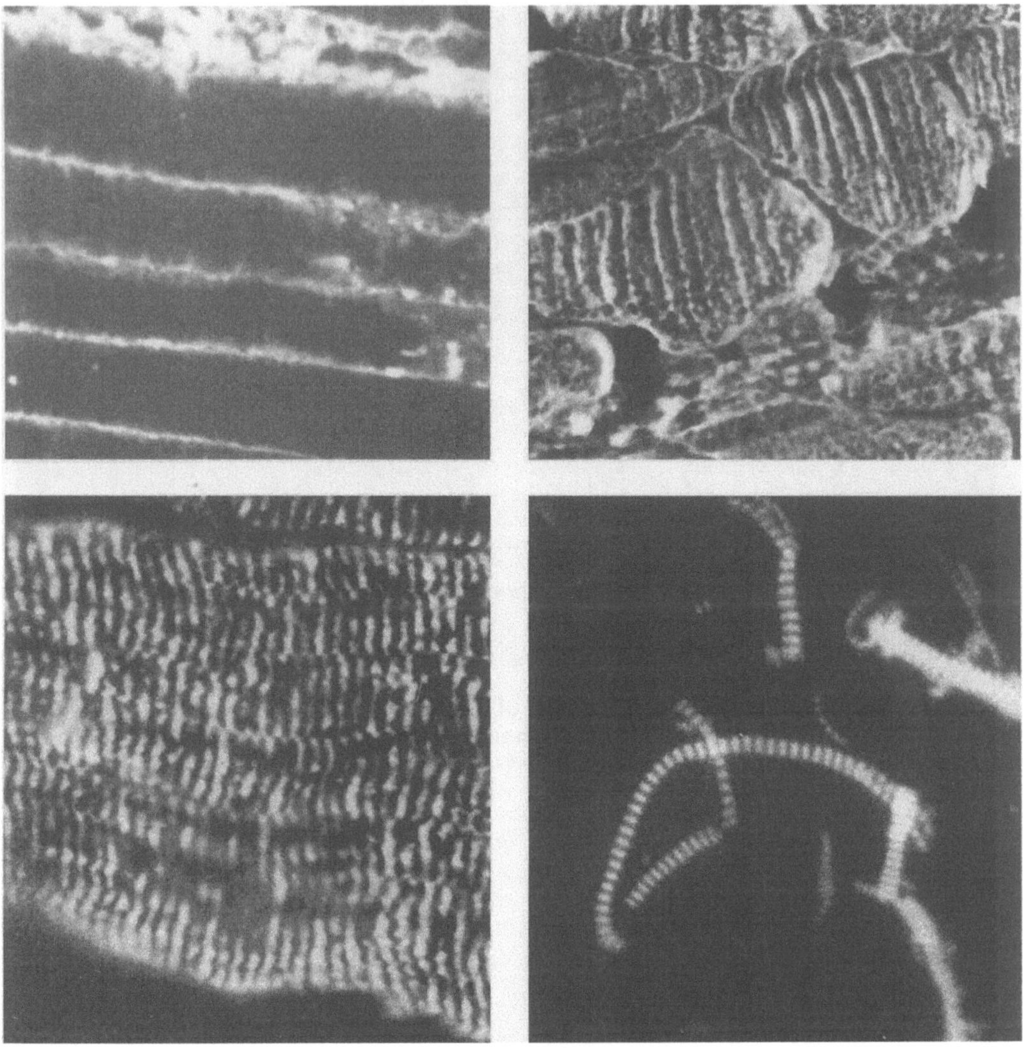

c

Figure 124. Antibodies against skeletal muscle: antigen, (a) and (b) human muscle, (c) and (d) bovine muscle – (a) conjugate control: cross-reaction of the conjugate (anti-human IgG) with the sarcolemma including vessel walls and perivascular connective tissue (objective ×25, ×216); (b) incubation with a positive control consisting of autoantibodies and conjugate: specific fluorescence of the cross-striations (objective ×25, ×300); (c) bovine muscle, otherwise as for (b): fluorescence of the cross-striations, no reaction with the sarcolemma (objective ×100, ×1200); (d) isolated myofibrils, incubated as in (b) (objective ×100, ×1456).

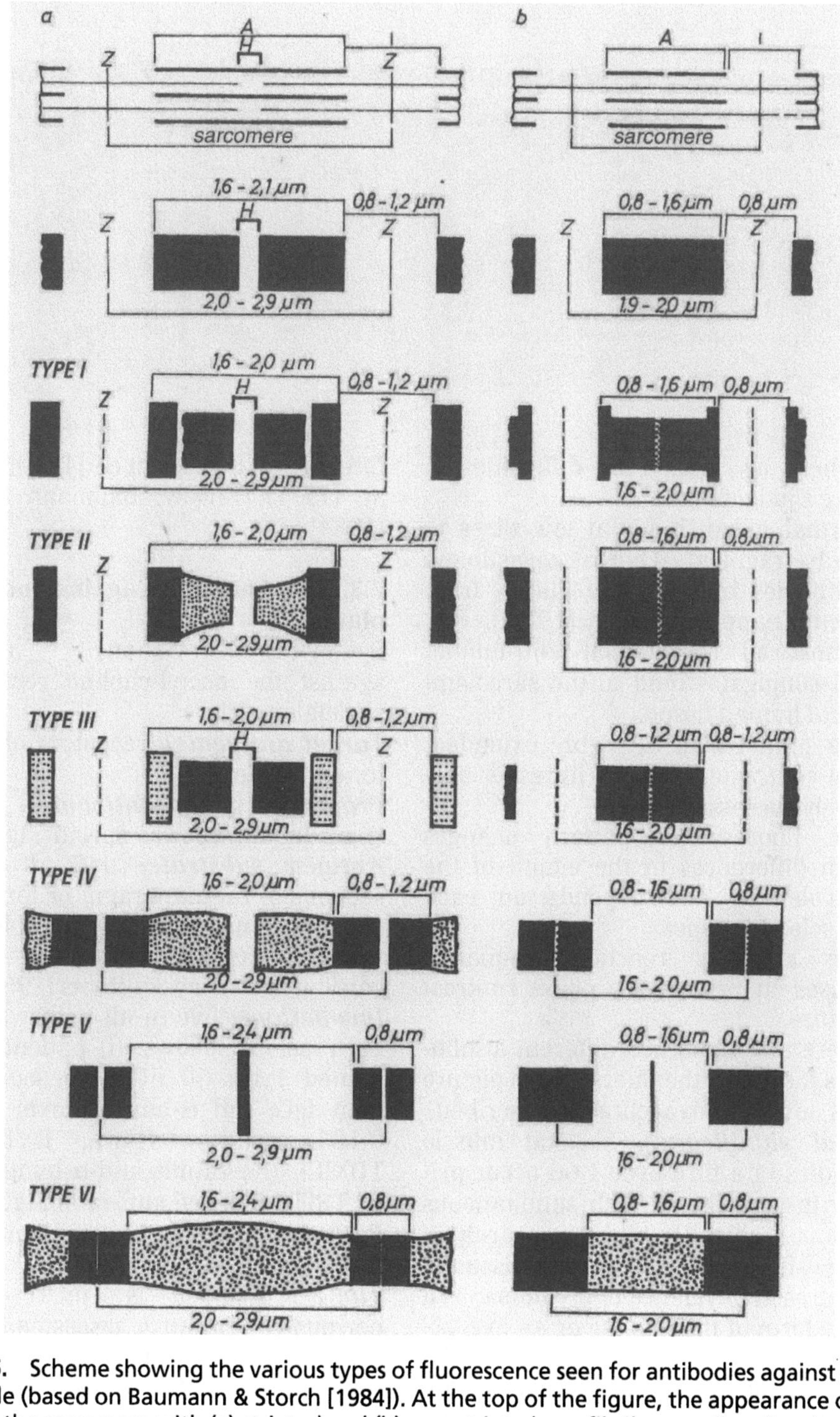

Figure 125. Scheme showing the various types of fluorescence seen for antibodies against skeletal muscle (based on Baumann & Storch [1984]). At the top of the figure, the appearance of regions of the sarcomere with (a) striated and (b) non-striated myofibrils, as well as the phase-contrast image. In fluorescence patterns 1–6, the black and punctate sarcomere regions appear as areas with different intensities of fluorescence. One must recognise that it is only possible in exceptional cases to differentiate the type of reaction for non-striated fibrils.

Table 29. Frequency of skeletal muscle antibodies with a titre over 1:60 (based on Strauss et al. [1966]).

Diagnosis	Number of cases	Number of positives	% positive
Myasthenia (all types)	336	99	32.6
Thymoma with myasthenia	20	19	95.0
Myasthenia without thymoma	316	80	25.3
Thymoma without myasthenia	129	31	24.0
Controls	803	2	0.3

Tip: there are numerous difficulties in making a judgement.

1. Normal serum binds at low titres to the F_C-fragment, whereas myasthenia antibodies bind via the $F(ab')_2$ fragment (Knapp & Paterisky [1972]).
2. Almost all commercial anti-human IgG conjugates bind to the sarcolemma of human tissue.
3. The preparation of highly extended, non-contracted muscle fibres is only rarely successful.
4. The fluorescence pattern changes with differences in the length of the sarcolemma and depends on each muscle fibre type.
5. The antibody reaction frequently causes changes in the phase contrast picture.
6. There are numerous different antibodies for which the fluorescence picture has not yet been accurately described.

Clinical significance: skeletal muscle antibodies in a titre over 1:60 occur primarily in myasthenia with simultaneous thymoma (Table 29), and more rarely in polymyositis and other muscle diseases. High titres are evidence of thymoma, as a serious form of the disease or an exacerbation. Constantly high titres can be the sign of a poor prognosis, whereas a drop in titre indicates an improvement. One frequently finds antibodies in younger patients at the beginning of the disease and in known HLA-antigen carriers (Feltkamp et al. [1974]). The titre need not

Literature: Strauss et al. [1960]; Peers et al. [1977]; review, Baumann & Storch [1984].

7.3.5.4.1 Antibodies against motor end-plates

Synonyms/abbreviations: antibodies against the acetyl-choline receptors of skeletal muscle.

Target antigen(s): receptors of the motor end-plate.

Properties of the antibodies: generally tissue but not species-specific; IgG type.

Antigen substrate: unfixed cryotome sections of rat diaphragm or tongue with as many end-plates as possible, determined by staining for cholinesterase according to Karnovsky-Roots [1964].

Incubations: five in all with washing in between, as follows: (i) patient's serum diluted 1:10; (ii) FITC-labelled anti-human IgG; (iii) α-bungarotoxin solution, 0.125 μg/ml – CARE, EXTREMELY TOXIC!; (iv) rabbit anti-α-bungarotoxin; (v) TRITC-labelled anti-rabbit IgG.

Pattern of fluorescence: staining of the motor end-plates (Fig. 126).

Tip: a prerequisite is a microscope that permits comparative assessment of both FITC and TRITC-labelled deposits. Both the antibody against acetyl-choline esterase and the α-bungarotoxin bind to motor end-plates. Detection can be successfully achieved through the use of double-labelling even in the presence of "normal" skeletal muscle antibodies.

Clinical significance: still not clear, antibodies are probably typical for the disease in myasthenia gravis, especially in the presence of hyperplasia of the thymus (Storch et al. [1985]), because negative results are seen in other neuromuscular diseases and in healthy patients. The method has a clearly reduced sensitivity, however, compared to radio-immunoassay methods.
Literature: Sondag-Tschroots et al. [1977]; Storch & Trautmann [1981b].

7.3.5.5 Antibodies against cardiac muscle

Synonyms/abbreviations: myocardial antibodies.
Target antigen(s): as for antibodies against smooth muscle, numerous antigens have to be considered; descriptively one can assign these to the sarcolemma and the myolemma, to fibrils, e.g., actin, myosin and Z-striae, to mitochondria and to the sarcoplasmic reticulum, and in addition to fibres of the bundle of His (conducting tissue) and to the endocardium. At a molecular level the following antigens are recognised: α- and ß-receptors, calcium channels, sarcosine dehydrogenase, creatine kinase, dihydrolipoamide dehydrogenase and ATPase (for a review, see Maisch et al. [1995]).
Properties of the antibodies: partly organ-specific, not species or disease-specific; cross-reactions with components of *Streptococci*, of smooth muscle and striated muscle and thymocytes; predominantly IgG, sometimes IgM.
Antigen substrate: fresh heart tissue, also biopsy material.
Conjugate: anti-human IgG.
Pattern of fluorescence: one can distinguish at least eight types: (i) diffuse cyto- or sarcoplasmic fluorescence; (ii) an inter-myofibrillar type; (iii) a myolemmal type; (iv) sarcolemmal or sub-sarcolemmal types (Fig. 127a); (v) affecting transverse striations; (vi) involving the intercalated disks (Fig. 127b); (vii) fluorescence

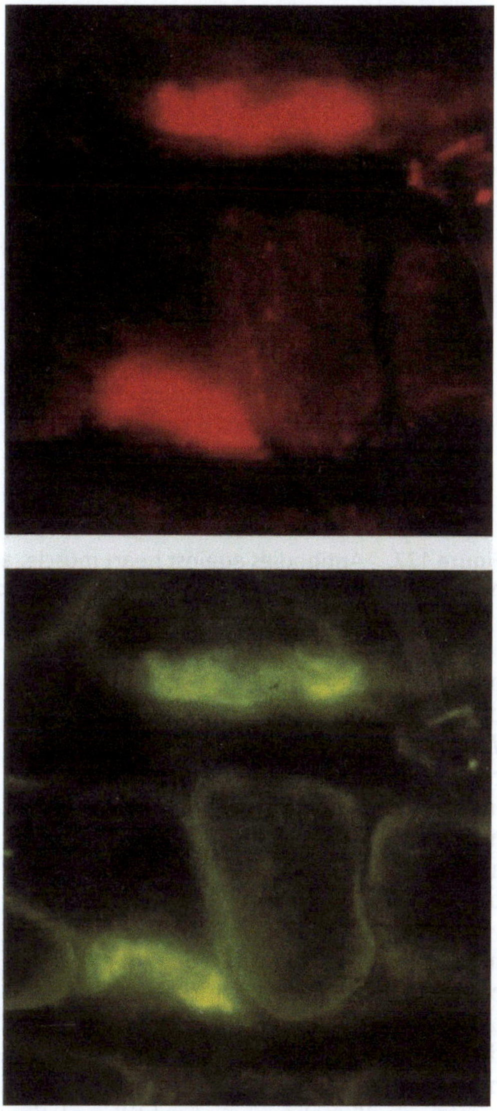

a

b

Figure 126. Antibodies against the motor end-plate in skeletal muscle: antigen, rat tongue – five layer immunofluorescence assay using human serum, FITC-labelled goat anti-human IgG, α-bungarotoxin, rabbit anti-α-bungarotoxin serum, and goat TRITC-labelled anti-rabbit IgG; one sees a red fluorescence exclusively affecting the motor end-plates on excitation in the green, with a green fluorescence of the motor end-plates using human antibodies and, in addition, fluorescence of the sarcolemma brought about by simultaneously present antibodies (objective ×25, ×500).

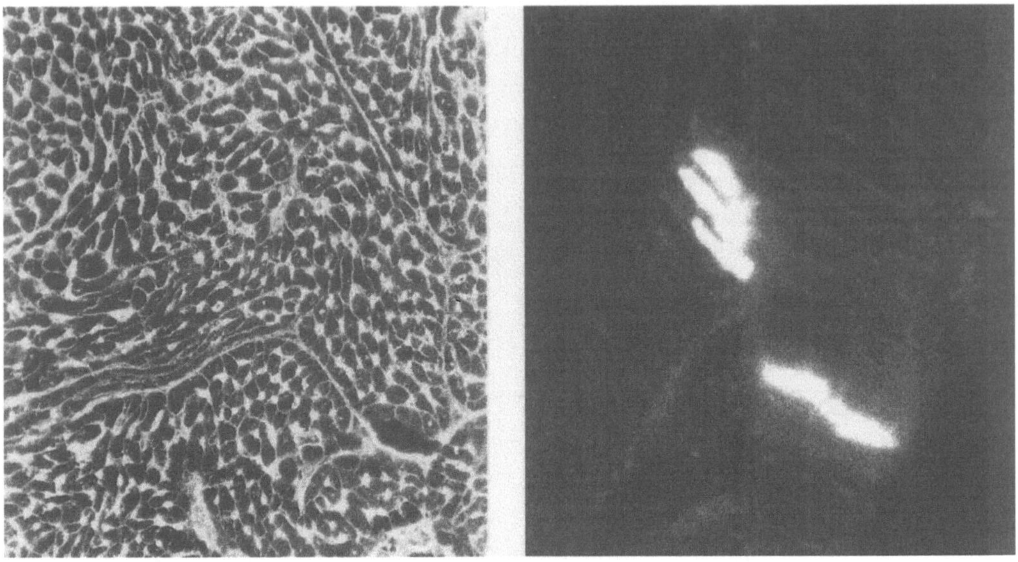

a

Figure 127. Antibodies against heart muscle: antigen, rat heart – (a) sarcolemmal-subsarcolemmal type (objective ×16, ×330); (b) intercalated disk type (objective ×40, ×800).

of the fibres forming the conducting bundle; and (viii) affecting the endocardium.

Tip: watch out for non-specific staining resulting from the use of unsuitable conjugates! In order to recognise cross-reacting antibodies against skeletal muscle and smooth muscle, a combined tissue block containing these tissues should always be used, or these tissues examined in addition. In order to recognise or exclude without doubt antibodies directed against connective tissue, vascular endothelium, mitochondria and nuclear material, amongst others which are not organ-specific, the use of additional tissues such as kidney, stomach and liver, is to be recommended.

Clinical significance: because the antibodies occur to some extent in healthy patients, after common streptococcal infections and also completely physiologically as "clearance" antibodies after myocardial infarction, a decision is only possible when, in individual cases, the type, persistence, titre and results of absorption studies are available, and the clinical history is considered. The sarcolemmal / sub-sarcolemmal type occurs primarily as a sequel to a streptococcal infection. The titres are high in patients with active rheumatic fever (also in remission), in whom the antibodies can be absorbed with Group A *Streptococci*. No significance can be attached to the diffuse type of reaction, in which streptococcal absorption is negative, and which occurs physiologically as the result of the presence of "clearance" antibodies in healthy patients and after various forms of myocardial damage, e.g., after myocardial infarction or cardiotomy. A rare type of fluorescence affecting the transverse striations is valuable for excluding skeletal muscle antibodies in the diagnosis of rheumatic fever. Recently, the occurrence of antibodies in viral myocarditis and pericarditis, idiopathic cardiomyopathy and left ventricular hypertrophy, has been repeatedly pointed out. Frequencies have been variously cited: in acute pericarditis these lie around 10%.

Literature: Kaplan et al. [1961]; Danilova & Lyampert [1972]; Caforio et al. [1992] Boudonas et al. [1994]; review, Maisch et al. [1995].

7.3.5.6 Antibodies against vascular endothelium

Target antigen(s): still insufficient details known; they appear to be related to blood-group substance A.

Properties of the antibodies: exclusively tissue but not species or disease-specific, IgG type, fix complement.

Antigen substrate: where possible human tissue of blood-group A_1 (kidney, liver and stomach), also rat tissue.

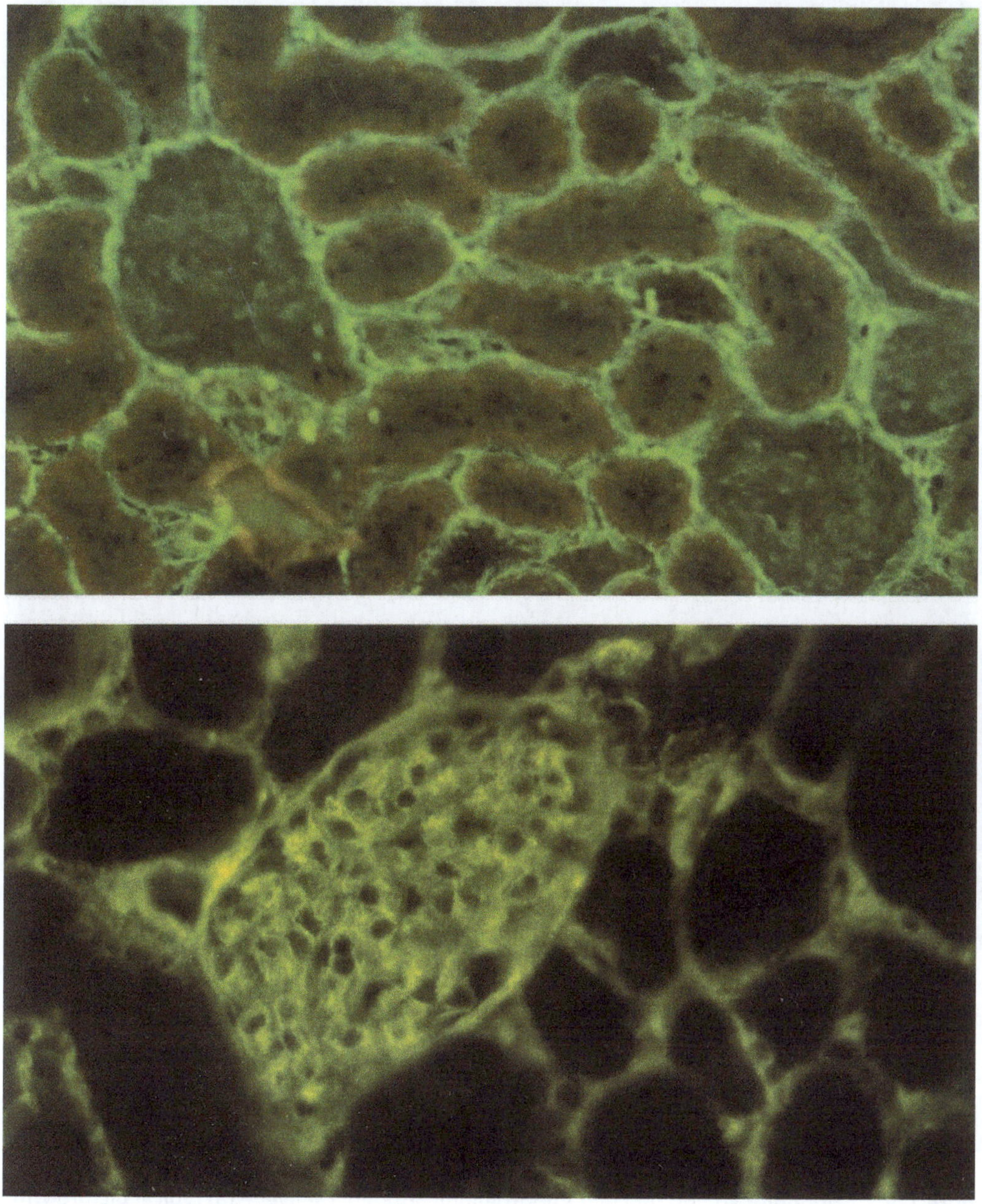

a

b

Figure 128. Antibodies against vascular endothelium: antigen, rat kidney – (a) typical picture, with fluorescence of the peritubular capillaries and with a suggestion of those of the glomeruli (objective ×20, ×250); (b) atypical picture, with intense fluorescence of the capillaries and of the

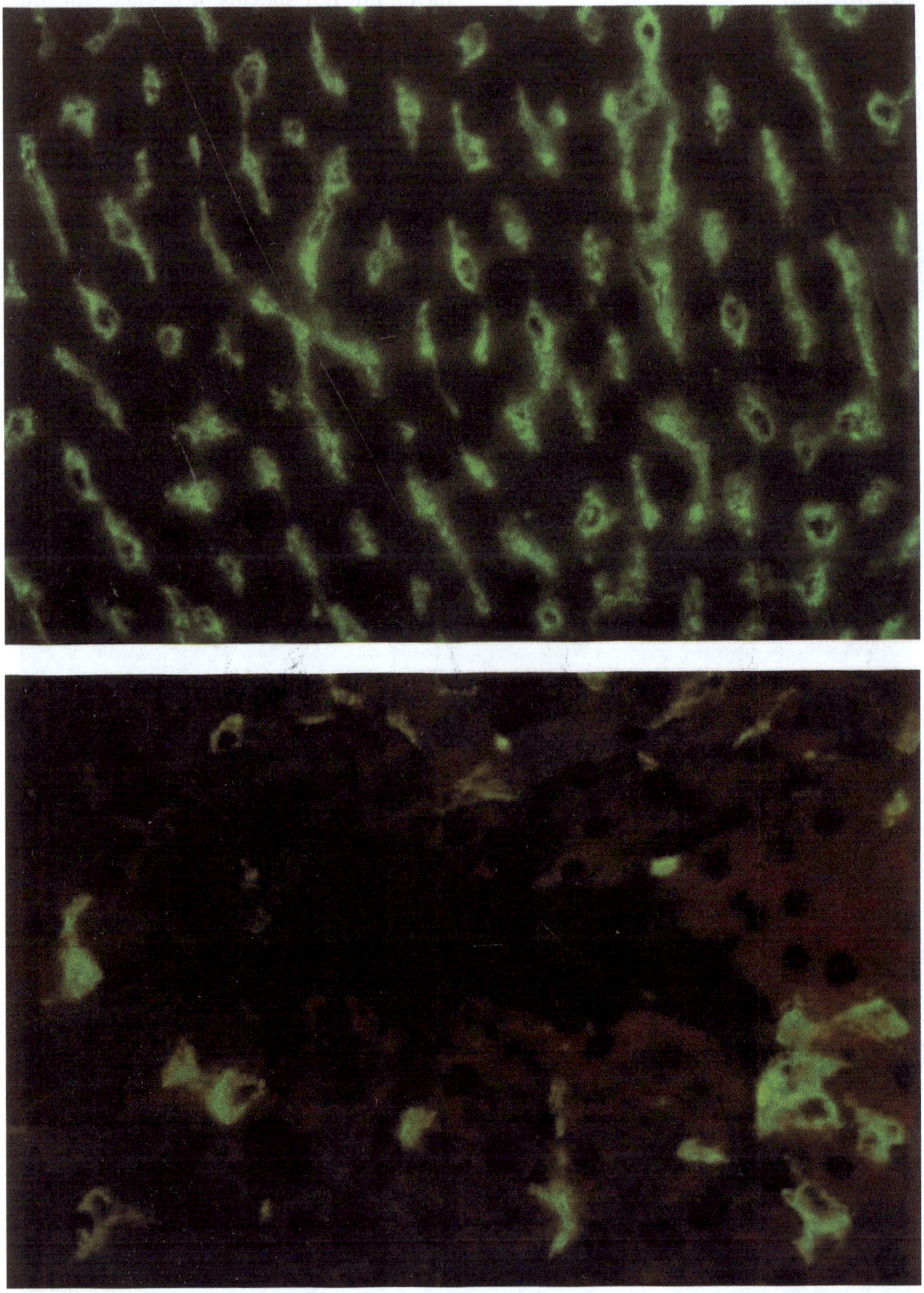

a

b

Figure 129. Antibodies against vascular endothelium: antigen, rat liver – (a) typical picture with diffuse fluorescence of all the sinusoids and veins; (b) atypical picture with hearth-like, i.e., focal, fluorescence of some sinusoids (objective ×40, ×500).

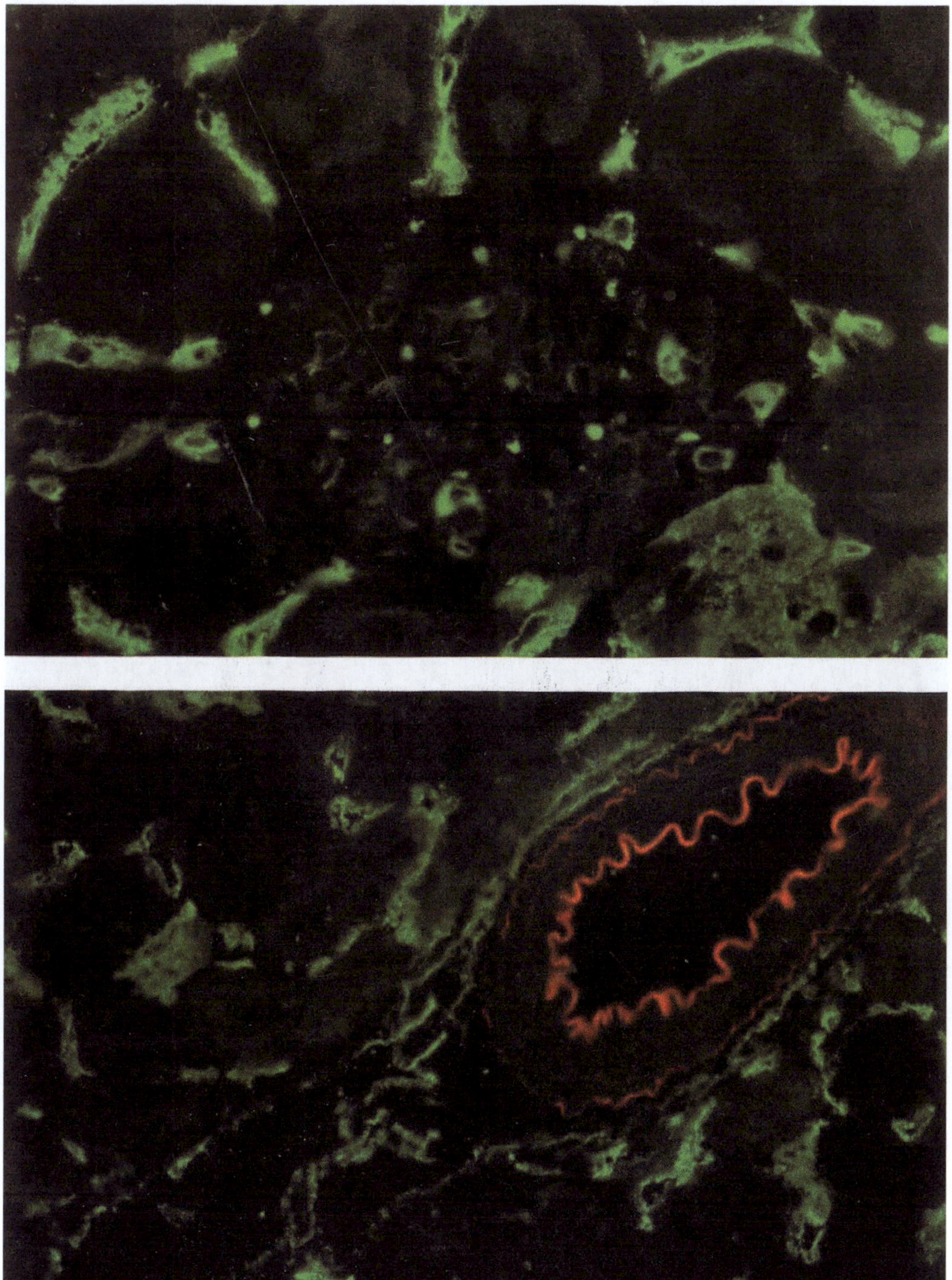

a

b

Figure 130. Antibodies against vascular endothelium (atypical picture): antigen, rat kidney – (a) fluorescence of the peritubular capillaries and hearth-like fluorescence of the glomeruli, as well as finely granular fluorescence of individual distal tubules (lower right); (b) fluorescence of the peritubular capillaries and of the endothelium of the veins but not of the arteries (objective ×40, ×500).

a

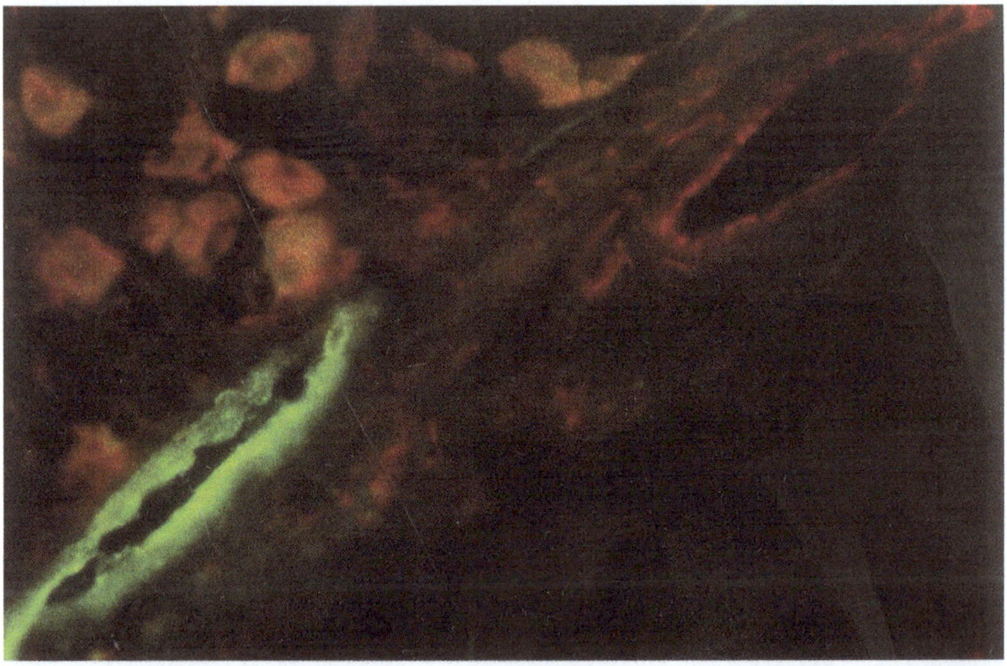

b

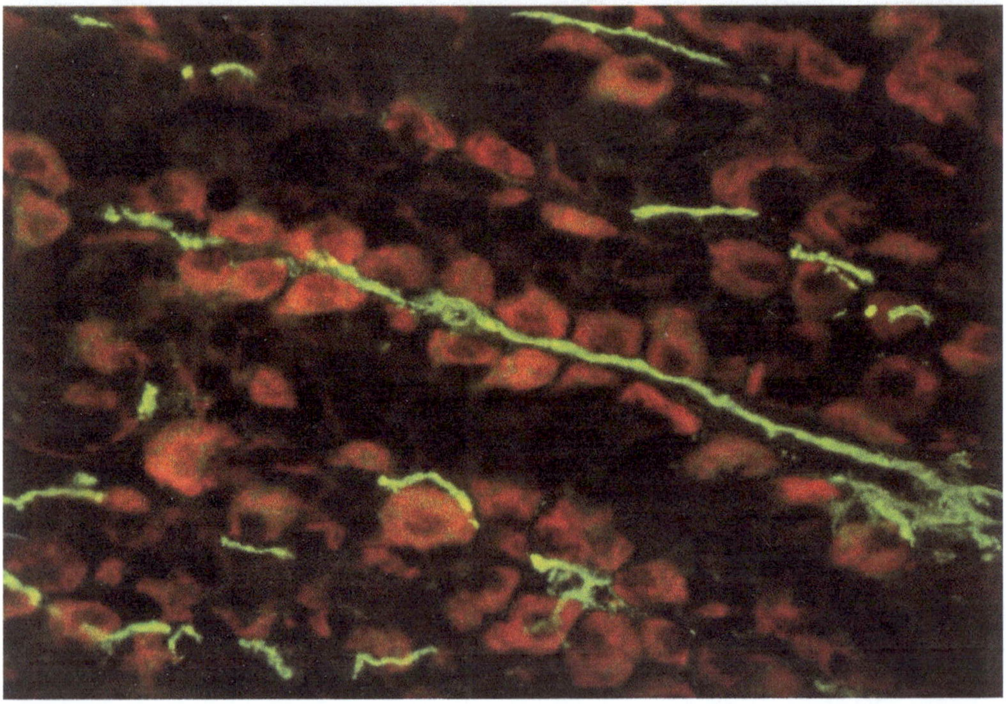

Figure 131. Antibodies against vascular endothelium (atypical picture): antigen, rat stomach – (a) widespread fluorescence of the endothelium and wall of a venule but not of the arterioles (upper right); (b) fluorescence of the mucosal capillaries along the septa between the glands (ob jective ×40, ×500).

Conjugate: anti-human IgG.
Pattern of fluorescence: *kidney* – fluorescence of the endothelium of peritubular vessels, of the glomeruli and of individual distal tubules; *liver* – fluorescence of the sinusoids; *stomach* – fluorescence of the endothelium of small vessels, especially well seen in the submucosa (Figs 128–131).
Tip: the fluorescence picture of tissues reacting with antibodies directed against smooth muscle, connective tissue and bile canaliculi, must be distinguished.
Clinical significance: antibodies against homologous vascular endothelium appear to be, quite generally, markers for viral infections, particularly if antibodies against smooth muscle and connective tissue can be detected simultaneously.
Literature: Piomelli et al. [1959]; Berg et al. [1973]; Berg [175]; Lindqvist & Osterland [1971]; Storch [1975].

7.3.5.7 Antibodies against connective tissue

Synonyms/abbreviations: reticulin antibodies. CTA, especially CTA-3, see below.
Target antigen(s): still insufficiently investigated; apart from those against collagen and reticulin, numerous others.
Properties of the antibodies: enormously heterogeneous, partly tissue-specific, neither species nor disease-specific; predominantly IgG, also some IgA, to some extent able to fix complement. Numerous cross-reactions with antibodies against bacteria (primarily *Streptococci*), smooth muscle, vascular endothelium, the brush-border of the proximal tubules of the kidney, bile canaliculi, and so on.
Antigen substrate: a block containing stomach, liver and kidney, if possible thyroid, cardiac and skeletal muscle of human or animal origin.
Conjugate: anti-human IgG with a low F/P ratio, approx. 1:1, and 25–50 µg antibody protein at the working dilution.
Pattern of fluorescence: there are numerous different antibodies; these are relatively easy to identify when they occur in isolation, for example, antibodies which react only with the sarcolemma of skeletal muscle, only with the membranes of Bowman's capsule or the basement membrane of the renal tubules, or only with the basement membrane of gut epithelium. Most exist as intermediate or mixed types, and amongst these one can distinguish three types relatively clearly:

- type 1 leads, in kidney and liver, simply to a fluorescence involving perivascular connective tissue, apart from a weak fluorescence affecting endothelium, mostly in the form of short, narrow bands; the sinusoids are negative (Fig. 132).
- type 2 produces mostly a widespread fluorescence of the sinusoids and periportal areas of the liver, in the absence of perivascular involvement; in kidney there is characteristic fluorescence affecting the vascular endothelium, the tubule basement membranes and peritubular structures (Fig. 133).
- type 3 is characterised by a predominantly endothelial fluorescence of the renal peritubular vessels and Bowman's capsule, with spotty periarterial fluorescence. The fluorescence associated with the tubular and glomerular basement membranes is only very limited. The spotty staining of perivascular connective tissue stands out particularly in liver; reaction with the *endomysium* is typical for stomach and oesophagus (Figs 134 & 135).

Tip: the conjugate should have a low F/P quotient because of the danger of non-specific staining. Cross-reactions, primarily with antibodies against smooth muscle, vascular endothelium and bacterial antigens, must also be excluded because sometimes these can produce virtually identical results.
Clinical significance: types 1 and 2 have, quite probably, no clinical significance; they occur commonly in healthy patients, especially in the old. Type 3 counts as a characteristic marker for pediatric coeliac disease (non-tropical

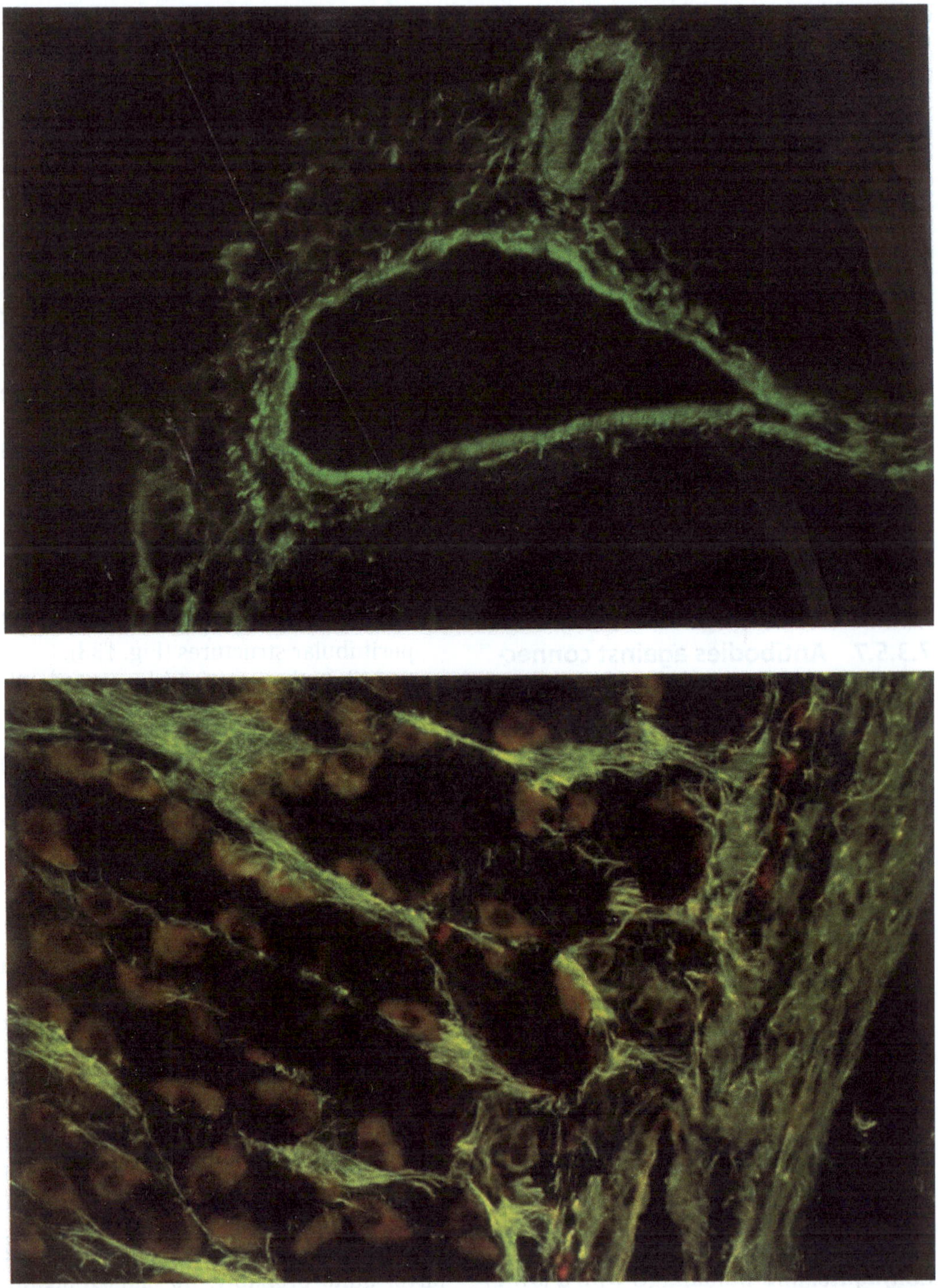

a

b

Figure 132. Antibodies against connective tissue type 1: antigen, block containing rat stomach, liver and kidney – (a) *liver*: perivascular fluorescence, in this instance with, additionally, antibodies against smooth muscle; (b) *gastric mucosa*: fine fibres (objective ×40, ×400).

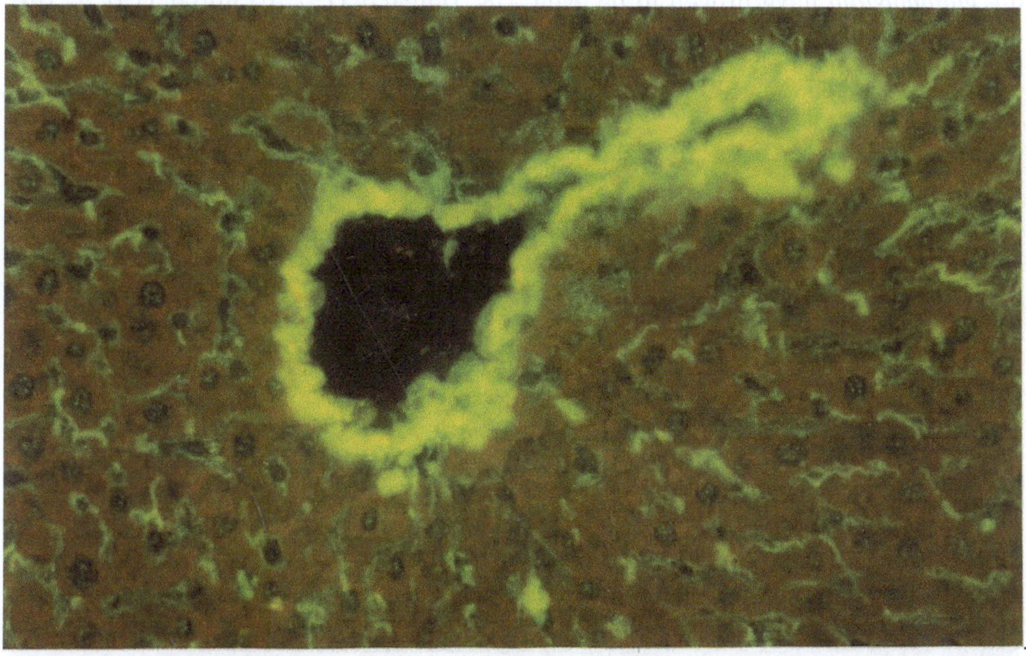

Figure 133. Antibodies against connective tissue type 2: antigen, as for Figure 132 – *liver*: widespread fluorescence of the vein-walls and sinusoids, in this case also ANF of the stippled type (objective ×40, ×500).

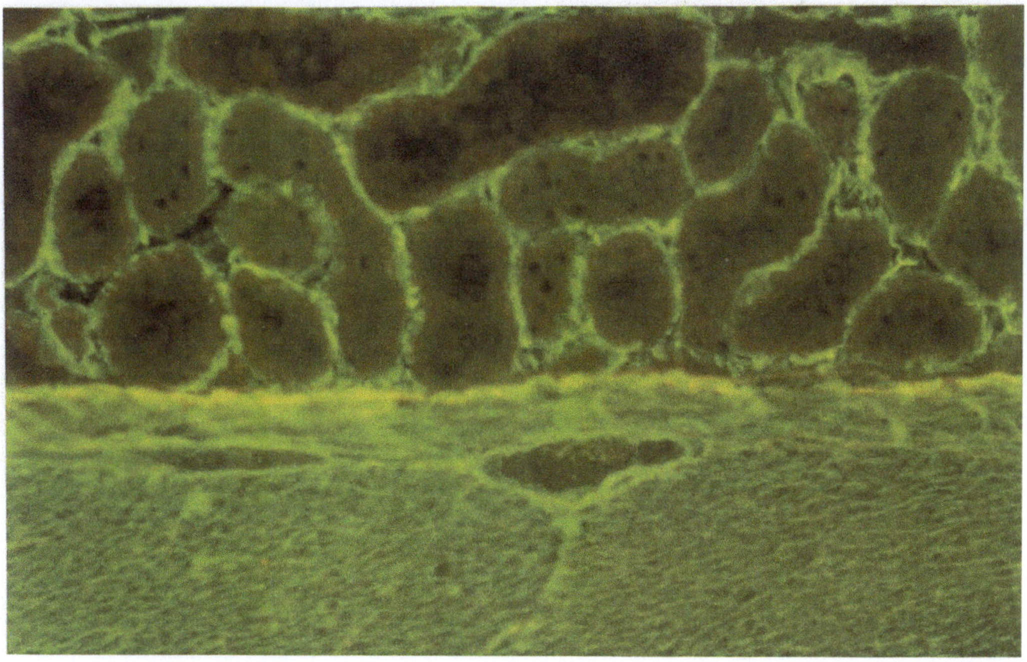

Figure 134. Antibodies against connective tissue type 3: antigen, as for Figure 132 – border between kidney and stomach tissue; fluorescence of the peritubular capillaries and the *endomysium* of the gastric *muscularis* (objective ×40, ×500).

sprue) and for herpetiform dermatitis (Storch [1979]); IgA antibodies against the *endomysium* have proven to be even more typical of this diagnosis. They will be discussed separately.

Literature: Kingston & Glynn [1971, 1976]; Seah et al. [1971, 1973]; Rizzetto & Doniach [1973]; Storch [1979]; Unsworth et al. [1987].

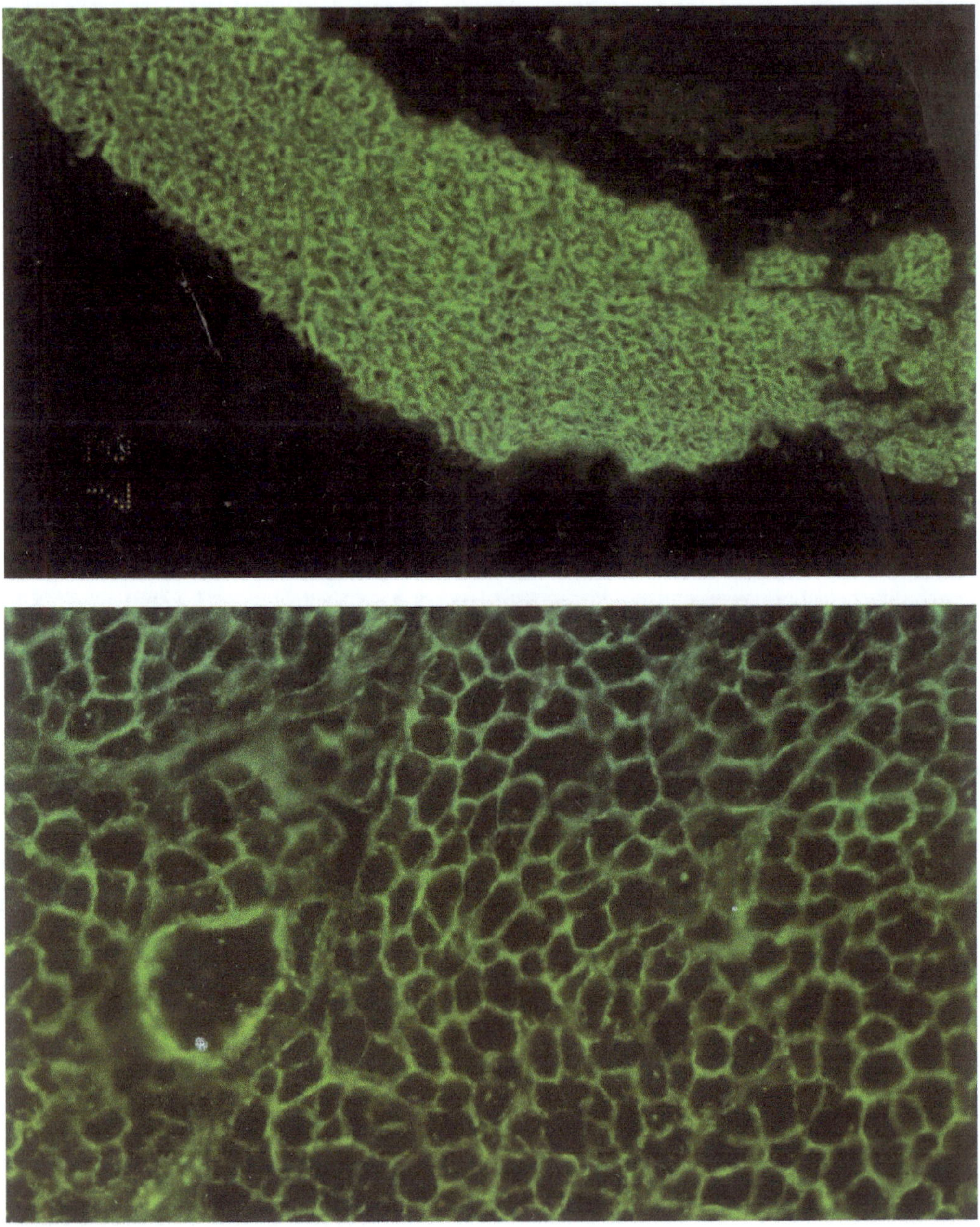

a

b

Figure 135. IgA antibodies against connective tissue type 3 (endomysium): antigen, monkey oesophagus – (a) objective ×20, ×250; (b) objective ×100, ×1250.

7.3.5.7.1 IgA antibodies against *endomysium*

Target antigen(s): *endomysium*, especially of smooth muscle.

Properties of the antibodies: predominantly IgA but also IgG.

Antigen substrate: the lower third of monkey oesophagus, also human umbilical cord (Ladinser et al. [1994]).

Conjugate: anti-human IgA.

Pattern of fluorescence: fluorescence of the *endomysium*, primarily of the *muscularis mucosae* and *muscularis* of the lower oesophagus, is typical (Fig. 135).

Clinical significance: IgA antibodies against *endomysium* count as characteristic markers for untreated or subclinical coeliac disease and for herpetiform dermatitis; the sensitivity and specificity lie between 60% and 100% depending on the literature source.

Literature: Chorzelski et al. [1979, 1983]; Calvani et al. [1994]; Lindqvist et al. [1994]; Valentini et al. [1994].

7.3.5.8 Antibodies against renal glomerular basement membrane

Synonyms/abbreviations: GBM antibodies.

Target antigen(s): type IV collagen - C-terminal non-collagenous globular NC–1 domains of the α–3-(IV)-chain (Saus et al. [1988]).

Properties of the antibodies: predominantly IgG, widely typical of the organ for kidney and lung.

Antigen substrate: fresh human or monkey kidney, also EHS mouse sarcoma which produces type IV collagen (Wick et al. [1983]); kidney from rodents is not suitable. Human kidney should not contain a raised IgG content in the basement membranes. Initial serum dilution, 1:2.

Conjugate: anti-human IgG, where possible consisting of the $F(ab')_2$ fragment, without cross-reacting antibodies directed against the glomerular basement membrane or against monkey IgG.

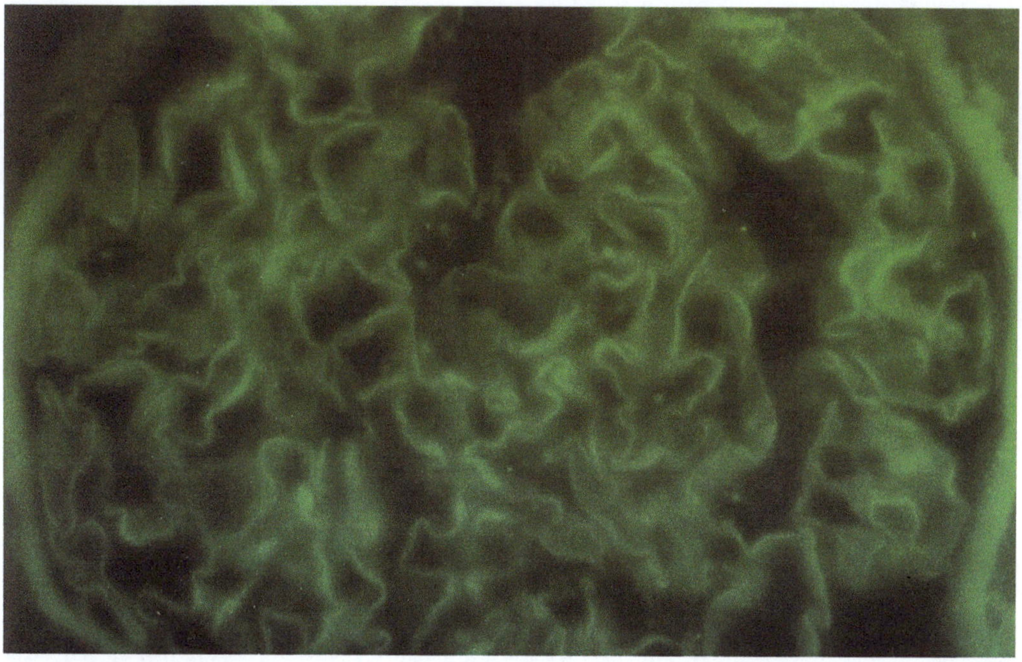

Figure 136. Antibodies against the glomerular basement membrane: antigen, monkey kidney (objective ×100, ×825).

Pattern of fluorescence: linear fluorescence of the glomerular basement membrane (Fig. 136). High titre antisera also produce a fluorescence affecting the basement membrane of proximal tubules.

Tip: there are numerous pictures that can lead to misinterpretation, especially where there are other antibodies present simultaneously, for example, against vascular endothelium.

Clinical significance: a positive result points to a diagnosis of Goodpasture's syndrome or a special form of glomerulonephritis, which is possibly still in the primary phase. Antibodies are found in 70% of cases of Goodpasture's syndrome (involvement of the lungs and kidneys), and are only found in approx. 30% of patients with glomerulonephritis. Titres are frequently low (1:2 to 1:16). In the fully developed phase, one mostly does not find GBM antibodies in the serum but only in the kidneys that are involved.

Literature: Scheer & Grossman [1964]; Pasternak et al. [1965]; review, Weber et al. [1995].

7.3.5.9 Antibodies against renal tubule basement membranes

Properties of the antibodies: predominantly IgG.

Antigen substrate: fresh human kidney, also rat kidney.

Conjugate: anti-human IgG, F/P approx. 1:1.

Pattern of fluorescence: fluorescence strictly limited to the tubule basement membranes; at least three types can be distinguished (Fig. 137).

Tip: confusion with other antibodies directed against connective tissue and vascular endothelium is possible.

Clinical significance: still not clear; when antibodies against the brush-border and vascular endothelium occur simultaneously, pyelonephritis should be suspected.

Literature: Klassen et al. [1973]; Storch [1973, 1975].

a

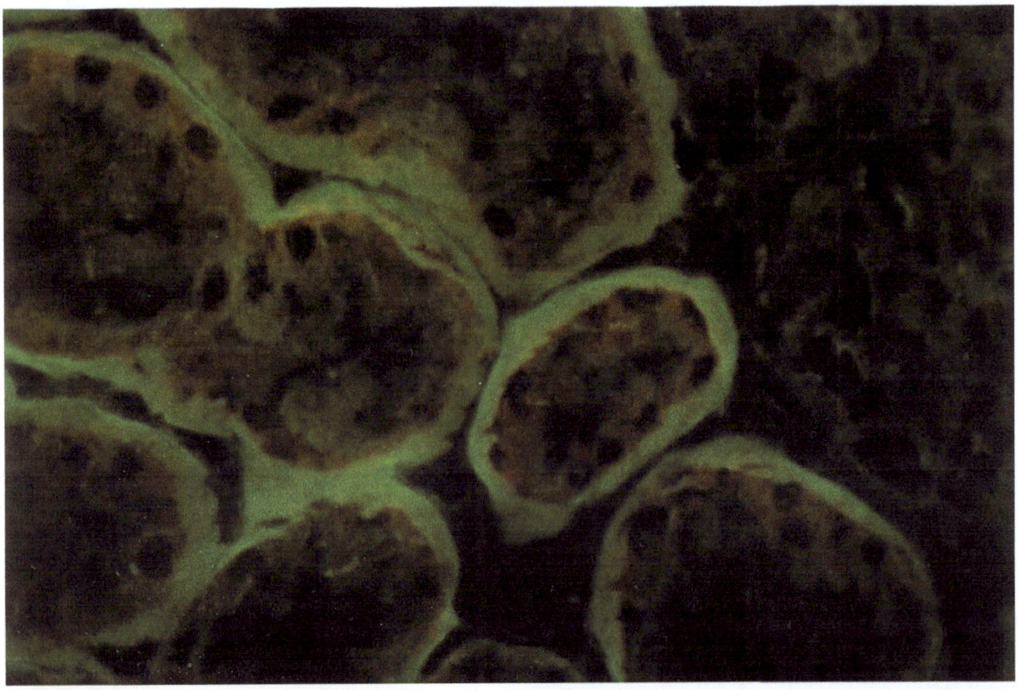

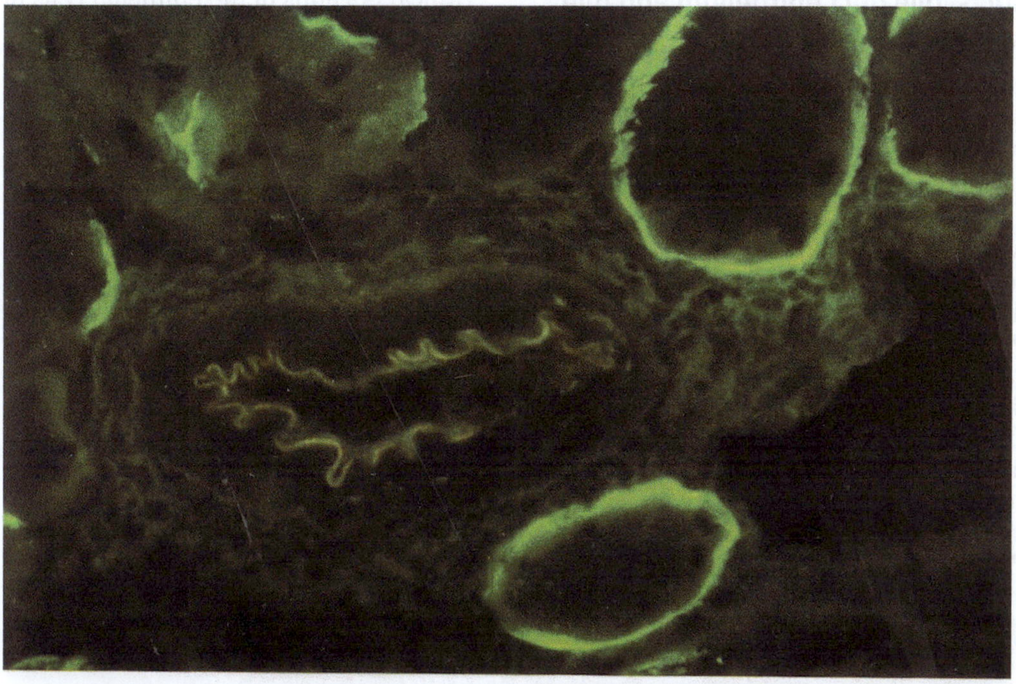

b

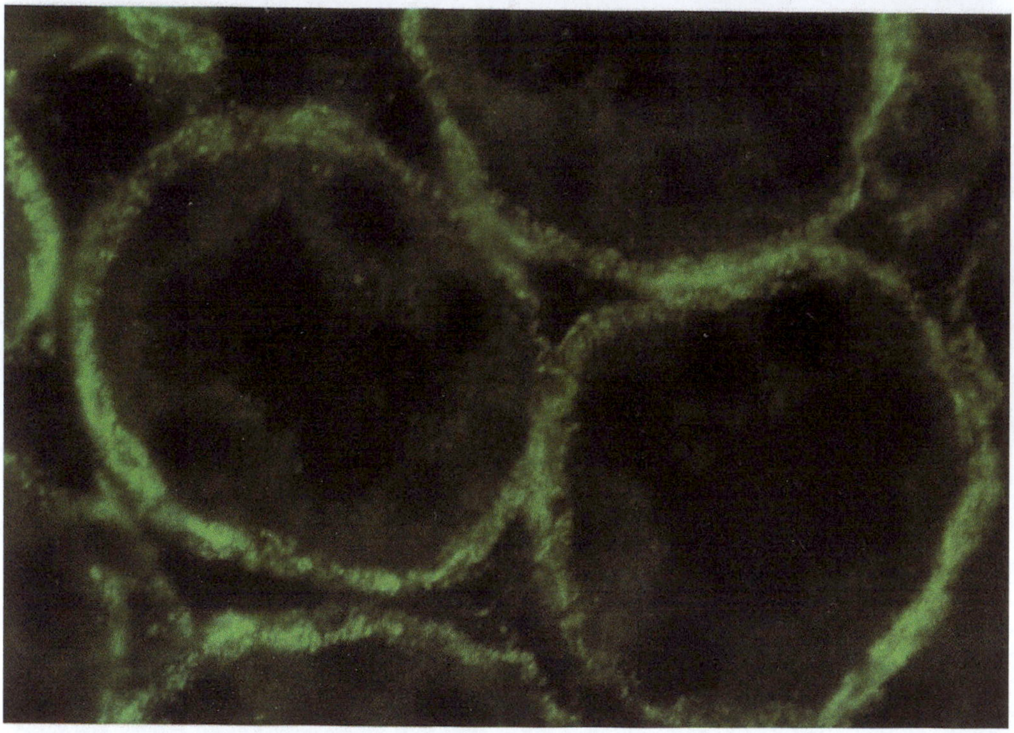

c

Figure 137. Antibodies against the basement membrane of the tubules: antigen, monkey kidney – (a) diffuse widespread appearance; (b) hearth-like linear appearance; (c) diffuse granular appearance (objective ×20, ×250 (a) and (b), ×40, ×500 (c)).

7.3.5.10 Antibodies against bile ducts, bile duct epithelium and bile canaliculi

Synonyms/abbreviations: bile canaliculi antibody (BCA).

Target antigen(s): still insufficiently investigated; at least three locations for antigens should be considered: (i) in the wall of large bile ducts, (ii) in the apical region of the bile duct epithelial cells, and (iii) in the bile canaliculi.

Properties of the antibodies: neither organ nor species-specific but typical for the disease; they belong to the IgG class and sometimes fix complement.

Antigen substrate: bovine or rabbit liver.

Conjugate: anti-human IgG.

Pattern of fluorescence:

1. fluorescence of the wall of the larger bile ducts;
2. fluorescence affecting especially the apical portions of the epithelial cells of the smaller bile ducts (Fig. 138);
3. dense staining of the bile canaliculi (Figs 139 & 140);
4. double-line (tram-line) fluorescence;

(5. "polygonal" hepatocyte fluorescence).

Tip: the antibodies against bile canaliculi, which are densely stained, can be especially easily confused with discontinuous double-linear or pericellular "polygonal" hepatocyte fluorescence, which can be caused by a particular antibody against smooth muscle and/or an antibody against the hepatocyte membrane. BCA almost always give rise to "prozone" phenomena.

Clinical significance: still not clear; antibodies directed against the bile ducts and bile duct epithelium are of fundamental importance. Antibodies against bile canaliculi are found predominantly in chronic

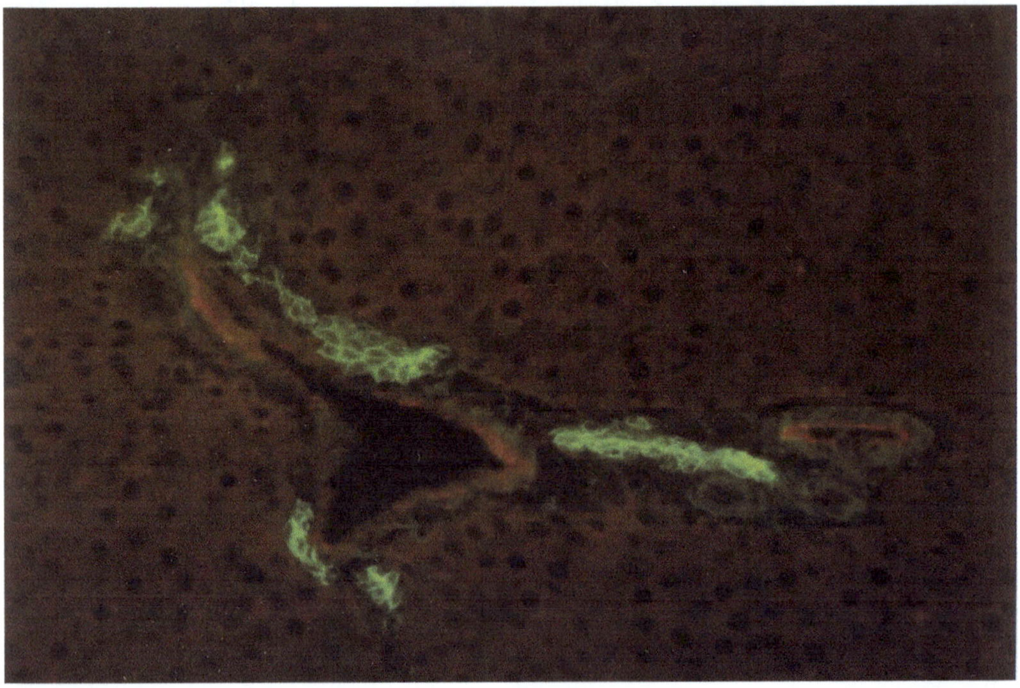

a

Figure 138a, b. Antibodies against the bile duct epithelium: antigen, rat liver – apically accentuated fluorescence (objective ×20, ×250 **(a)**; ×100, ×1250 **(b)**).

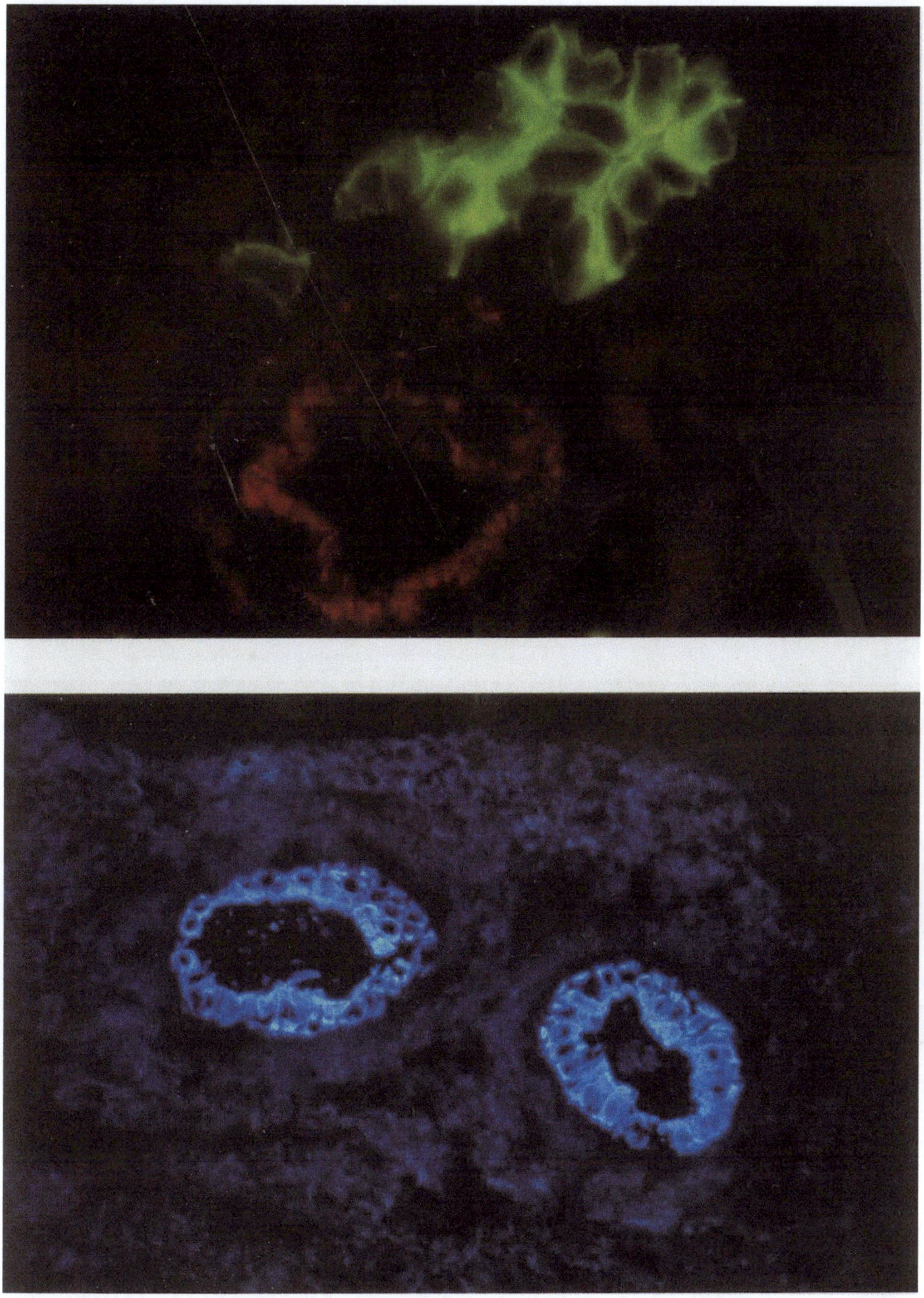

b

c

Figure 138c. Human liver: appearance of cytokeratin 19 in the cytoplasm of the epithelial cells of the bile duct, demonstrated using animal-experiment antibodies against cytokeratin 19 and AMCA-conjugated secondary antibodies (objective ×25, ×330; Mrs. Kuhn and Prof. Franke, Heidelberg).

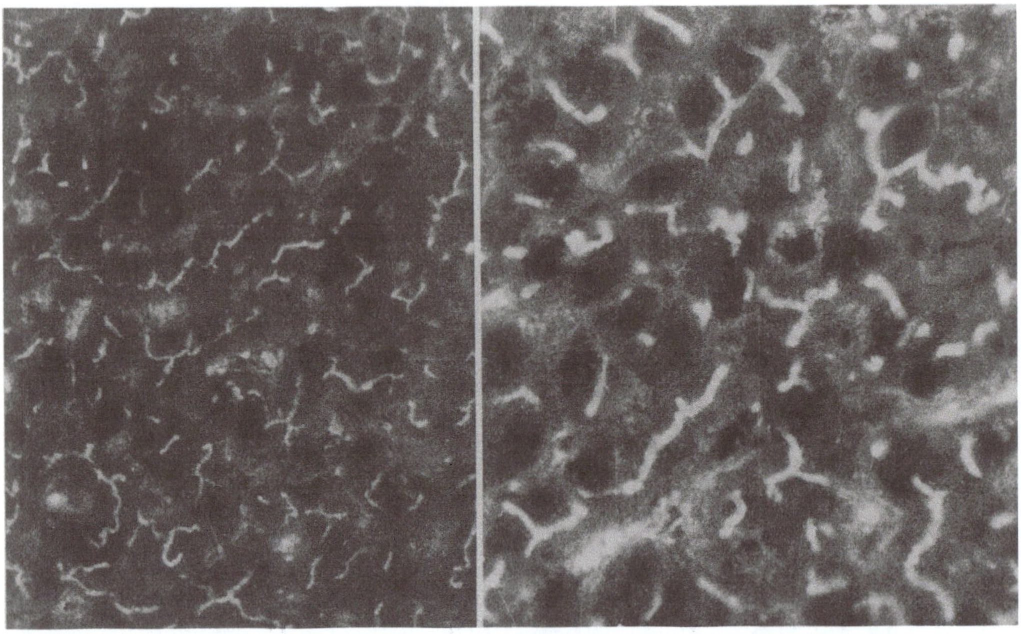

a

Figure 139. Antibodies against bile canaliculi: antigen, rabbit liver (objective ×16, ×196 (a), ×40, ×480 (b)).

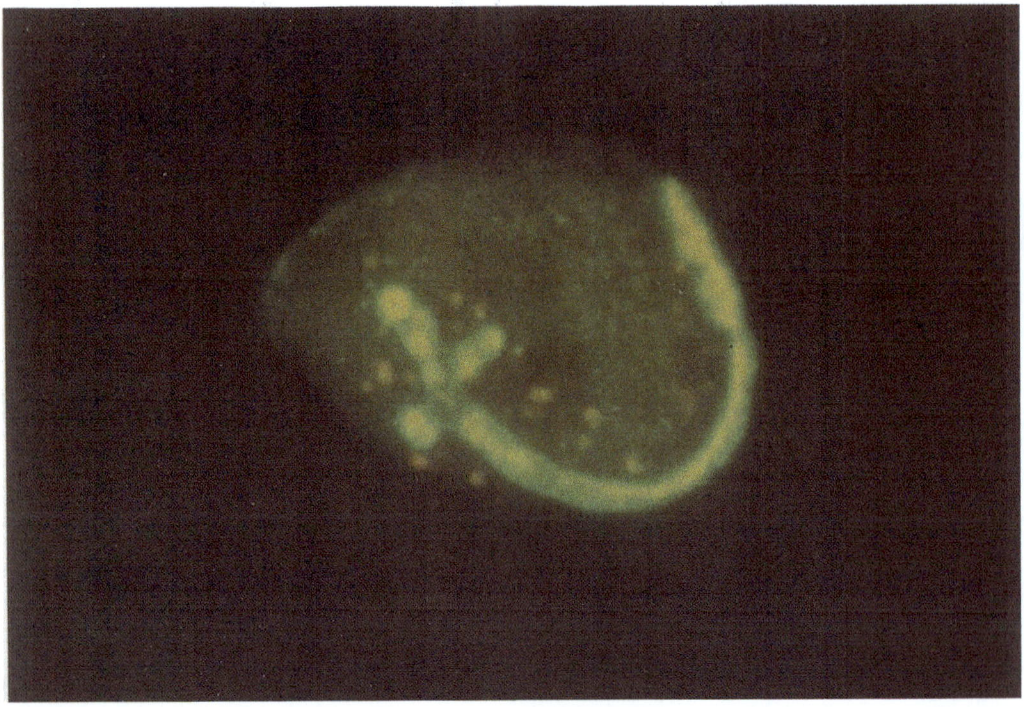

a

Figure 140. Antibodies against bile canaliculi: (a) antigen, isolated rabbit liver cells (objective ×100, ×1400; Prof. Diederichsen, Odense); (b) antigen, Vero cells (objective ×40, ×600).

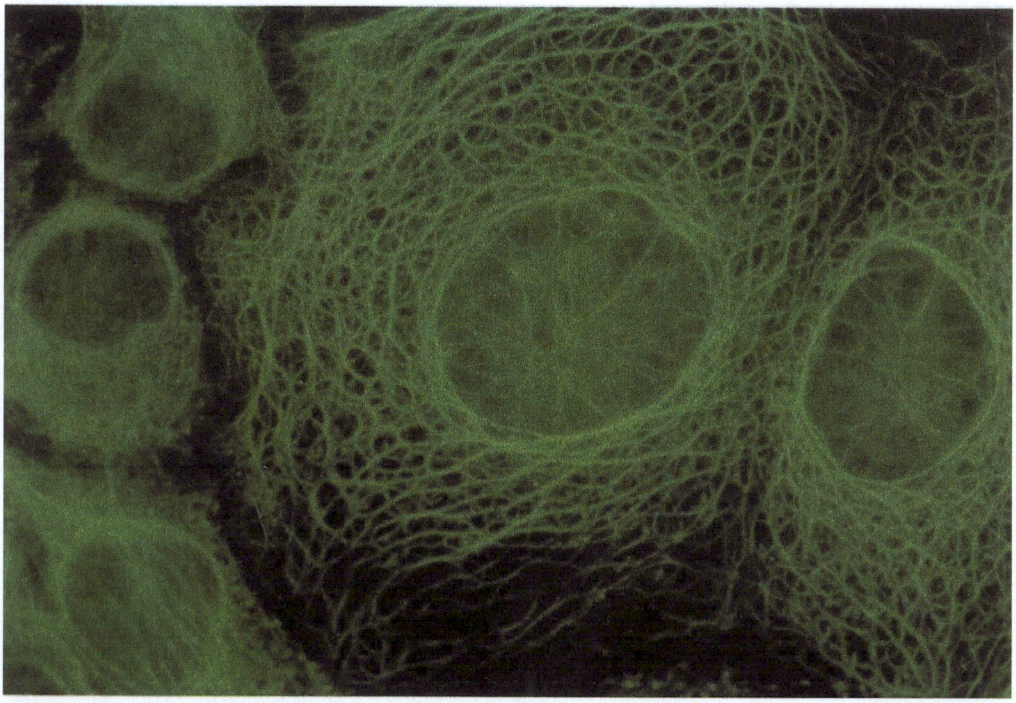

Abb. 140b

liver disease but also occasionally in malignancy as well as in healthy patients.
Literature: Johnson et al. [1966]; Diederichsen [1969, 1975]; Diederichsen et al. [1975]; Ablin et al. [1974]; Storch [1975].

7.4 Detection of immunoglobulin receptors on lymphocytes

The appropriate methodology has already been described in section 4.3.1.

Pattern of fluorescence: the following are regarded as specific in the sense of membrane immunofluorescence (Fig. 141):

(i) a fine dot-like fluorescence of the cell membrane, which frequently appears in the form of a ring of fluorescence when the field of view is in the equatorial plane;
(ii) a cap-like fluorescence ("capping") of the cell membrane.

Tip: the following are regarded as nonspecific in the sense of membrane immunofluorescence (Fig. 142):

opposite page:
Figure 141a–f. Specific types of fluorescence and patterns of distribution, respectively, for immunoglobulin determinants on human lymphocytes, using a comparison of indirect membrane fluorescence (right) and phase-contrast microscopy (left). Rabbit anti-human IgM and FITC-labelled anti-rabbit immunoglobulin were used, showing (a, b) fine, partly interrupted rings of fluorescence, (c, d) larger granules arranged in the form of rings, and (e, f) cap-like fluorescence (objective ×100, ×1200; Prof. F. Emmrich, Leipzig).

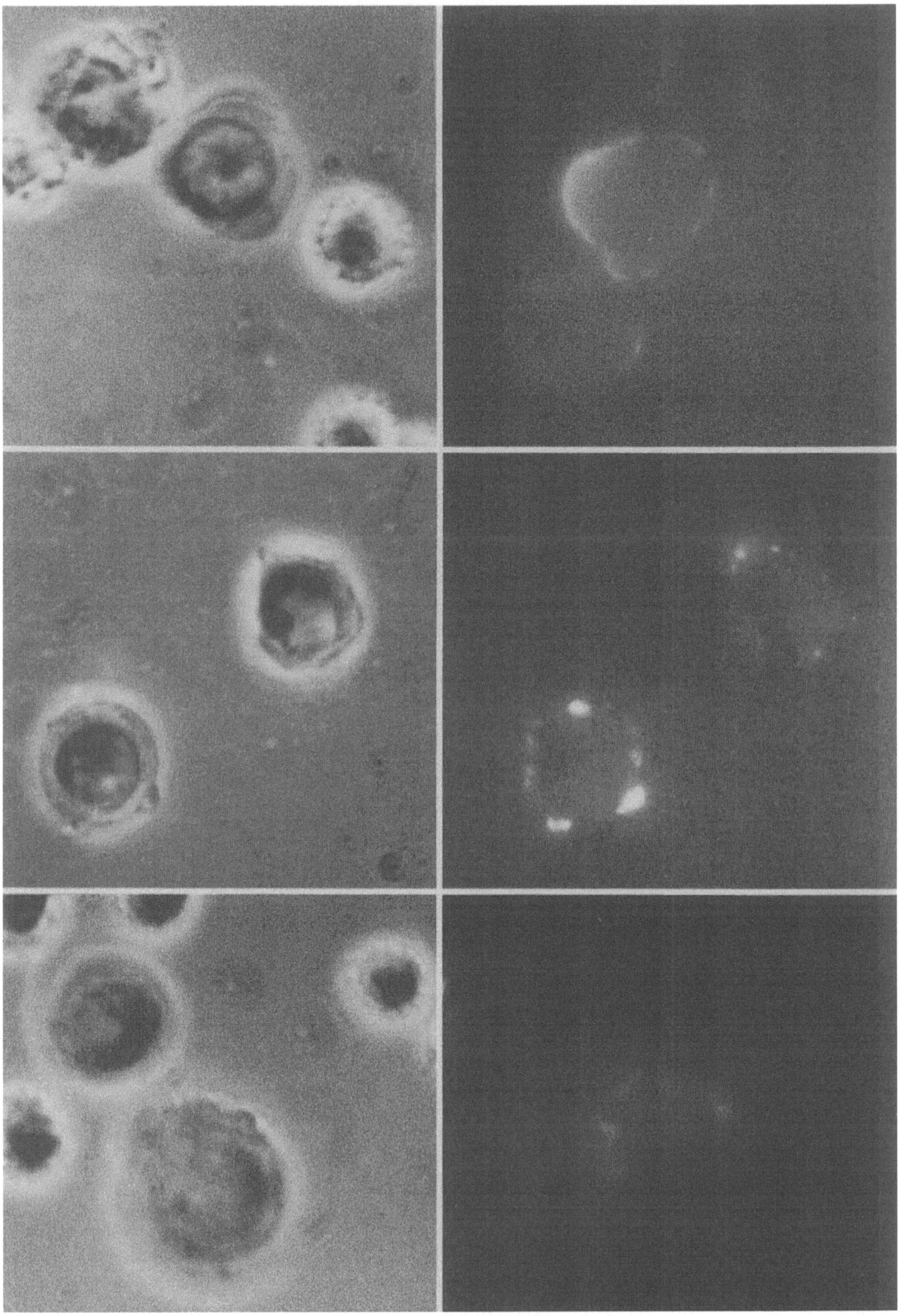

a

c

e

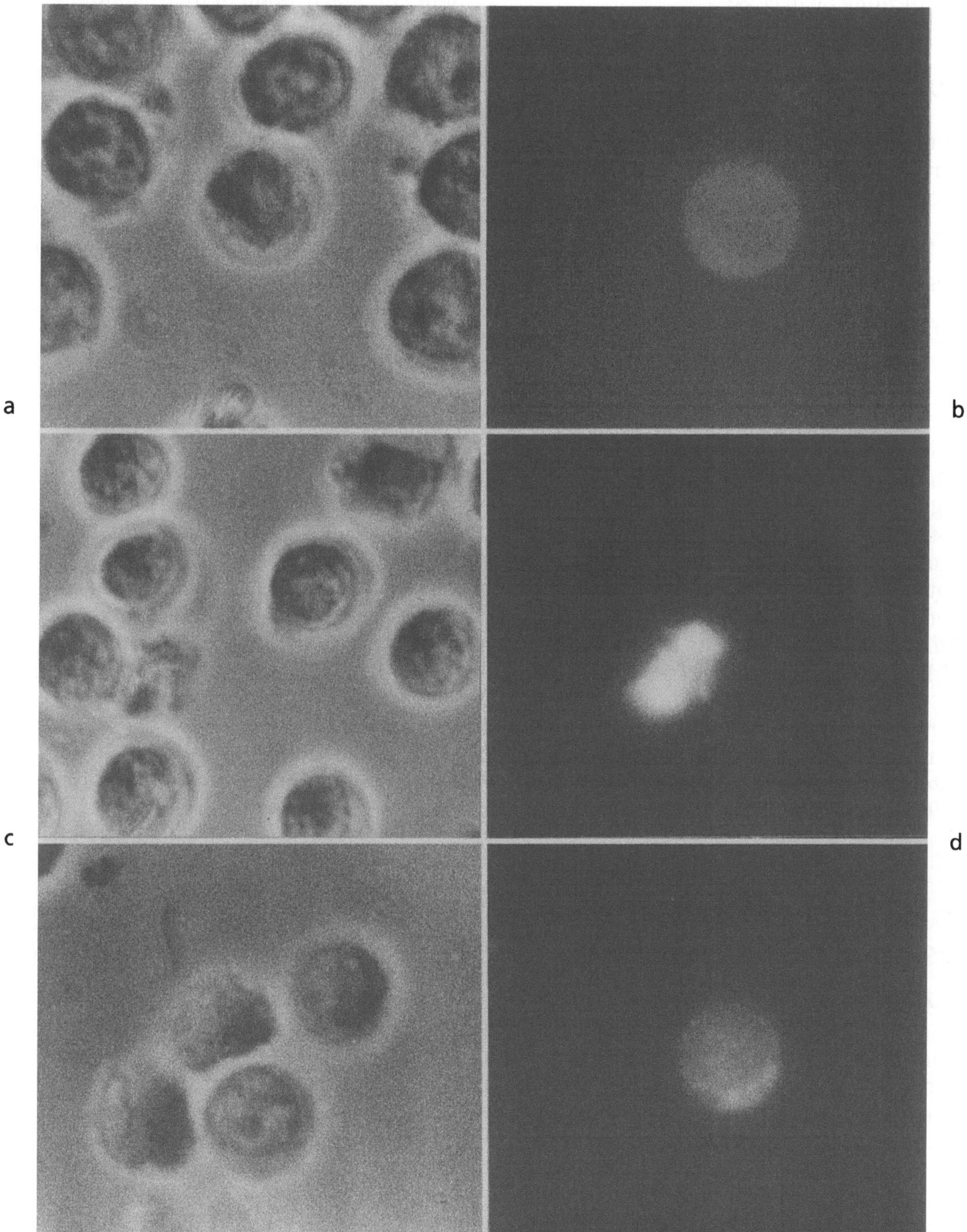

Figure 142a–f. Non-specific types of fluorescence, in the sense of membrane fluorescence, for lymphocytes; comparison shown between phase-contrast (left) and fluorescence (right) microscopy as in figure 141. (a, b) diffuse fluorescence; (c, d) coarse deposition of fluorescent particles; (e, f) cytoplasmic fluorescence (objective ×100, ×1200; Prof. F. Emmrich, Leipzig).

(i) diffuse intracellular fluorescence, indicating dead cells;
(ii) coarse deposition of fluorescent particles on the cell membrane;
(iii) cytoplasmic fluorescence.

In order to determine the number of cells carrying immunoglobulin, preparations should be examined, for comparison, under phase contrast.

Result =

$$\frac{\text{Number of fluorescent lymphocytes}}{\text{Total number of lymphocytes}} \cdot 100$$

False positive results can result, for example, from non-specific binding of conjugates caused by a high fluorochrome content, by interfering accompanying proteins or through the F_c fragment; auto-antibodies and cross-reacting antibodies can also be present.
Clinical significance: the determination of B cells in the periphery is, at present, of greatest importance; the method is generally carried out in most large laboratories using a fluorescent-cell sorter.
Literature: Möller [1961]; Emmrich [1975]; Emmrich et al. [1974, 1975].

7.5 Immunofluorescence using artificial spherical substrates

The so-called DASS or "defined antigen substrate spheres" method has found particularly widespread favour (van Dalen et al. [1973], see 3.2.7). This method makes quantitative immunofluorescence possible, is sensitive and reproducible (Deelder & Ploem [1974]; Capel [1975]). Sepharose 2B, 4B, 6B and Sephadex G25 (Pharmacia, Uppsala, Sweden) have been employed predominantly as the carriers, as well as Ultrogel AcA44 and Biogel P300. The preparation of the spherules encompasses their activation, coupling to protein and inactivation of the remaining reactive groups. Cyanogen bromide (CNBr) has been used in most cases for activation but so have benzoquinone and sodium *meta*-periodate (Wilson & Nakane [1976]; Raybould & Chantler [1978]).

The incubation volume, its duration, the temperature and the pH, as well as the number and diameter of the spherules, are important variables during incubation (Schauenstein & Wick [1980]). The number of spherules must be kept absolutely constant because increasing it appears to reduce the sensitivity of the method (Capel [1975]). A ring of fluorescence must be visible on focussing in the equatorial plane during surface marking. In assessing the fluorescence, the integrity and size of the spherules must be considered; the spheres should not vary in size.

Defective surfaces on the spherules lead to general staining, similar to that seen for dead cells in immunofluorescence. The fluorescence intensity for a sphere can be determined independently of its size by setting the aperture to be less than the diameter of the sphere. According to the methodology as originally reported, the reactions should be carried out in small tubes. The carrier spheres are washed by centrifugation and the supernatant removed with a pipette. Finally, the prepared suspension is placed on a microscope slide and visually or photometrically assessed. In order to simplify the procedure and to save on the use of antiserum, the use of slides already coated with spherules has been recommended (Streefkerk et al. [1975]).

The use of micro-assay plates, for example, micro-titre, Terasaki or Takatsy plates, together with an inverted microscope and objectives with large working distances, allows for much wider application of the method as well as for automation.

Fluoro-immuno-cytoadherence (Richter & Wick [1977]) represents a special method. With its help one can recognise antigen-binding lymphocytes. The antigen, as an example, thyroglobulin, is linked to a carrier such as Sephadex G25, which is itself labelled with a fluorescent dye. The spherules are then incubated with the cells to be investigated. Cells with antigen-specific receptors bind to the spherules in the form of rosettes. Spherules with more than two cells attached are determined by fluorescence microscopy. Uncoupled spherules, or spherules linked to another antigen or labelled with a different fluorochrome, are incubated as controls. This method has been employed, amongst others, in the diagnosis of thyroid disorders; patients with Hashimoto's thyroiditis possess more thyroglobulin-binding cells than patients with other thyroid pathologies, or none at all (Richter et al. [1978]).

7.5.1 Detection of antibodies against D-penicillamine using hapten-coupled Sephadex spherules – according to Storch [1984]

Principle: D-penicillamine is coupled to Sephadex and used as the antigen substrate for the detection of the antibody.

Tip: the conjugates and sera being investigated should be absorbed onto Sephadex immediately before the assay is carried out, so as to avoid undesirable binding to the Sephadex. Immunofluorescence is carried out, in a somewhat modified way, according to the procedure of Berger-Hernandez et al. [1977] in pointed centrifuge tubes.

Coupling to Sephadex: Sephadex G25 Fine (Pharmacia, Uppsala, Sweden) is soaked overnight in distilled water. 4 ml is suspended in 100 ml 0.1 mM $NaIO_4$ and allowed to stand at room temperature for one hour. The oxidised Sephadex particles are washed three times with water. 500 mg D-penicillamine (Homburg, Germany) are dissolved in 50 ml PBS (10 mM "PO_4", 145 mM NaCl, pH 7.1–7.2) and adjusted to pH 8.0 by the addition of 1 M NaOH. 2.5 ml oxidised Sephadex bead-sediment are suspended in this solution and lightly agitated for 20 hours at room temperature. The sediment is then removed, suspended in 50 ml freshly prepared 0.26 M $NaBH_4$ (sodium borohydride) solution and, after 15 minutes at room temperature, washed with PBS. The DPA-Sephadex particles can be kept in PBS at +4°C for about four days.

Carrying out immunofluorescence:

(i) **Absorption** – 200 µl Sephadex sediment, washed with PBS, and 400 µl of diluted serum or FITC-conjugate, are allowed to stand for one hour at room temperature with frequent shaking.

(ii) **Incubation** – 50 µl of the DPA-coupled Sephadex sediment are mixed with 200 µl of the diluted serum which is to be investigated (starting dilution 1:4) and incubated for one hour at room temperature with occasional shaking. The sediment is then washed twice with PBS. 200 µl of a 1:20 dilution of FITC-labelled anti-guinea pig immunoglobulin (Nordic Immunology, the Netherlands, Batch No. 9–1274) are added to the sediment and the mixture allowed to stand for a further hour at room temperature with periodic shaking. The Sephadex particles are then washed three times with PBS, mixed with buffered glycerol, pH 8.6, placed dropwise on a microscope slide and covered with a cover-slip. Assessment should be carried out using fluorescence microscopy with FITC-narrow band excitation and ×80 magnification. A positive result is a fluorescence of the surface of the gel beads which, on focussing in the equatorial plane, appears as ring-shaped (Fig. 143).

Literature: Storch [1986, 1990]; Trautmann & Storch [1986].

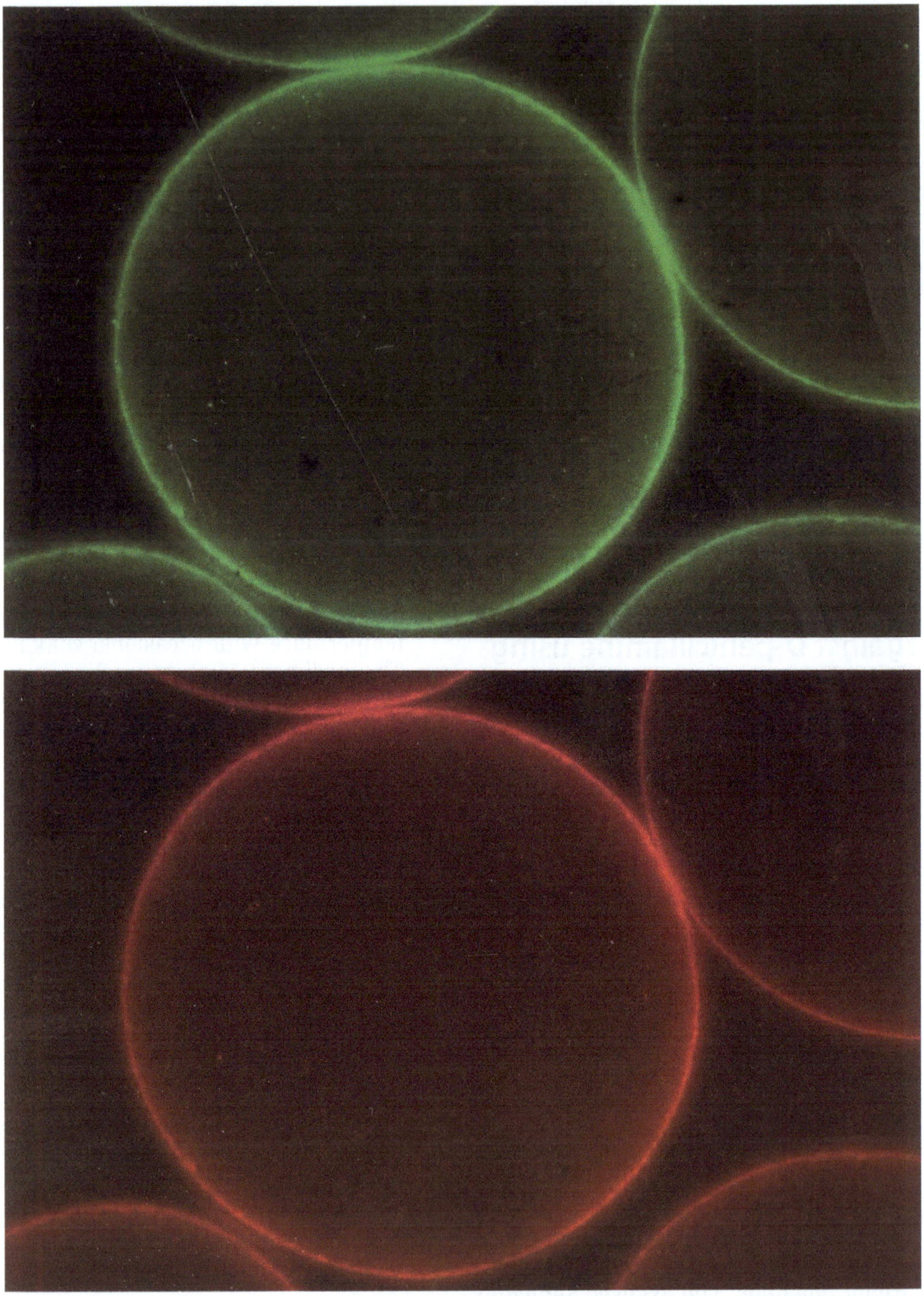

a

b

Figure 143. Antibodies against D-penicillamine: antigen, Sephadex G25 to which D-penicillamine has been coupled; four layer immunofluorescence assay for the detection of complement-fixing antibodies; incubated with heat-inactivated human serum and FITC-labelled anti-human IgG – (a) with fresh human serum as the source of complement and TRITC-labelled anti-human

7.6 Detection of complement activation by C-reactive protein

Synonyms/abbreviations: C3/C4 test.
Principle: C-reactive protein (CRP – an acute-phase protein) probably mediates the binding of complement components to renal vessels, analogous to the activation of complement described by Linder et al. [1981].
Antigen substrate: female Wistar rat kidney.
Conjugate: anti-human C3 or C4.
Pattern of fluorescence: fluorescence, presumably in the first instance of the capillaries, primarily affecting the renal medulla (Fig. 144).
Tip: an initial serum dilution of 1:10 should be employed.
Clinical significance: a positive result in this assay is regarded as a reliable criterion for active, untreated rheumatic polymyalgia and/or giant-cell arteritis, although it is not specific for it.
Literature: Vaith et al. [1986, 1988].

As a summary the essential antibodies are again shown in tabular form (Tab. 30).

Acknowledgements

I should like to thank the following people for provision of various reference sera, in particular Professor Humbel, Luxembourg (Figs 81, 84, 87, 91, 99, 101, 103 and 107a), Professor Bautz, Heidelberg (Figs 92 and 96), Privatdozent Conrad, Dresden (Fig. 96), and Professor Vaith, Freiburg (Fig. 97a).

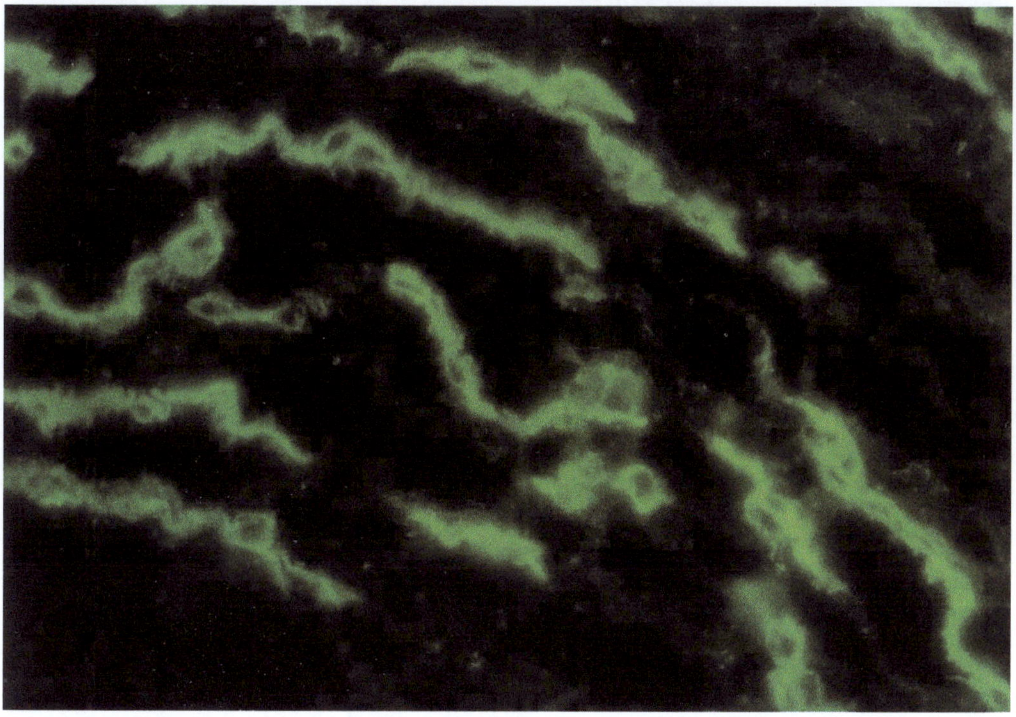

Figure 144. Detection of complement activation by C-reactive protein (CRP): antigen, rat kidney – incubation with human serum and with FITC-labelled anti-human C3; fluorescence of the capillaries of the renal medulla (objective ×40, ×500).

Table 30. Detection of auto-antibodies by immunofluorescence and associated diagnoses (a selection).

Antibody against	Ag substrate for detection	Pattern of fluorescence	Occurs most frequently in
a. Hepatocyte membrane	self or rabbit hepatocytes	granular	chronic aggressive hepatitis
b. Thyroid cell membranes	human thyroid	linear	thyrotoxicosis
c. Lymphocytic membrane Ag	autogenous lymphocytes	linear	viral disease SLE
d. Tumour cell membrane Ag	autogenous or cultivated tumour cells	linear or granular	melanoblast tumours mammary ca.
e. Thyroid epithelium microsomes	human thyrotoxic thyroid (unfixed)	cytoplasmic	Hashimoto's primary myoedema thyrotoxicosis focal thyroiditis
f. Thyroglobulin	fixed sections	globular/fluffy	
g. 2. colloid Ag	fixed sections	cloudy	
h. Parietal cell microsomes	human fundus post-operative for duodenal ulcer, or organ block (v.i.)	cytoplasmic	pernicious anaemia; chronic atrophy of the stomach; thyroid disease healthy patients > 60 yr
i. Pancreas	human pancreas	islet cells	juvenile insulin-requiring diabetes
j. Hormone-producing cells of the adrenal	human adrenal where possible post-operative in Cushing's syndrome	cytoplasmic especially of the zona glomerulosa	idiopathic Addison's; idiopathic hypoparathyroidism
k. Parathyroid	human parathyroid	cytoplasm of oxyphil and principal cells	idiopathic hypoparathyroidism; idiopathic Addison's; Hashimoto's
l. Salivary gland efferent ducts	human parotid	cytoplasm of efferent duct epithelium	Sjögren's syndrome
m. Thrombocytes	autogenous megakaryocytes (sternal marrow)	cytoplasmic	idiopathic thrombocytopenic purpura
n. Mitochondria	block containing rat kidney, rabbit liver and mouse stomach	**kidney**: cytoplasmic proximal and distal tubule. **liver**: cytoplasmic, perinuclear emphasis. **stomach**: parietal cells.	primary biliary cirrhosis; pseudo-LE syndrome; chronic aggressive hepatitis
o. Endoplasmic reticulum	as above	**kidney**: cytoplasmic prox. tubule (P3). **liver**: diffuse cytoplasmic. **stomach**: none.	chronic aggressive hepatitis
p. Ribosomes	as above	**kidney**: glomeruli, cytoplasmic of prox. and distal tubule. **liver**: cytoplasmic perinuclear emphasis **stomach**: chief cells.	active SLE; CAH (chronic aggressive hepatitis)
q. Nuclear material	as above	homogen., peripheral, speckled, punctate	SLE; Scleroderma; Sjögren's; rheumatoid arthritis; PAN; Myasthenia gravis; CAH; healthy > 60 yr.

Antibody against	Ag substrate for detection	Pattern of fluorescence	Occurs most frequently in
r. Smooth muscle	as above	muscularis & muscularis mucosae, septa of the mucosa, arterioles, mesangium of the glomeruli	CAH; acute viral hepatitis; viral infections; malignancy
s. Skeletal muscle	human or bovine skeletal muscle human thymus	striae; motor end-plates	Myasthenia gravis; Dermato-myositis
t. Cardiac muscle	rat heart	(sub)-sarcolemmal, intermyo-fibrillar, diffuse	acute rheumatic fever; streptococcal infections; post-infarction; myocar-ditis; cardiotomy
u. Spermatozoa	human sperm	head mid-portion tail	Infertility – male and female
v. Squamous epithelium basement membrane	rabbit tongue or oesophagus guinea-pig lip	basement membrane linear	bullous pemphigoid
w. Squamous epithelium – intercellular region	as above	inter-cellular substance	Pemphigus vulgaris
x. Glomerulus – basement membrane	monkey or human kidney	basement membrane linear	Goodpasture's syn-drome; autoimmune glomerulonephritis
y. Auerbach's plexus	rat stomach	cytoplasmic	Achalasia Chagas' disease
z. Podocytes	rat kidney	glomerular podocytes	glomerulo-nephritis

SLE = systemic lupus erythematosus
CAH = chronic aggressive hepatitis
PAN = polyarteriitis nodosa or panarteriitis nodosa

Bibliography

AARDEN, L. A., CROOT, E. R. DE and FELTKAMP, T. E. W.: Immunology of DNA. III. *Crithidia luciliae*, a simple substrate for the determination of anti-ds-DNA with the immunofluorescence technique. Ann. N.Y. Acad. Sci. 254 (1975), 505–515.

AARLI, J. A. and CLOSS, O.: Myasthenia gravis: cross-striational fluorescence of muscle fibers mediated by F_c fragments. J. Immunol. 109 (1972), 271 –274.

AARLI, J. A., APARICIO, S. R. LUMSDEN, C. E. and TÖNDER O.: Binding of normal human IgG to myelin sheaths, glia and neurons. Immunology 28 (1975), 171–185.

AARLI, J. A., STEFANSSON, K., MARTON, L. S. G. and WOLLMANN, R. L.: Patients with myasthenia gravis and thymoma have in their sera IgG autoantibodies against Titin. Clin. Exp. Immunol. 82 (1990), 284–288.

ABUAF, N., JOHANET, C. and HOMBERG, J.-C.: Autoantibodies in autoimmune chronic active hepatitis. In: KRAWITT, E. L. and WIESNER, R. H. (Eds.): Autoimmune liver diseases. Raven Press, New York (1991), 93–109.

ABUAF, N., JOHANET, C., SOULIE, E., LOEPER, J. and HOMBERG, J.-C.: Anti-liver cytosol antibodies in hepatology: autoimmune hepatitis, viral hepatitis C and graft-versus-host disease. In: MEYER zum BÜSCHENFELDE, K.-H., HOOFNAGLE, J. and MANNS, M. (Eds.): Immunology and liver (Proceedings of the 70th Falk symposium held in Basel, October 18–20 1992), Kluwer Acad. Publ., Dordrecht (1993), 215–226.

AHO, K., PALUSO, T. and KURKI, P.: Marker antibodies of rheumatoid arthritis: diagnostic and pathogenetic implications. Seminars Arthr. Rheum. 23 (1994), 379–387.

ALBERTS, B., BRAY, D., LEWIS, J., RAFF, M., ROBERTS, K. and WATSON, J. D.: Molecular biology of the cell. 3.Edition, Garland Publ., New York, London (1994).

ALSPAUGH, M. A. and TAN, E. M.: Antibodies to cellular antigens in Sjögren's syndrome. J. Clin. Invest. 55 (1975), 1067–1073.

ALTERMATTT, H. J., WILLIAMS, C. L. and LENNON, V. A.: Paraneoplastic cerebellar autoantibodies associated with gynaecological cancer bind to myenteric plexus neurons. Ann. Neurol. 29 (1991), 687–688.

ALTERMATTT, H. J., GERBER, H. A. and SCHEITHAUER, B. W.: Neurologische and paraneoplastische Syndrome. Anti-Purkinje-Zell-Antikörper und Anti-Neuronal-Nukleäre-Antikörper als serologische Marker. Dtsch. med. Wschr. 118 (1993), 1735–1741.

AMAGAI, M.: Adhesion molecules. I: Keratinocyte-keratinocyte interactions; Cadherins and Pemphigus. J. Invest. Dermatol. 104 (1995),146–152.

AMANTE, L., ANCONA, A. and FORNI, L.: The conjugation of immunoglobulins with tetramethyl-rhodamine isothiocyanate: a comparison between the amorphous and the crystalline fluorochrome. J. Immunol. Meth. I (1972), 289–301.

AMBROSIUS, H.: Immunologische Grundlagen der Immunhistochemie. In: GRAUMANN, W. and NEUMANN, K. (Eds.): Handbuch der Histochemie Vol. IV, Proteine Teil 2 (Immunreaktionen in der Histochemie, bearb. von Luppa, H., Ambrosius, H. and Storch, W.), G. Fischer, Stuttgart (1986), 1–81.

ANDERSON, N. E., ROSENBLUM, M. K. and GRAUS, F.: Autoantibodies in paraneoplastic syndrome associated with small-cell lung cancer. Neurology 38 (1988), 1391–1398.

ANDRADE, L. E. C., CHAN, E. K. L. RASKA, I., PEEBLES, C. L. ROOS, G. and TAN, E. M.: Human autoantibody to a novel protein of the nuclear coiled body: immunological characterization and cDNA cloning of p 80-coilin. J. Exp. Med. 173 (1991), 1407–1419.

ANDRÉ, C., ABUAF, N. and HOMBERG, J.-C.: Un nouvel anticorps antimicrosome de foie et de rein marqueur de l'hépatite induite par l'acide tiénilique. Gastroenterology 5 (1981), 90A.

ANHALT, G. J., KIM, S. C., STANLEY, J. R., KORMAN, N. J., JABS, D. A., KORY, M., IZUMI, H., RATRIE, H., MUTASIM, D., ARISS-ABDO, L. and LABIB, R. S.: Paraneoplastic pemphigus. An autoimmune mucocutaneous disease associated with neoplasia. N. Engl. J. Med. 323 (1990), 1729–1735.

AUER-GRUMBACH, P. and ARCHLEITNER, B.: Epidemiology and clinical associations of NuMA (nuclear mitotic apparatus protein) autoantibodies. J. Rheumatol. 21 (1994), 1779–1781.

BAEKKESKOV, S., WARNOCK, G., CHRISTIE, M., RAJOTTE, R. V., LARSEN, P. M. and FEY, S.: Revelation of specificity of 64 k autoantibodies in IDDM serums by high-resolution 2-D gel electrophoresis. Unambiguous identification of 64 k target antigen. Diabetes 38 (1989), 1133–1141.

BALFOUR, B. M., DONIACH, D., ROITT, I. M. and COUCHMAN, K. G.: Fluorescent antibody studies in human thyroiditis: autoantibodies to an antigen of the thyroid colloid distinct from thyroglobulin. Brit. J. Exp. Path. 12 (1961), 307–316.

BALLARDINI, G. GROFF, P., GIOSTRA, F., FRANCESCONI, R., MINIERO, R., GHETTI, S., ZAULI, D. and BIANCHI, F. B.: Hepatocellular co-distribution of c 100, c 33, c 22, and NS 5 hepatitis C virus antigens detected by using immunopurified polyclonal spontaneous human antibodies. Hepatology 21(1995), 730–734.

BANNASCH, P., ZERBAN, H., SCHMID, E. and FRANKE, W. W.: Characterization of cytoskeletal components in epithelial and mesenchymal liver tumors by electron and immunofluorescence microscopy. Virchows Arch. 36 (1981), 139–158.

BARTHELMES, H. and SÖNNICHSEN, N.: Die Bedeutung der Immunfluoreszenzverfahren für die Dermatologie (Technik and klinische Anwendung). Derm. Wschr. 156 (1970), 184–211.

BARTHELMES, H., STORCH, W. and RACKOW, S.: Nachweis von Antikörpern gegen doppelsträngige Desoxyribonucleinsäure im Serum (Indirekte Immunfluoreszenz). Vorschläge zum Arzneibuch der DDR, Diagnostische Laboratoriumsmethoden AB (D.L.)-DDR 85. Zbl. Pharm. 122 (1983), 653–657.

BAUMANN, I. and STORCH, W.: Zur Heterogenität von Antikörpern gegen Skelettmuskulatur. Dt. Gesundh.-wesen 39(1984), 1294–1299.

BAUTZ, F. and BLÜTHNER, M.: The PM/Scl antigens. In: VENROOIJ, VAN, W. J. and MAINI, R. N. (Eds.): Manual of biological markers of disease. Kluwer Acad. Publ., Dordrecht (1994), B 5.4,1–17.

BEAUNE, Ph., DANSETTE, P. M., MANSUY, D., KIFFEL, L., FINCK, M., AMAR, C., LEROUX, J. P. and HOMBERG, J.-C.: Human anti-endoplasmic reticulum autoantibodies appearing in a drug-induced hepatitis are directed against a human liver cytochrome P–450 that hydroxylates the drug. Proc. Natl. Acad. Sci. USA 84 (1987), 551–555.

BECKER, H. and MAYERSBACH, H. V.: Histochemische Blockade unspezifischer Fluoreszenz in der Immunhistologie. Acta Histochem. 24 (1969), 10–25.

BEESLEY, J. E. (Ed.): Immunocytochemistry: a practical approach. Oxford Univ. Press, Oxford (1993).

BELVISI, L., BOMBELLI, F., SIRONI, L. and DOLDI, N.: Organ-specific autoimmunity in patients with premature ovarian failure. J. Endocrinol. Invest. 16 (1993), 889–892.

BERG, P. A., BRANDT, H. and MARKER, A.: Nachweis and Bedeutung von Antikörpern gegen homologes Gefäßendothel bei Leber- und anderen Krankheiten. Verh. dt. Ges. inn. Med. 79 (1973), 647–650.

BERG, P. A.: Humorale Immunphänomene bei akuter and chronischer Virushepatitis. Immunität u. Infekt. 3(1975), 172–181.

BERG, P. A.: Chronisch-aktive (aggressive) Hepatitis. Zur Differentialdiagnose der lupuiden (ANA-positiven) and cholestatisch verlaufenden (AMA-positiven) Hepatitis. Dtsch. Med. Wochenschr. 101 (1976), 1536–1543.

BERG., P. A. and KLEIN, R.: Mitochondrial antigen/antibody systems in primary biliary cirrhosis: revisited. Liver 15(1995), 281–292.

BERGER-HERNANDES, R., BÜRGIN-WOLFF, A. and JUST, M.: The use of antigen-coated ultragel polyacrylamide-agarose in an immunofluorescence test for antibodies to food proteins. Science Tools 25 (1977), 56–59.

BERGQVIST, N. R. and NILSSON, P.: The conjugation of immunoglobulins with tetramethylrhodamine isothiocyanate by utilization of dimethylsulfoxide (DMSO) as a solvent. J. Immunol. Meth. 5 (1974), 189–198.

BERNSTEIN, R. M., NEUBERGER, J. M. and BUNN, C. C.: Diversity of autoantibodies in primary biliary cirrhosis and chronic active hepatitis. Clin. Exp. Immunol. 55 (1984), 553–560.

BERTHELOT, J.-M., VINCENT, CH., SERRE, G. and YOUINOU, P.: The antiperinuclear factor. In: VENROOIJ, VAN, W. J. and MAINI, R. N. (Eds.): Manual of biological markers of disease. Kluwer Acad. Publ., Dordrecht (1994), B 1.2, 1–9.

BERTRAM, K. and HALBERG, P.: A specific antibody against the epithelium of the salivary ducts in sera from patients with Sjögren's syndrome. Acta Allergol. 19 (1964), 458–478.

BETTERLE, C, ZANETTE, F., ZANCHETTA, R., PEDINI, B., TREVISA, A., MANTERO, F. and RIGON, F.: Complement-fixing adrenal autoantibodies as a marker for predicting onset of idiopathic Addison's disease. Lancet (1983), 1238- 1241.

BEUTNER, E. H., WITEBSKY, E., RICKEN, D. and ADLER, R. H.: Studies on autoantibodies in myasthenia gravis. J. Amer. Med. Assoc. 182 (1962), 46–58.

BEUTNER, E. H. and JORDON, R. E.: Demonstration of skin antibodies in sera of pemphigus vulgaris patients by indirect immunofluorescent staining. Proc. Soc. Exp. Biol. (N.Y.) 117 (1964), 505–510.

BEUTNER, E. H., HOLBOROW, E. J. and JOHNSON, G. D.: A new fluorescent antibody method: mixed antiglobulin immunofluorescence or labelled antigen indirect immunofluorescence staining. Nature (London) 208 (1965), 353–355.

BEUTNER, E. H., LEVER, W. F., WITEBSKY, E., JORDON, R. and CHERTOCK, B.: Autoantibodies in pemphigus vulgaris. J. Amer. Med. Assoc. 192 (1965), 98–104.

BEUTNER, E. H., HOLBOROW, E. J. and JOHNSON, G. D.: Quantitative studies of immunofluorescent staining. – I. Analysis of mixed immunofluorescence. Immunology 12 (1967), 327–337.

BEUTNER, E. H., SEPULVEDA, M. R. and BARNETT, E. V.: Quantitative studies of immunofluorescent staining. Bull. World Health Organisation 39 (1968), 587–606.

BEUTNER, E. H., WICK, G., SEPULVEDA, M. R. and MONIN, K.: A reverse immunodiffusion assay for antibody protein concentration in antisera or conjugates to human IgG. In: HOLBOROW, E. J. (Eds.) Standardization in immunofluorescence. Blackwell, Oxford, Edinburgh (1970), 165–169.

BEUTNER, E. H. CHORZELSKI, T. D., BEAN, S. F. and JORDON, R. E. (Eds.): Immunopathology of the skin. Labeled antibody studies. Dowden, Hutchinson and Ross, Stroudsbourg (1973).

BEUTNER, E. H., CHORZELSKI, T. P. and KUMAR, V. (Eds.): Immunopathology of the skin. 3. Edition, Churchill Livingstone, New York (1987a).

BEUTNER, E. H., KUMAR, V., JABLONSKA, S. and JONES., P. C.: Nature and function of stratum corneum/keratinocyte cytoplasm autoantibodies and their antigens. In: BEUTNER, E. H., CHORZELSKI, T. P. and KUMAR, V. (Eds.): Immunopathology of the skin. 3. Edition, Churchill Livingstone, New York (1987b), 681–702.

BINGLEY, P. J., CHRISTIE, M. R., BONIFACIO, E., BONFANTI, R., SHATTOCK, M., FONTE, M.-T., BOTTAZZO, G.-F. and GALE, E. A. M.: Combined analysis of autoantibodies improves prediction of IDDM islet cell antibody-positive relatives. Diabetes 43 (1994), 1304–1310.

BLAKSLEE, D. and BAINES, M.: Immunofluorescence using dichlorotriazinyl-aminofluorescein (DTAF). I. Preparation and fractionation of labelled IgG. J. Immun. Methods 13 (1976), 305–320.

BLASCHEK, M. A.: Some new autoantibodies in connective tissue diseases. Med. Diss., Tübingen (1988).

BLASCHEK, M., MULLER, S. and YOUINOU, P.: Anti-"dividing-cell antigen" autoantibody: a novel antinuclear antibody pattern related to histones in systemic lupus erythematosus. J. Clin. Immunol. 13 (1993), 329–338.

BLIZZARD, R. M., CHANDLER, R. W., KYLE, M. A. and HUNG, W.: Adrenal antibodies in Addison's disease. Lancet (1962) II, 901–903.

BLIZZARD, R. M., CHEE, D. and DAVIS, W.: The incidence of parathyroid and other antibodies in the sera of patients with idiopathic hypoparathyroidism. Clin. Exp. Immunol. 1 (1966), 119–128.

BLIZZARD, R. M., CHEE, D. and DAVIS, W.: The incidence of adrenal and other antibodies in the sera of patients with idiopathic adrenal insufficiency (Addison's disease). Clin. Exp. Immunol. 2 (1967), 19–30.

BLÜTHNER, M. and BAUTZ, F. A.: Cloning and characterization of the cDNA coding for a polymyositis-scleroderma overlap syndrome related nucleolar 100-kd protein. J. Exp. Med. 176 (1992), 973–980.

BONFA, E. and ELKON, K. B.: Clinical and serological associations of the antiribosomal P protein antibody. Arthr. Rheum. 29 (1986), 981 -985.

BORNE, A. E. G. K. VON DEM, VERHEUGT, F. W. A., OOSTERHOF, F., VON RIESZ, E., BRUTEL de la RIVIERE, A. and ENGELFRIET, C. P.: A simple immunofluorescence test for the detection of platelet antibodies. Brit. J. Haematol. 39 (1978), 195–207.

BOTTAZZO, G.-F., FLORIN-CHRISTENSEN, A. and DONIACH, D.: Islet-cell antibodies in diabetes mellitus with autoimmune polyendocrine deficiencies. Lancet (1974) II, 1297–1282.

BOTTAZZO, G.-F., FLORIN-CHRISTENSEN, A., FAIRFAX, A., SWANA, G., DONIACH, D. and GRÖSCHELSTEWART, U.: Classification of smooth muscle autoantibodies detected by immunofluorescence. J. Clin. Pathol. 29(1976), 403–410.

BOTTAZZO, G.-F. and LENDRUM, R.: Separate autoantibodies to human pancreatic glucagon and somatostatin cells. Lancet (1976), 873–876.

BOTTAZZO, G.-F., MCINTOSH, C. W., STANFORD,W. and PREECE, M.: Growth-hormone cell antibodies and partial growth-hormone deficiency in a girl with Turner's syndrome. Clin. Endocrinol. 12 (1980a), 1–9.

BOTTAZZO, G.-F., DEAN, B. M., GORSUCH, A. N., CUDWORTH, A. G. and DONIACH, D.: Complement-fixing islet-cell antibodies in type 1 diabetes: possible monitors of active beta cell damage. Lancet (1980b), 668–672.

BOUDONAS, G., BOURA, P., LEFKOS, N., ZACHARIOYDAKI, E. and EFTHYMIADIS, A.: A possible role for autoantibodies in left ventricular hypertrophy. Cardiology 84 (1994), 278–283.

BOURDI, M., LARREY, D., NATAF, J., BERNUAU, J., PESAYRE, D., IWASAKI, M., GUENGERICH, F. P. and BEAUNE, P. H.: Anti-liver endoplasmic reticulum autoantibodies are directed against human cytochrome P–450/A 2. J. Clin. Invest. 85 (1990), 1967–1973.

BOYUM, A.: Isolation of lymphocytes, granulocytes and macrophages. Scand. J. Immunol. 5, Suppl. 5 (1976), 9–15.

BRADFORD, M. M.: A rapid and sensitive method for the quantitation of microgram quantities of protein utilizing the principle of protein-dye binding. Anal. Biochem. 72 (1976), 248–254.

BRANDTZAEG, P.: Rhodamine conjugates. Specific and non-specific binding properties in immunohistochemistry. Ann. N.Y. Acad. Sci. 254 (1975), 35–53.

BRAVENY, I. and DISKO, R.: Der Toxoplasma-Immunfluoreszenztest: Standardisierung und Bedeutung. Arztl. Lab. 16 (1976), 16–20.

BRENNER, S., PEPPER, O. and BERNS, M. W.: Autoantibodies in human serum selectively bind to the centriole region in cultured cells. J. Cell. Biol. 87 (1980) MI 1826, p. 240 abstract book.

BRINKLEY, B. R., OUSPENSKI, I. and ZINKOWSKI, R. P.: Structure and molecular organization of the centromeric kinetochore complex. Trends in Cell Biol. 2 (1992), 15–21.

BURLINGAME, R. W. and RUBIN, R. L.: Histones. In: VENROOIJ, VAN, W. J. and MAINI, R. N. (Eds.): Manual of biological markers of disease. Kluwer Acad. Publ., Dordrecht (1994), B2.2, 1–28.

BYSTRYN, J.-C.: Clinical significance of basal cell layer antibodies. Arch. Dermatol. 113 (1977), 1380–1382.

CAFORIO, A. L., MARTINETTI, M., SCHWARZ, G., BONIFACIO, E., GAVAZZI, A., GRAZIANO, G., LORINI, R., CUCCIA, M., McKENNA, W. J. and BOTTAZZO, G.-F.: Idiopathic dilated cardiomyopathy: lack of association between circulating organ-specific cardiac antibodies and HLA-DR antigens. Tissue Antigens 39 (1992), 236–240.

CALARCO-GILLAM, P. D., SIEBERT, M. C., HUBBLE, R., MITCHISON, T. and KIRSCHNER, M.: Centrosome development in early mouse embryos as defined by an autoantibody against pericentriolar material. Cell 35 (1983), 621–629.

CALVANI, M., PARISI, G. GIANELLI, C., CERI, E. and GRAZIANI, M. G.: Rub degli anticorpi endomisio nella diagnosi e nel monitoraggio della malatti celica. Recenti Prog. Med. 85 (1994), 318–322.

CAPEL, P. J. A.: A quantitative immunofluorescence method based on the covalent coupling of protein to Sepharose beads. J. Immunol. Meth. 5 (1974), 165–178.

CAPEL, P. J. A.: The defined antigen substrate spheres (DASS) system and some of its applications. Ann N.Y. Acad. Sci. 254 (1975), 108–118.

CARMO-FONSECA, M., BOHMANN, K., CARVALHO, M. T. and LAMOND, A. I.: P 80 coilin. In: VENROOIJ, VAN, W. J. and MAINI, R. N. (Eds.): Manual of biological markers of disease. Kluwer Acad. Publ., Dordrecht (1994), B8.2, 1–11.

CASIANO, C. A., LANDBERG, G., OCHS, L. R. and TAN, E. M.: Autoantibodies to a novel cell cycle-regulated protein that accumulates in the nuclear matrix during S phase and is localized in the kinetochores and the spindle midzone during mitosis. J. Cell Sci. 106 (1993), 1045–1056.

CASIANO, C. A., HUMBEL, R. L., PEEBLES, C., CORINI, G. and TAN, E. M.: Autoimmunity to the cell cycle-dependent centromere protein p330d/ CENP-F in disorders associated with cell proliferation. J. Autoimmunity 8 (1995), 575–586.

CEBRA, J. J. and GOLDSTEIN, G.: Chromatographic purification of tetra methyl-rhodamine-immune globulin conjugates and their use in the cellular localization of rabbit-γ-globulin polypeptide chains. J. Immunol. 95 (1965), 230–245.

CELlS, J. E. (Eds.): Cell biology: a laboratory handbook, Vol. 2, Academic Press, London (1994).

CELlS, A., DEJGAARD, K., CELlS, J. E.: Production of mouse monoclonal antibodies. In: CELlS, J. E. (Eds.): Cell biology: a laboratory handbook, Vol. 2, Academic Press, London (1994), 269–275.

CHADWICK, C. S., McENTEGART, M. G. and NAIRN, R. C.: Fluorescent protein tracers. A trial of new fluorochromes and the development of an alternative to fluorescein. Immunology 1 (1958), 315–327.

CHAN, E. K. L. and BUYON, J. P.: The SS-A/Ro antigen. In: VENROOIJ, VAN, W. J. and MAINI, R. N. (Eds.): Manual of biological markers of disease. Kluwer Acad. Publ., Dordrecht (1994), B4.1, 1–118.

CHAN, H. L., LEE, Y. S., HONG, H. S. and KUO, T. T.: Anticentromere antibodies (ACA): clinical distribution and disease specificity. Clin. Exp. Dermatol. 19 (1994), 298–302.

CHANARIN, I., LOEWI, G., TAVILL, A. S., SWAIN, C. P. and TIDMARSH, E.: Defect of renal tubular acidification with antibody to loop of Henle. Lancet 1 (1974), 317–318.

CHANARIN, I.: The stomach in allergic diseases. In: GELL, P. G. H., COOMBS, R. R. A. and LACHMANN, P. J.: 3rd Edition, Blackwell Oxford, London u. a. (1975), pp. 1429–1440.

CHERRY, W. B., GOLDMAN, M. and CARSKI, T. R.: Fluorescent antibody techniques in the diagnosis of communicable diseases. (U.S. Dept. of Health, Communicable Disease Center). Atlanta, Georgia (1960).

CHORZELSKI, T. P., JABLONSKA, S. and BEUTNER, E. H.: Linear IgA dermatoses. In: BEUTNER, E. H., CHORZELSKI, T. P. and BEAN, S. F. (Eds.): Immunopathology of the skin. 2nd Edition, Wiley, New York (1979), 315–323.

CHORZELSKI, T. P., SULEJ, J. and TCHORZEWSKA, H.: IgA class endomysium antibodies in dermatitis herpetiformis and celiac disease. Ann. N.Y. Acad. Sci. 420 (1983), 325–334.

CHURO, J. BERNSTEIN, J. and GLASSOCK JR. (EDS.): Renal disease: Classification and atlas of glomerular diseases. 2nd Edition. Igaku Shoin, New York (1995).

CLARK, G., REICHLIN, M. and TOMASI, T. B., JR.: Characterization of a soluble cytoplasmic antigen reactive with sera from patients with systemic lupus erythematosus. J. Immunol. 102 (1969), 117–124.

COMPTON, D. A. and CLEVELAND, D. W.: NuMA, a nuclear protein involved in mitosis and nuclear reformation. Curr. Opin. Cell Biol. 6 (1994),343–346.

CONRAD, K., DEPARADE, M. and SOKOLOWSKI, G.: Autoantikörper. Teil 1, Nichtorganspezifische Autoantikörper. Ges. f. Immunchemie and Immunbiologie. dr-Design, Hamburg (1993).

COONS, A. H., CREECH, R. J. and JONES, R. N.: Immunological properties of an antibody containing a fluorescent group. Proc. Soc. Exp. Biol. (N.Y.) 47 (1941), 200–202.

COONS, A. H., CREECH, R. J., JONES, R. N. and BERLINER, E.: The demonstration of pneumococcal antigen in tissues by the use of fluorescent antibody. J. Immunol. 45 (1942), 159–170.

COONS, A. R. and KAPLAN, M. H.: Localization of antigen in tissue cells. II. Improvements in a method for the detection of antigen by means of fluorescent antibody. J. Exp. Med. 91 (1950), 1–13.

COONS, A. H., LEDUC, E. R. and KAPLAN, M. H.: Localization of antigen in tissue cells. VI. The fate of injected foreign protein in the mouse. J. Exp. Med. 93 (1951), 153–188.

CORAZZA, G. R., BIAGI, F., VOLTA, U., ANDREANI, M. L., DE FRANCESCHI, L. and GASBARRINI, G.: Autoimmune enteropathy and villous atrophy in adults. Lancet 350 (1997), 106–109.

COURVALIN, J. C., HERNANDEZ-VERDUN, D., McCARTY, M. C., MAUNOURI, R. and BORNENS, M.: A 80 kD protein is associated with the nucleolus organizer (NOR) of human cell lines. Protides of the biological fluid, Proc. 35th. Colloquium. R. Peeters Ed. Pergamon Press Ltd, (1985), 221–224.

CRIVELLI, O., LAVARINI, C., CHIABERGE, E., AMOROSO, A., FARCI, P., NEGRO, F. and RIZZETTO, M.: Microsomal autoantibodies in chronic infection with the HBsAg associated delta (6) agent. Clin. Exp. Immunol. 54 (1983), 232–238.

CRUICKSHANK, B. and STUART-SMITH, D. A.: Orchitis associated with sperm-agglutinating antibodies. Lancet (1959), I, 708.

DALEN, J. P. R. VAN, KNAPP, W. and PLOEM, J. S.: Microfluorometry on antigen-antibody interaction in immunofluorescence using antigens covalently bound to agarose beads. J. Immunol. Meth. 2 (1973), 383–392.

DALMAU, J., FURNEAUX, H. M., GRALLA, R. J., KRIS, M. G; and POSNER, J. B.: Detection of the anti-Hu antibody in the serum of patients with small cell lung cancer. A quantitative western blot analysis. Ann. Neurol. 27 (1990), 544–552.

DALMAU, J. and POSNER, J. B.: Neurological paraneoplastic syndromes. Springer Semin. Immunopathol. 18 (1996), 85–95.

DANILOVA, T. A. and LYAMPERT, l. M.: An immunofluorescence study of antibodies to components of the connective tissue and muscle fiber of the myocardium in the sera of patients with rheumatism. Bull. Exp. Biol. Med. 73 (1972), 68–71.

DARTSCH, P. C.: Darstellung von Zytoskelett- and Zelloberflächenproteinen in kultivierten Gefäßwandzellen and Gewebsschnitten durch indirekte Immunfluoreszenz-Mikroskopie. Lab. Med. 12 (1989), 565–577.

DAVIES, D. J., MORAN, J. E., NIALL, J. F. and RYAN, G. B.: Segmental necrotizing glomerulonephritis with anti-neutrophil antibody: possible arbovirus aetiology? Brit. Med. J. 285 (1982), 606.

DAWSON, R. M. C., ELLIOTT, D. C., ELLIOTT, W. H. and JONES, K. H.: Data for Biochemical Research. 3rd Edition, Clarendon Press, Oxford, (1986), pp. 541–543.

DEELDER, A. M. and PLOEM, J. S.: An immunofluorescence reaction for Schistosoma mansoni using the defined antigen substrate spheres-DASS-system. J. Immunol. Methods 4 (1974), 239–251.

DIEDERICHSEN, H.: Hetero-antibody against bile canaliculi in patients with chronic, clinically active hepatitis. Acta Med. Scand. 186 (1969), 299–302.

DIEDERICHSEN, H.: Bile canalicular antibodies in non-hepatic diseases. Lancet 1 (1975), 1142–1143.

DIEDERICHSEN, H., MÖLLER-NIELSEN, P. and LINDE, N. C.: Antikörper gegen Gallencanalicul. Probleme des Nachweises and klinische Bedeutung. Z. Gastroenterol. 13 (1975), 271–277.

DIETERICH, W., EHNIS, T., BAUER, M., DONNER, P., VOLTA, U., RIECKEN, E. O. and SCHUPPAN, D.: Identification of tissue transglutaminase as the autoantigen of coeliac disease. Nature Med. 3 (1997), 797–801.

DONIACH, D., LINDQVIST, H. L. and BERG, P. A.: Non-organ-specific cytoplasmic antibodies detected by immunofluorescence. Internat. Arch. Allergy & Appl. Immunol. 41 (1971), 501–513.

DOUVAS, A. S., ACHTEN, M. and TAN, E. M.: Identification of a nuclear protein (Scl–70) as a unique target of human antinuclear antibodies in scleroderma. J. Biol. Chem. 254 (1979), 10514–10522.

ELKON, K. B.: Ribosomal RNP. In: VENROOIJ, VAN, W. J. and MAINI, R. N. (Eds.): Manual of biological markers of disease. Kluwer Acad. Publ., Dordrecht (1994), B 2.5, 1–11.

ELSON, C. O.: The immunology of inflammatory bowel disease. In: KIRSNER, J. and SHORTER, R. G. (Eds.): Inflammatory bowel disease. 3rd Edition, Lea & Febiger, Philadelphia (1989), 97–164.

EMERY, D. J., DIAZ, L. A., FAIRLEY, J. A., LOPEZ, A., TAYLOR, A. F. and GIUDICE, G. J.: Pemphigus foliaceus and pemphigus vulgaris autoantibodies react with the extracellular domain of desmoglein–1. J. Invest. Dermatol. 105 (1995), 323–328.

EMMRICH, F., STORCH, W. and BAEHR, R. VON: Immunfluoreszenzmuster an Zellmembranen menschlicher Lymphozyten. Folia Haematol. 101 (1974), 523–531.

EMMRICH, F.: Untersuchungen an Membranimmunglobulin menschlicher Lymphozyten. Promotion A, Leipzig (1975).

EMMRICH, F., STORCH, W., BAEHR, R. VON and SCHWENKE, H.: Demonstration of membrane-bound immunoglobulin determinants and the significance of a quantitative characterization of the antisera used. Folia Biol. (Praha) 21 (1975), 155–159.

ENGELBERTH, O. and JEZKOVA, Z.: Autoantibodies in Sheehan's syndrome. Lancet (1965), 1075.

ENGELHARDT, J. and KNEBEL, W.: Leica TCS – das konfokale Laserscanning-Mikroskop der neuesten Generation; Technik and Anwendung. Leitz Mitt. Wissenschaft Technik 10 (1993), 159–168.

ERNST, L. A., GUPTA, R. K., MUJUMDAR, R. B. and WAGGONER, A. S.: Cyanine dye labelling reagents for sulphydryl groups. Cytometry 10 (1989), 3–10.

FALK, R. J. and JENNETTE, J. C.: Anti-neutrophil cytoplasmic autoantibodies with specificity for myeloperoxidase in patients with systemic vasculitis and idiopathic necrotising and crescentic glomerulonephritis. N. Engl. J. Med. 318 (1988), 1651–1657.

FELTKAMP, T. E. W., BERO-LOONEN, P. M. VAN DEN, NIJENHUIS, L. E., ENGELFRIET, C. P., ROSSUM, A. L. VAN, LOCHEM, J. J. VAN and OOSTERHUIS, H. J. G. H.: Myasthenia gravis, autoantibodies, and HL-A Antigens. Brit. Med. J. 1 (1974), 131–133.

FELTKAMP, T. E. W.: Standards and reference preparations. In: VENROOIJ, VAN, W. J. and MAINI, R. N. (EDS.): Manual of biological markers of disease. Kluwer Acad. Publ., Dordrecht (1993), A 11, 1–12.

FEY, H.: Eriochrome black, a means for reduction of non-specificity in immunofluorescence. Path. Microbiol. 38 (1972), 271–277.

FORD, P. M.: A naturally occurring human antibody to loops of Henle. Clin. Exp. Immunol. 14 (1973), 569–572.

FRASER, K. B.: Immunological tracing: viruses and rickettsias. In: NAIRN, R. C.: Fluorescent protein tracing, 4. Edition. , Churchill Livingston, Edinburgh, London, New York (1976), 225–277.

FRITZLER, M. J., KINSELLA, T. D. and GARBUTT, E.: The CREST syndrome: A distinct serologic entity with anticentromeric antibodies. Am. J. Med. 69 (1980), 520–526.

FRITZLER, M. J., ALI, R. and TAN, E. M.: Antibodies from patients with mixed connective tissue disease react with heterogenous nuclear ribonucleoproteins or ribonucleic acid (hn RNP/RNA) of the nuclear matrix. J. Immunol. 132 (1984), 1216–1222.

FRITZLER, M. J., McCARTY, G. A. and VALENCIA, D. W. (Eds.): Antinuclear antibodies: Contemporary techniques and clinical application to connective tissue disease. Univ. Press, N.Y. Oxford, (1984).

FRITZLER, M. J., AYER, L. M., GOHILL, J., O'CONNOR, C., LAXER, R. M. and HUMBEL, R.-L.: An antigen in metaphase chromatin and the midbody of mammalian cells bind to scleroderma sera. J. Rheumatol. 14 (1987), 291–294.

FURNEAUX, H. DROPCHO, E. J., BARBUT, D., CHEN, Y. T., ROSENBLUM, M. K., OLD, L. J. and POSNER, J. B.: Characterization of a cDNA encoding a 34-kDa Purkinje neuron protein recognized by sera from patients with paraneoplastic cerebellar degeneration. Proc. Natl. Acad. Sci. (Wash.) 86 (1989), 28–73.

FURUKAWA, F., TAKIGAWA, M., MATSUYOSHI, N., SHIRAHAMA, S., WAKITA, H., FUJITA, M., HORIGUCHI, Y. and IMAMURA, S.: Cadherins in cutaneous biology. J. Dermatol. 21 (1994), 802–813.

FUSCONI, M., CASSANI, F., GOVONI, M., CASELLI, A., FARABEGOLI, F., LENZI, M. and BALLARDINI, G.: Anti-nuclear antibodies of primary biliary cirrhosis recognize 78–92-kD and 96–100-kD proteins of nuclear bodies. Clin. Exp. Immunol. 83 (1991), 291–297.

GAARDER, P. I. and HEIER, H. E.: A human autoantibody to renal collecting duct. Clin. Exp. Immunol. 51 (1983), 29–37.

GABBIANI, G., RYAN, G. B., LAMELIN, J.-P., VASSALI, P., MAJNO, G., BOUVIER, C. A., CRUCHAUD, A. and LOSCHER, E. F.: Human smooth muscle autoantibody. Its identification as an anti-actin antibody and a study of its binding to non-muscular cells. Amer. J. Path. 3 (1973), 473–488.

GAMMON, W. R., BRIGGAMAN, R. A., INMAN, A. O., QUEEN, L. L. and WHEELER, C. E.: Differentiating anti-lamina lucida and anti-sublamina densa anti-BMZ antibodies by indirect immunofluorescence on 1.0 M sodium chloride-separated skin. J. Invest. Dermatol. 82 (1984), 139–144.

GAMMON, W. R., FINE, J. D., FORBES, M. and BRIGGAMAN, R. A.: Immunofluorescence on split skin for the detection and differentiation of basement membrane zone autoantibodies. J. Am. Acad. Dermatol. 27 (1992), 79–87.

GARREIS, J. l. and GIBSON, W.: Identification and characterization of multiple forms of actin. Cell 9 (1976), 793–805.

GATELY, L. E. and NESBITT, L. T.: Update on immunofluorescent testing in bullous diseases and lupus erythematosus. Dermatol. Clinics 12 (1994), 133–142.

GERMUTH, F. G. and RODRIGUEZ, E.: Immunopathology of the renal glomerulus immune complex deposit and anti-basement membrane disease. Little Brown & Co., Boston (1973).

GERNAND, K., HOST, D., HANKE, E. and BENT, H.: Einfache Fixierung von Antigensubstraten für den Autoantikörpernachweis. Acta Histochem. Suppl. 22 (1980), 99–100.

GITLIN, D., LANDIG, B. H. and WHIPPLE, A.: The localization of homologous plasma proteins in the tissues of young human beings as demonstrated with fluorescent antibodies. J. Exp. Med. 97 (1953), 163–176.

GLEICHER, N. (Eds.): Autoimmunity in reproduction. Immunology and allergy clinics of North America. 14 (1994), 701–909.

GLÜCK, M., SCHRELL, U. and SCHERBAUM, W. A.: Reactivity and intracellular location of the ACTH cell autoantigen in human fetal and adult anterior pituitary tissue. Autoimmunity 14 (1993), 299–305.

GODING, J. W..: Conjugation of antibodies with fluorochromes: modifications to the standard methods. J. Immunol. Meth. 13 (1976), 215–226.

GOLDSCHMEDING, R. VAN DER SCHROOT, C. E., TEN BOKKEL HUININK, D., HACK, C. E., VAN DEN ENDE, M. E., KALLENBERG, C. G. M. and VON DEM BORNE, A. E. G. K. R.: Wegener's granulomatosis autoantibodies identify a novel di-*iso*-propyl-fluorphosphate-binding in the lysosomes of normal human neutrophils. J. Clin. Invest. 84 (1989), 1577–1587.

GOLDWASSER, R. A. and SHEPARD, C. C.: Fluorescent antibody methods in the differentiation of murine and epidemic typhus sera; specificity changes resulting from previous immunization. J. Immunol. 82 (1959), 373–380.

GOMÉS-DAUDRIX, V., SEBBAG, M., GIRBAL, E., VINCENT, CH., SIMON, M., RAKOTOARIVONY, J., ABBAL, M. FOURNIÉ, B. and SERRE, G.: Immunoblotting detection of so-called "anti-keratin antibodies": a new assay for the diagnosis of rheumatoid arthritis. Ann. Rheum. Dis. 53 (1994), 735–742.

GORNALL, A. G., BARDAWILL, C. J. and DAVID, M. M.: Determination of serum proteins by means of the biuret reaction. J. Biol. Chem. 177 (1949), 751–766.

GÖTZ, H. and BÜCHERL, E. S.: Applied tumor immunology. De Gruyter, Berlin (1975).

GRAUS, F., CORDON-CARDO, C. and POSNER, J. B.: Neuronal antinuclear antibody in sensory neuronopathy from lung cancer. Neurology 35 (1985), 538–543.

GRÖSCHEL-STEWART, U.: Herstellung and Reinigung von Antikörpern für die Immunzytochemie am Beispiel des Aktin-Myosin-Systems. Acta Histochem. Suppl. 25 (1982), 37–41.

GROSS, W. L., SCHMITT, W. H. and CSERNOK, E.: ANCA and associated diseases: immunodiagnostic and pathogenetic aspects. Clin. Exp. Immunol. (1993), 1–12.

GROSSL, C. E. and MAYERSBACH, H. VON: Technical factors effecting immunohistological methods. l. The influence of different pH and buffer salts for unspecific reactions. Acta Histochem. 19 (1964), 382–398.

GUDAT, F. and BIANCHI. L.: Immunpathologie der Virus-hepatitis. Therapiewoche 30 (1980), 7717–7727.

GULDNER, H. H., SZOSTECKI, C. and VOSBERG, H. P.: Scl–70 autoantibodies from scleroderma patients recognize a 95 kD protein identified as DNA topoisomerase l. Chromosoma 94 (1986), 132–138.

HAAIJMAN, J. J. and BLOEMMEN, F. J.: Microfluorometric immunoassays with antigens bound to Sepharose beads. Ann. N.Y. Acad. Sci. 254 (1975), 137–150.

HADI, S. M., BARNETSON, R. ST. C. and GAWKRODGER, D. J.: Clinical, histological, and immunological studies in 50 patients with bullous pemphigoid. Dermatologica 176 (1988), 6–17.

HAITINGER, M.: Fluoreszenzmikroskopie, 2. Edition. Akad. Verlagsges. Leipzig (1959).

HALSTENSEN, T. S., DAS, K. M. and BRANDTZAEG, P.: Epithelial deposits of immunoglobulin G 1 and activated complement co-localise with Mr 40 kD putative autoantigen in ulcerative colitis. Gut 34 (1993), 650–657.

HAMPERL, H.: Die Fluoreszenzmikroskopie menschlicher Gewebe. Virchows Arch. Path. Anat. 292 (1934), 1–51.

HANLY, J. G., WALSH, N. M. G. and SANGALANG, V.: Brain pathology in systemic lupus erythematosus. J. Rheumatol. 19 (1992), 732–741.

HEBERT, G. A., PITTMAN, B., McKINNEY, R. M. and CHERRY, W. B.: The preparation and physicochemical characterization of fluorescent antibody reagents. U.S. Dept. Health, education and welfare. Publ. Health Serv. Atlanta (1972).

HERKNER, K. R., HEINZ, M., KOHLROSS, CH., SWOBODA, W. and BAYER, P. M.: Labordiagnostische Immunologie. Autoimmunologisches Screening auf HEp–2 mittels indirekter Immunfluoreszenz. Facultas-Universitätsverlag, Wien (1994).

HERON, I.: A paraffin embedding method in kidney immunofluorescent studies. Acta Path. Microbiol. Scand. Sekt. B 78 (1970), 444–450.

HERTZ, K. C., GAZZE,L. A., KIRKPATRICK, CH. and KATZ, S. I.: Autoimmune vitiligo: detection of antibodies to melanin-producing cells. N. Engl. J. Med. 297 (1977), 634–637.

HESS, G. and ARNOLD, W.: Methodische Probleme des Hepatitis-B-Surface-Antigen (HB_sAg)- and Hepatitis-B-Core-Antigen (HB_cAg)-Nachweises auf Gewebsebene. Acta Histochem. 58 (1977), 47–55.

HICKS, J. D. and MATTHAEI, E.: Fluorescence in histology. J. Path. Bact. 70 (1955), 1–12.

HIDA, C., TSKAMOTO, T., AWANO, H. and YAMAMOTO, T.: Ultrastructural localization of anti-Purkinje cell antibody-binding sites in paraneoplastic cerebellar degeneration. Arch. Neurol. 51 (1994), 555–558.

HIJMANS, W., SCHUIT, H. R. E. and KLEIN, F.: An immunofluorescence procedure for the detection of intracellular immunoglobulins. Clin. Exp. Immunol. 4 (1969), 457–472.

HOCH, S. O.: The Sm antigens. In: VENROOIJ, VAN W. J. and MAINI, R. N. (Eds.): Manual of biological markers of disease. Kluwer Acad. Publ., Dordrecht (1994), B2.4, 1–29.

HOET, R. M. A.: Detection of antiperinuclear factor and anti-keratin antibodies. In: VENROOIJ, VAN, W. J. and MAIN, R. N. (Eds.): Manual of biological markers of disease. Kluwer Acad. Publ., Dordrecht (1993), A10, 1–8.

HOLZ, H. M.: Was man von der Fluoreszenzmikroskopie wissen sollte. Fa. Opton, Oberkochen (1975), 5.

HOMBERG, J.-C., MICOUIN, C., FELTIER, A., SALMON. CH. and CAROLI, J.: Un nouvel anticorps non-spécifique d'organe au cours d'hépatite chronique. Méd. Chir. Dig. 3 (1974a), 85–92.

HOMBERG, J.-C., RIZZETTO, M. and DONIACH, D.: Ribosomal antibodies detected by immunofluorescence in systemic lupus erythematosus and other collagenoses. Clin. Exp. Immunol. 17 (1974b), 617–628.

HONG, H. S., MORSHED, S. A., TANAKA, S., FUJIWARA, T., IKEHARA, Y. and NISHIOKA, M.: Anti-Golgi antibody in rheumatoid arthritis patients recognizes a novel antigen of 79 kDa (doublet) by Western Blot. Scand. J. Immunol. 36 (1992), 785–792.

HOPF, U., MEYER zum BÜSCHENFELDE, K. H. and FREUDENBERG, J.: Liver-specific antigens of different species. II. Localization of a membrane antigen at cell surface of isolated hepatocytes. Clin. Exp. Immunol. 16 (1974), 117–124.

HOPF, U., MEYER ZUM BÜSCHENFELDE, K. H. and ARNOLD, W.: Detection of a liver-membrane autoantibody in HB_{Ag} negative chronic active hepatitis. New. Engl. J. Med. 294 (1976a),578–582.

HOPF, U., MEYER zum BÜSCHENFELDE, K. H. and DIERICH. K. H.: Demonstration of binding sites for IgG. F_c and the third complement component (C3) on isolated hepatocytes. J. Immunol. 117 (1976b), 639–645.

HOPF, U., JAHN, H-U., MÖLLER, B., STEMEROWICZ, R., WITTENBRINK, C., KLEIN, R. and BERG, P. A.: Liver membrane antibodies (LMA) recognize a 26-kD protein on the hepatocellular surface. Clin. Exp. Immunol. 79 (1990), 54–61.

HUANG, S. N.: Immunohistochemical demonstration of hepatitis B core and surface antigens in paraffin sections. Lab. Invest. 33 (1975), 88–95.

HUET, Ch.: Production of polyclonal antibodies in rabbits. In: CELIS, J. E. (Eds.): Cell biology: a laboratory handbook, Vol. 2, Academic Press, London (1994a), 257–268.

HUET, Ch.: Purification of immunoglobulins. In: CELIS, J. E. (Eds.): Cell biology: a laboratory handbook, Vol. 2, Academic Press, London (1994b), 291–296.

HUMBEL, R.-L.: Autoantibodies to the cellular mitotic apparatus. Immunobiology 173 (1986), 211.

HUMBEL, R.-L.: Detection of antinuclear antibodies by immunofluorescence. In: VENROOIJ, VAN, W. J. and MAINI, R. N. (Eds.): Manual of biological markers of disease. Kluwer Acad. Publ., Dordrecht (1993a), A 2, 1–16.

HUMBEL, R -L.: Utilisation des cellules HEp2 pour la recherche et l'identification des anticorps antinucléaires et anticytoplasmiques. Option/Bio Suppl. 88 (1993 b), 1–10.

IRETON, H. J. C., MULLER, H. K. and McGIVEN, A. R.: Human antibody against rat gastric parietal cells and kidney brush border. Clin. Exp. Immunol. 8 (1971), 783–789.

IRVINE, W. J. and BARNES, E. W.: Addison's disease and associated conditions. In: Clinical aspects of immunology. Ed. by GELL, P. G. H., COOMBs, R. R. A. and LACHMANN, P. J., 3rd Edition, Blackwell Sci. Publ. Oxford, London u. a. (1975), 1301–1354.

ISENBERO, D. A., EHRENSTEIN, M. R., LONGHURST, C. and KALSI, J. K.: The origin, sequence, structure and consequences of developing anti-DNA antibodies. Arthr. Rheum. 37 (1994), 169–180.

JAECKLE, K. A., GRAUS, F., HOUGHTON, A., CARDON-CARDO, C., NIELSEN, S. L. and POSNER, J. B.: Autoimmune response of patients with paraneoplastic cerebellar degeneration to a Purkinje cell cytoplasmic antigen. Ann. Neurol. 18 (1985), 592–600.

JENKINS, R. E., RODENAS, J., BHOGAL, B. S. and BLACK, M. M.: Optimal conditions of 1 M NaCl splitting technique to demonstrate basement zone antigens in bullous pemphigoid, epidermolysis bullosa aquisita and linear bullous dermatoses. Dermatology 189 (Suppl. 1) (1994), 133–134.

JENTZSCH, K. -D.: Immunfluoreszenz in der medizinischen Mikrobiologie. Barth, Leipzig (1967).

JOHNSON, G. D., HOLBOROW, E. J. and GLYNN, L. E.: Antibody to smooth muscle in patients with liver disease. Lancet (1965)11, 878–879.

JOHNSON, G. D., HOLBOROW, E. J. and GLYNN, L. E.: Antibody to liver in lupoid hepatitis. Lancet (1966)11, 416–418.

JOHNSON, G. D., HOLBOROW, E. J. and DORLING, J.: Immunofluonescence and immunoenzyme techniques. In: WEIR, D. M. (Eds.): Handbook of experimental immunology. 3rd Edition, Vol. 1, Blackwell Sc. Publ. Oxford (1978), 15.1–15.30.

JOHNSON, G. D., DAVIDSON, R. S., McNAMEE, K. C. RUSSELL, G., GOODWIN, D. and HOLBOROW, E. 1.: Fading of immunofluorescence during microscopy: a study of the phenomenon and its remedy. J. Immunol. Methods 55 (1982), 231–241.

JOHNSON, E. F.: The cytochrome P 450 supergene family: a class of autoantigens. In: MEYER zum BÜSCHENFELDE, K. H., HOOFNAGLE, J. and MANNS, M. (Eds.): Immunology and liver. (Proceedings of the 70th Falk symposium held in Basel, October 18–20 1992), Kluwer Acad. Publ., Dordrecht (1993), 259–268.

JONES, G. L., HEBERT, G. A. and CHERRY, W. B.: Fluorescent antibody techniques and bacterial applications. US Dept. of Public Health Education and Welfare. Center for Disease Control. Atlanta, Georgia, USA, (1978).

JORDON, R. E., BEUTNER, E. H. and WITEBSKY, E.: Basement zone antibodies in bullous pemphigoid. 200 J. Amer. Med. Assoc. (1967), 751–756.

JORDON, R. E., XIA, P. and GEOGHEGAN, W. D.: Bullous pemphigoid autoantibodies reactive with intracellular basal keratinocyte antigens: studies of subclass distribution and complement activation. J. Clin. Immunol. 12 (1992), 163–169.

KALLENBERG, C. G. M., BROUWER, E., WEENING, J. J. and COHEN TERVAERT, J. W.: Anti-neutrophil cytoplasmic antibodies: current diagnostic and pathophysiological potential. Kidney Int. 46 (1994), 1–15.

KAPLAN, M. H., MEYESERIAN, M. and KUSHNER, I.: Immunologic studies of heart tissue. – IV. Serologic reactions with human heart tissue as revealed by immunofluorescent methods; isoimmune, Wassermann and autoimmune reactions. J. Exp. Med. 113 (1961), 17–36.

KARLSSON, F. A., BURMAN, P., LÖÖF, L. and MARDH, S.: Major parietal cell antigen in autoimmune gastritis with pernicious anemia is the acid-producing H^+-K^+-adenosine triphosphatase of the stomach. J. Clin. Invest. 81 (1988), 475–479.

KARNOVSKY, M. J. and ROOTS, L.: A "direct-colouring" thiocholine method for cholinesterases. J. Histochem. Cytochem. 12 (1964), 219–221.

KARTENBECK, J. and FRANKE, W. W.: Desmoplakins. In: KREIS, TH. and VALE, R. (Eds.): Guidebook to the cytoskeletal and motor proteins. Oxford Univ. Press Oxford, (1993), pp. 228–230.

KHALFAN, H., ABUKNESHA, R., RAND-WEAVER, M., PRICE, R. G. and ROBINSON, D.: Amino-methyl coumanin acetic acid: a new fluorescent labelling agent for protein. Histochemical J. 18 (1986), 497–499.

KIEFEL, V., KROLL, H. and MUELLER-ECKHARDT, Ch.: Neonatale Allo-immunthrombozytopenie. Dtsch. Med. Wschr. 119 (1994), 1512-1517.

KIFFEL, L., LOEPER, J., FLINOLS, J. P., CELIER, C., HOMBERG, J.-C. and LEROUX, J. P.: Does the anti-liver-kidney microsome antibody (anti LKM 1) recognize an isozyme of hepatic cytochrome P–450? 10th European drug metabolism workshop, Guilford, UK, July 6–7, (1986), abstract P 1/17.

KINGSTON, D. and GLYNN, L. E.: A cross-reaction between *Streptococcus pyogenes* and human fibroblast, endothelial cells and astrocytes. Immunology 21 (1971), 1003–1016.

KINGSTON, D. and GLYNN, L. E.: Anti-streptococcal antibodies reacting with brain tissue. I. Immunofluorescent studies. Brit. J. Exp. Path. 57 (1976), 114–128.

KITMIMA, Y., HIRAKO, Y., KATSUSHI, O., MORI, S. and YAOITA, H.: Antibody-binding to the 180kD bullous pemphigoid antigens at the lateral cell surface causes their internalization and inhibits their assembly at the basal cell surface in cultured keratinocytes. J. Dermatol. 21 (1994), 838–846.

KLASSEN, J., KANO, K., MILGROM, F., MENNO, A. B., ANTHONE, S., ANTHONE, H., SEPULVEDA, M., ELWOOD, C. M. and ANDRES, G. A.: Tubular lesions produced by autoantibodies to tubular basement membrane in human renal allografts. Internat. Arch. Allergy & Appl. Immunol. 45 (1973), 675–689.

KLEIN, P. and BURKHOLDER, P.: Die Darstellung von fixiertem Komplement mit markiertem Antikomplement. Schweiz. Z. Alg. Path. 22 (1959), 729–731.

KLEIN, R. and BERG, P. A.: Anti-M 9 antibodies in sera from patients with primary biliary cirrhosis recognize an epitope of glycogen phosphorylase. Clin. Exp. Immunol. 81 (1990a), 65–71.

KLEIN, R. and BERG, P. A.: Anti-mitochondrial antibodies (anti-M 7) in heart disease recognize epitopes on bacterial and mammalian sarcosine dehydrogenase. Clin. Exp. Immunol. 82 (1990b), 289–293.

KLEIN, R. and BERG, P. A.: Anti-M 4 antibodies in primary biliary cirrhosis react with sulfite oxidase, an enzyme of the mitochondrial intermembrane space. Clin. Exp. Immunol. 84 (1991), 445–448.

KLEIN, R., GUNNEWIEK, J. M. T. and VAN VENROOIJ, W. J.: Autoantigens contained in the Ul small nuclear ribonucleoprotein complex. In: VENROOIJ, VAN, W. J. and MAINI, R. N. (Eds.): Manual of biological markers and disease. Kluwer Acad. Publ., Dordrecht (1994), B3.1, 1–20.

KNAPP, W. and PATEISKY, K.: Muskelantikörper bei Myasthenia gravis, Nachweis ihrer spezifischen Reaktion am $F(ab')_2$-Stück. Z. Immun.-Forsch. 143, (1972), 477–482.

KNAPP, W., HAAIJMAN, J. J., SCHUIT, H. R. E., RADL, J., BERG, P., VAN DEN PLOEM, J. S. and HIJMANS, W.: Microfluorometric evaluation of conjugate specificity with the defined antigen substrate spheres (DASS) system. Ann. N.Y. Acad. Sci. 254 (1975), 94–107.

KNAPP, W. and LUDWIG, H.: Entwicklungen auf dem Gebiet der Immunfluoreszenztechnik. Wien. Klin. Wochenschr. 88 (1976), 1–5.

KOBAYASHI, I., IMAMURA, K., KUBOTA, M., ISHIKAWA, S., YAMADA, M., TONOKI, H., OKANO, M., STORCH, W. B., MORIUCHI, T., SAKIYAMA, Y. and KOBAYASHI, K.: Identification of an autoimmune enteropathy-related 75 kDa antigen. Gastroenterology, 117 (1999), 823–830.

KONISHI, K., HAYASHI, M. and SARUTA, T.: Renal tubular acidosis with autoantibody directed to renal collecting-duct cells. N. Engl. J. Med. 331 (1994), 1593–1594.

KORMAN, N. J.: Bullous pemphigoid. Dermatol. Clinics 11 (1993), 483–498.

KREIS, T. and VALE, R. (Eds.): Guidebook to the cytoskeletal and motor proteins. Oxford, New York (1993).

KRIETE, A. (Eds.): Visualization in biomedical microscopies. 3-D imaging and computer applications. VCH Verlagsgesellschaft, Weinheim (1992).

KROGH, H. K.: Antibodies in human sera to stratum corneum. Int. Arch. Allergy Appl. Immunol. 36 (1969), 415–426.

KURKI, P., AHO, K., PALOSUO, T. and HELIÖVAARA, M.: Immunopathology of rheumatoid arthritis: anti-keratin antibodies precede the clinical disease. Arthritis Rheum. 35 (1992), 914–917.

LAAFF, H., WIEMERS, S., UEKI, H. and YOITA, H.: Dermato-Immunhistochemie. Atlas and Handbuch für Klinik and Praxis. Wiss. Verlagsgesellschaft, Stuttgart (1991).

LACHMANN, P. J., PETERS. K. and WALPORT, M. J. (Eds.): Clinical aspects of immunology. 5th Edition, Blackwell Sc. Publ., Boston (1993).

LADINSER, B., ROSSIPAL, E. and PITTSCHIELER, K.: Endomysium antibodies in coeliac disease: an improved method. Gut 35 (1994), 776–778.

LARISON, K. D., BREMILLER, R., SAM WELLS, K., CLEMENTS, I. and HAUGLAND, R. P.: Use of a new fluorogenic phosphatase substrate in immunohistochemical applications. J. Histochem. Cytochem. 43 (1995), 77–83.

LARSSON, L.-I.: Gastrointestinal cells producing endocrine, neurocrine and paracrine messengers. In: CREUTZFELD, W. (Eds.): Clinics in gastroenterology. Saunders, London (1980), 485–516.

LASSOUED, K., GUILLY, M. N., DANON, F., CHANTAL, A., DHUMEAUX, D., CLAUVEL, J.-P., BROUET, J.C., SELIGMANN, M. and COURVALIN. J.-C.: Antinuclear autoantibodies specific for lamins. Ann. Intern. Med. 108 (1988), 829–833.

LE PECQ, J. B. and PAOLETTI, C.: A fluorescent complex between ethidium bromide and nucleic acids. J. Mol. Biol. 27 (1967), 87–106.

LENDRUM, R. and WALKER, G.: Serum antibodies in human pancreatic disease. Gut 16 (1975), 365–371.

LENKEI, R., BIBERFELD, G., MAGNIUS, L. O., FAGRAEUS, A. and BIBERFELD, P.: Autoantibodies to the basal cells of squamous epithelium react with thymic epithelial cells. Clin. Immunol. Immunopathol. 34 (1985) 11–19.

LENNON, V. A.: Anti-Purkinje cell cytoplasmic and neuronal nuclear antibodies aid diagnosis of paraneoplastic autoimmune neurological disorders. J. Neurol. Neurosurg. Psychiat. 52 (1989), 1438–1439.

LERMA, B. DE: Die Anwendung von Fluoreszenzlicht in der Histochemie. In: GRAUMANN, W. and NEUMANN, K. H. (Eds.): Handbuch der Histochemie, Vol. I/I. Fischer, Stuttgart (1958), 78–159.

LEUNG, P. S. C., MACKAY, I. and GERSHWIN, M. E.: Mitochondrial autoantigens. In: VENROOIJ, VAN, W. J. and MAINI, R. N. (Eds.): Manual of biological markers of disease. Kluwer Acad. Publ., Dordrecht (1994), B8.1, 1–14.

LIN, S.-C., SUN, A., WU, W.-C. and CHIANG, C.-P.: Presence of anti-basal cell antibodies in oral lichen planus. J. Amer. Acad. Dermatol. 26 (1991), 943–947.

LINDER, E., RHEN, M. and MERI, S.: An immunofluorescence assay for complement activation by the classical pathway. J. Immunol. Methods 47 (1981), 49–59.

LINDERSTROM-LANG, K. and MOGENSEN, K. R.: Studies on enzymatic histochemistry. XXXI. Histological control of histochemical investigations. C. R. Tray. Labor. Carlsberg Sér. Chim. 23 (1938), 27–35.

LINDQUIST, B. L., ROGOZINSKI, T., MOL, H., DANIELSSON, D. and OLCEN, P.: Endomysium and gliadin IgA antibodies in children with coeliac disease. Scand. J. Gastroenterol. 29 (1994), 452–456.

LINDQVIST, K. J. and OSTERLAND, C. K.: Human antibodies to vascular endothelium. Clin. Exp. Immunol. 9 (1971), 753–760.

LIU, A. Y., VALENZUELA, H., HELM, T. N., CAMISA, C., MELTON, A. L. and BERGFELD, W. F.: Indirect immunofluorescence on rat bladder transitional epithelium: a test with high specificity for paraneoplastic pemphigus. J. Amer. Acad. Dermatol. 28 (1993), 696–699.

LOWRY, O. H., ROSENBROUGH, N. J., FARR, A. L. and RANDALL, R. J.: Protein measurement with the Folin phenol reagent. J. Biol. Chem. 193 (1951), 265–275.

LUCA, D., DE: Immunofluorescence analysis. In: MARCHALONIS, J. J. and WARR, G. W. (Eds.): Antibody as a tool. John Wiley and Sons. Chichester (1982), 189–231.

LÜDEMANN, J., UTECHT, B. and GROSS, W. L.: Anti-neutrophil cytoplasm antibodies in Wegener's granulomatosis recognize an elastinolytic enzyme. J. Exp. Med. 171 (1990), 357-362.

LUDWIG, H. and SCHERNTHANER, G.: Multiorganspezifische Autoimmunität bei idiopathischer Nebenniereninsuffizienz. Wien. Klin. Wschr. 90 (1978), 736–741.

LUNZENAUER, K., SEIDLER, E., BIESTER, l. und STEINMANN, CH.: Untersuchungen über die Spezifizität der im Gewebe auftretenden Fluoreszenz bei Anwendung fluoreszenzmarkierter Antikörper. Dtsch. Ges.wesen 21 (1966), 1500–1504.

LUQUE, A., FURNEAUX, H. M., WRAY, S., SCHOLD, C., GLANTZ, M., RIVER, M., LESSER, M., BIRAN, H. and POSNER, J. B.: Anti-Ri: an autoantibody associated with paraneoplastic opsoclonus. Ann. Neurol. 26 (1989), 178 (Abstract).

LUQUE, F. A., FURNEAUX, H. F., FERZIGER, R., ROSENBLUM, M. K., WRAY, S. H., SCHOLD, S. C., GLANTZ, M. J., JAECKLE, K. A., BIRAN, H., LESSER, M., PAULSEN, W. A., RIVER, M. E. and POSNER, J. B.: Anti-Ri. An antibody associated with paraneoplastic opsoclonus and breast cancer. Ann. Neurol. 29 (1991), 241-251.

LYDERSEN, B. K. and PETTIJOHN, D. E.: Human-specific nuclear protein that associates with the polar region of the mitotic apparatus: distribution in a human/hamster hybrid cell. Cell 22 (1980), 489–499.

MacCUISH, A. C., BARNES, E. W., IRVINE, W. J. and DUNCAN, L. J. P.: Antibodies to pancreatic islet cells in insulin-dependent diabetics with coexistent autoimmune disease. Lancet (1974), II, 1529–1531.

MacLAREN, N. K., HUANG, S. W. and FOGH, J.: Antibody to cultured insulinoma cells in insulin-dependent diabetes. Lancet (1975), 997–1000.

MacSWEEN, R. N. M., ANTHONY, P. P., SCHEUER, P. J., BURT, A. D. and PORTMANN, B. C. (Eds.): Pathology of the liver. Churchill Livingstone, Edinburgh (1994).

MAISCH, B., SCHÖNIAN, U., HERZUM M. and HUFNAGEL, G.: Immunserologische und immunhistologische Untersuchungen bei Myokarditis and Perikarditis. Internist 36 (1995), 448–457.

MANCINI, G., VAERMAN, J.-P., CARBONARA, A. O. and HEREMANS, J. F.: A single radial diffusion method for the immunological quantitation of proteins. In: PEETERS, H. (Ed.), Protides of the biological fluids. Elsevier Publishing Comp., Amsterdam 1964, 370–373.

MARSHALL, J. M.: Localization of adrenocorticotrophic hormone by histochemical and immunochemical methods. J. Exp. Med. 94 (1951), 21–30.

MARTIN, J. L., KENNA, J. G., MARTIN, B. M., THOMASSEN, D., REED, G. F. and POHL, L. R.: Halothane hepatitis patients have serum antibodies that react with protein disulfide isomerase. Hepatology 18 (1993), 858–863.

MARTIN-VILLA, J. M., REGUEIRO, J. R., DE JUAN, D., PÉREZ-ACIEGO, P., PÉRES BLAS, M., MANZANARES, J., VARELA, G. and ARNAIZ-VILLENA, A.: T-lymphocyte dysfunctions occurring together with apical gut epithelial cell autoantibodies. Gastroenterology 101 (1991), 390–397.

MARTINI, E., ABUAF,N., CAVALLI, F., DURAND, V., JOHANET, C. and HOMBERG, J.-C.: Antibody to liver cytosol (anti-LC 1) in patients with autoimmune chronic active hepatitis type. 2. Hepatology 8 (1988), 1662–1666.

MATHUR, S., PERESS, M. R., WILLIAMSON, H. O., YOUMANS, C. D., MANEY, S. A., GARVIN, A. J., RUST, P. F. and FUDENBERG, H. H.: Autoimmunity to endometrium and ovary in endometriosis. Clin. Exp. Immunol. 50 (1982), 259–266.

MATTIOLI, M. and REICHLIN, M.: Heterogeneity of RNA protein antigens reactive with sera of patients with systemic lupus erythematosus. Arthritis Rheum. 17 (1974), 421–429.

MATUHASI, T. and USUI, M.: Fluorescent complement fixation technique. Nature (Lond.) 212 (1966), 418–419.

MAUERHOFF, TH., MIRAKIAN, R. and BOTTAZZO, G.-F.: Autoimmunity and the pituitary. Baillére's Clin. Immunol. 1 (1987), 217–235.

MAUL, G. G., FRENCH, B. T., VAN VENROOIJ, W. J. and JIMENEZ, S. A.: Topoisomerase 1 identified by scleroderma anti-70 antisera: enrichment of topoisomerase 1 at the centromere in mouse mitotic cells before anaphase. Proc. Natl. Acad. Sci. USA 83 (1986), 5145–5149.

MAUNSBACH, A. B.: Observation on the segmentation of the proximal tubule in the rat kidney. Comparison of results from phase contrast, fluorescence and electron microscopy. J. Ultrastruct. Res. 16 (1966), 239–258.

MAYERSBACH, H. VON: Möglichkeiten der Artefaktbildung bei der Nachbehandlung gefroren getrockneten Zellmaterials. Acta Anat. (Basel) 30 (1957), 487–507.

MAYERSBACH, H. VON and SCHUBERT, G.: Immunhistologische Methoden. III. Die unspezifischen Reaktionen zwischen markierten Seren und Geweben bei der immunhistologischen Technik. Acta Histochem. 10 (1960), 44–82.

MAYERSBACH, H. VON: Immunhistologische Methoden in der Histochemie. In: GRAUMANN, W. and NEUMANN, K. (Eds.): Handbuch der Histochemie, Vol. l. T. 2. Fischer, Stuttgart (1966), 188–268.

MAYR, A., BACHMANN, P. A., BIBRACK, B. and WITTMAN, G.: Virologische Arbeitsmethoden, Vol. II. Fischer, Jena (1977), 535–656.

McCARTHY, D. M., KATZ, S. I., GAZZE, L., WALDMAN, T. A., NELSON. D. L. and STROBER, W.: Selective IgA deficiency associated with total villous atrophy of the small intestine and organ-specific anti-epithelial cell antibody. J. Immunol. 128 (1978), 932–938.

McCARTY, G. A., VALENCIA, D. W., FRITZLER, M. J. and BARADA, F. A.: A unique antinuclear antibody staining only the mitotic spindle apparatus. N. Engl. J. Med. 305 (1981), 703.

McCARTY, G. A., VALENCIA, D. W. and FRITZLER, M.: Antibody to the mitotic spindle apparatus: immunologic characteristics and cytologic studies. J. Rheumatol. 11 (1984), 213–218.

McENTEGART, M. G.: Immunological tracing: bacteria, protozoa, helminths and fungi. In: NAIRN, R. C. (1976), 170–224.

McHUGH, N. J. WHYTE, J. HARVEY, G. and HAUSTEIN, U. F.: Anti-topoisomerase 1 antibodies in silica-associated systemic sclerosis. Arthr. Rheum. 37 (1994), 1198–1205.

McKENNA, J. L. and PISCIOTTA, A. V.: Fluorescence of megakaryocytes in idiopathic thrombocytopenic purpura (ITP) stained with fluorescent antiglobulin serum. Blood 19 (1962), 664–675.

McKEON, F. D., TUFFANELLI, D. L., FUKUYAMA, K. and KIRSCHNER, M. W.: Autoimmune response directed against conserved determinants of nuclear envelope proteins in a patient with linear scleroderma. Proc. Natl. Acad. Sci. USA 80 (1983), 4374–4378.

McKEON, F. D., TUFFANELLI, D. L., KOBAYASHI, S. and KIRSCHNER, M. W.: The redistribution of a conserved nuclear envelope protein during the cell cycle suggests a pathway for chromosome condensation. Cell 36 (1984), 83–92.

McKINNEY, R. M., SPILLANE, J. T. and PEARCE, G. W.: Factors affecting the rate of reaction of fluorescein isothiocyanate with serum proteins. J. Immunol. 93 (1964a), 232–242.

McKINNEY, R. M., SPILLANE, J. T. and PEARCE, G. W.: Fluorescein diacetate as a reference color standard in fluorescent antibody studies. Analyt. Biochem. 9 (1964b), 474–476.

McKINNEY, R. M., SPILLANE, J. T. and PEARCE, G. W.: A simple method for determining the labeling efficiency of fluorescein isothiocyanate products. Analyt. Biochem. 14 (1966), 421–428.

McKINNEY, R. M. and SPILLANE, J. T.: An approach to quantitation in rhodamine isothiocyanate labeling. Ann. N.Y. Acad. Sci. 254 (1975), 55–64.

MEAD, G. M., COWIN, P. and WHITEHOUSE, J. M. A.: Anti-tubulin antibody in healthy adults and patients with infectious mononucleosis and its relationship to smooth muscle antibody (SMA). Clin. Exp. Immunol. 39 (1980), 328–336.

MELLBYE, O. J., MOLLNES, T. E. and SLETTEVOLL STEEN, L.: IgG subclass distribution and complement activation ability of autoantibodies to neutrophil cytoplasmic antigens (ANCA). Clin. Immunol. Immunopathol. 70 (1994), 32–39.

MINOTA, S., JABJOUR, W. N., SUZUKI, N., NOJIMA, Y., ROUBEY, R. A. S., MIMURA, T. YAMADA, A., HOSOYA, T., TAKAKU, F. and WINFIELD, J. B.: Autoantibodies to nucleolin in systemic lupus erythematosus and other diseases. J. Immunol. 146 (1991) 2249–2252.

MIRAKIAN, R., BOTTAZZO, G.-F. and DONIACH, D.: Autoantibodies to duodenal gastric-inhibitory peptide (GIP) cells and to secretin (S) cells in patients with coeliac disease, tropical sprue and maturity onset diabetes. Clin. Exp. Immunol. 41 (1980), 33–42.

MIRAKIAN, R., RICHARDSON, C. A., BOTTAZZO, G. F. and DONIACH, D.: Humoral autoimmunity to gut-related endocrine cells. Clin. Immunol. Newsletter 2 (1981), 161–167.

MIRAKIAN, R., RICHARDSON, C. A., MILLA, P. J., WALKER-SMITH, J. A., UNSWORTH, J., SAVAGE, M. O. and BOTTAZZO, G. F.: Protacted diarrhoea of infancy: evidence in support of an autoimmune variant. Brit. Med. J. 293 (1986), 1132–1136.

MIYACHI, K., FRITZLER, M. J. and TAN, E. M.: Autoantibody to a nuclear antigen in proliferating cells. J. Immunol. 121 (1978), 2228–2234.

MOLL, R., FRANKE, W. W. and SCHILLER, D. L.: The catalog of human cytokeratins: pattern of expression in normal epithelia, tumors and cultured cells. Cell 31 (1982), 11–24.

MÖLLER, G.: Demonstration of mouse isoantigens on the cellular level by the fluorescent antibody technique. J. Exp. Med. 114 (1961), 415–435.

MOODY, M. D., GOLDMAN, M. and THOMASON, B. M.: Staining bacterial smears with fluorescent antibody. J. Bact. 72 (1956), 357–361.

MOROI, Y., PEEBLES, C., FRITZLER, M. J., STEIGERWALD, J. and TAN, E. M.: Autoantibody to centrosome (kinetochore) in scleroderma sera. Proc. Natl. Acad. Sci. USA 77 (1980), 1627–1631.

MUELLER-ECHARDT, Ch.: Idiopathic thrombocytopenic purpura (ITP): Clinical and immunologic considerations. Seminars in Thrombosis and Hemostasis 3 (1977), 125–159.

MUJUMDAR, R. B., ERNST, L. A., MUJUMDAR, S. R. and WAGGONER, A. S.: Cyanine dye labeling reagents containing isothiocyanate groups. Cytometry 10 (1989), 11–19.

MULLER, H. K., MCGIVEN, A. R. and NAIRN, R. C.: Immunofluorescent staining of rat gastric parietal cells by human antibody unrelated to pernicious anaemia. J. Clin. Path. 24 (1971), 13–14.

MÜLLER, F.: Direkter and indirekter Nachweis der Komplementbindung an Virus-Antikörperkomplex in der Gewebekultur. Path. Microbiol. 23 (1960), 712–714.

MURO, Y and TAN, E. M.: PCNA. In: VENROOIJ, VAN, W. J. and MAINI, R. N. (Eds.): Manual of biological markers and disease. Kluwer Acad. Publ., Dordrecht (1994), B2.7, 1–12.

NAIRN, R. C., ROLLAND, J. M., WARD, H. A., MATTHEWS, N. and CHALMERS, P. J.: Immunofluorescence in cancer investigation and research. Ann. N.Y. Acad. Sci. 254 (1975), 523–527.

NAIRN, R. C. (Eds.): Fluorescent protein tracing. 4th Edition, Livingstone, Edinburgh, London (1976).

NATAF, J., BERNUAU, J., LARREY, D., GUILLIN, M. C., RUEFF, B. and BENHAMOU, J. P.: A new anti-liver microsome antibody: a specific marker of dihydralazine-induced hepatitis. Gastroenterology 90 (1986), 1751.

NAUGHTON, G. K., EISINGER, M. and BYSTRYN, J.-C.: Antibodies to normal human melanocytes in vitiligo. J. Exp. Med. 158 (1983), 246–251.

NERUP, J., LINDHOLM, J., SOBORG, M., HALBERO, P.: Organ-specific antibodies in idiopathic panhypopituitarism, primary thyroid and adrenal insufficiency. Acta Med. Scand. 185 (1969), 293–296.

NGUYEN, B. T. and MARKOVITS, A.: ANA atlas: antinuclear antibodies. Detection, identification and clinical relevance. Mardx Diagn. Inc, (1991).

NICHOLS, R. L. and McCOMB, D. E.: Immunofluorescent studies with trachoma and related antigens. J. Immunol. 89 (1962), 545–554.

NIENHUIS, R. L. F. and MANDEMA, E.: A new serum factor in patients with rheumatoid arthritis. The antiperinuclear factor. Ann. Rheum. Dis. 23 (1964), 302–305.

NILES, J. L., McCLUSKEY, R. T. AHMAD, M. F. and ARNAOUT, M. A.: Wegener's granulomatosis autoantigen is a novel neutrophil serine proteinase. Blood 74(6) (1989), 1888–1893.

NISHIKAI, M. and REICHLIN, M.: Heterogeneity of precipitating antibodies in polymyositis and dermatomyositis. Characterization of the Jo-i antibody system. Arthritis Rheum. 23 (1980), 881–888.

NÖLLE, B. and MÜLLER-RUCHHOLTZ, W.: Anti-retinal antibodies in retinitis pigmentosa: appearance of human anti-retinal antibodies in patients suffering from retinitis pigmentosa, uveitis or various autoimmune diseases. In: ZRENNER, E., KRASTEL, H. and GOEBEL, H-H. (Eds.): Research in retinitis pigmentosa. Adv. Biosciences 62 (1987), 497–505.

NÖLLE, B.: Antikörper gegen menschliche Retina. Untersuchungen mit Kaninchen and Humanseren. Med. Diss., Kiel (1992).

NÖLLE, B.: Humorale Immunreaktivität bei entzündlichen Augenerkrankungen – Eine Übersicht. Z. ärztl. Fortb. 88 (1994), 597–602.

NOUSARI, H. C. and ANHALT, G. J.: Bullous skin diseases. Curr. Opinion Immunol. 5 (1995), 844–852.

NOWOSLAWSKI, A., KRAWCZYNSKI, K, BRZOSKO, W. J. and MADALINSKI, K.: Tissue localization of Australia antigen immune complexes in acute and chronic hepatitis and liver cirrhosis. Amer. J. Path. 68 (1972), 31–56.

NOZICKA, Z.: Protilatka proti bazalnim bunkam vyvodu parotis. Cas. Lek. ces. 128 (1989), 917–918.

OCHS, R.L, LISCHWE, M. A., SPOHN, W. H. and BUSCH, H.: Fibrillarin: a new protein of the nucleolus identified by autoimmune sera. Biol. Cell. 54 (1985), 123–134.

OKANO, Y., STEEN, V. D. and MEDSGER, T. A. JR.: Autoantibody to U 3 nucleolar ribonucleoprotein (fibrillarin) in patients with systemic sclerosis. Arthritis Rheum. 35 (1992), 95–100.

OUCHTERLONY, O.: Diffusion in gel methods for immunological analysis. Progr. Allergy 5 (1958), 1–78.

PARODI, A. and REBORA, A.: Serum IgE antibodies bind to the epidermal side of the basement membrane zone splits in bullous pemphigoid. Br. J. Dermatol. 126 (1992), 526–527.

PASTERNAK, A., LINDER, E. and KUHLBÄCK, B.: Glomerulonephritis with initial pulmonary haemorrhage. Acta Med. Scand. 177 (1965), 601–605.

PASTERNAK, G. and SCHNEEWEISS, U.: Transplantations- and Tumorimmunologie. Fischer, Jena (1973).

PAWLEY, J. B. (Eds.): Handbook of biological confocal microscopy. 2. Edition., Plenum Press, New York (1995).

PEARSE, A. G. E.: Histochemistry – Theoretical and Applied. 3. Ed. Churchill, London (1968).

PEERS, J. M., McDONALD, B. L. and DAWKINS, R. L.: The reactivity of the anti-striational antibodies associated with thymoma and myasthenia gravis. Clin. Exp. Immunol. 27 (1977), 66–73.

PETER, H. H. and PICHLER, W. J. (EDS.): Klinische Immunologie. 2nd Edition, Urban & Schwarzenberg, München (1995).

PHILIPP, T., DURAZZO, M., TRAUTWEIN, C., ALEX, B., STRAUB, P., LAMB, J. G., JOHNSON, E. F., TUKEY, R. H. and MANNS, M. P.: Recognition of uridine diphosphate glucuronosyl transferase by LKM–3 antibodies in chronic hepatitis D. Lancet 344 (1994), 578–581.

PIOMELLI, S. STEFANINI, M. and MELE, R. H.: Antigenicity of human vascular endothelium: lack of relationship to the pathogenesis of vasculitis. J. Lab. Clin. Med. 54 (1959), 241–256.

PIRMOHAMED, M., KITTERINGUAM, N. R., BRECKENRIDGE, A. M. and PARK, B. K.: Detection of an autoantibody directed against human liver microsomal protein in a patient with carbamazepine hypersensitivity. Brit. J. Clin. Pharmacol. 33 (1992), 183–186.

PLOEM, J. S.: Immunofluorescence microscopy of conventional systems with interference filters and epi-illumination. In: BEUTNER, E. H. et al. (1973), 248–270.

POETSCHKE, G.: Spezifität and Reproduzierbarkeit der Immunfluoreszenz-Teste. Arztl. Lab. 22 (1976), 10–15.

PONS, C., DANSETTE, P. M., AMAR, C., JAOUEN, M., WOLF, C. R., GREGEOIS, J., HOMBERG, J.-C. and MANSUY, D.: Detection of human hepatitis anti-liver kidney microsomes (LKM 2) autoantibodies on rat liver sections is predominantly due to reactivity with rat liver P–450 II C11. J. Pharmacol. 259 (1991), 1328–1334.

POTWOROWSKI, E. F. and NAIRN, R. C.: Origin and fate of a thymocyte-specific antigen. Immunology 13 (1967), 597–602.

POUPLARD-BARTHELAIX, A., LEPINARD, V., LUXEMBOURGER, L., ROHMER, V., BERTHELOT, J. and BIGORGNE, J. C.: Circulating pituitary autoantibodies against cells secreting luteinising and follicle stimulating hormones in children with cryptorchism. Lancet (1984), 631–632.

PRICE, C. M., McCARTY, G. A. and PETTIJOHN, D. E.: NuMA protein is a human autoantigen. Arthr. Rheum. 27 (1984), 774–779.

PRUIJN, G. J. M.: The La (SS-B) antigen. In: VENROOIJ, VAN, W. J. and MAINI, R. N. (Eds.): Manual of biological markers of disease. Kluwer Acad. Publ., Dordrecht (1994), B4.2, 1–14.

PRYZWANSKY, K. B., MARTIN, L. E. and SPITZNAGEL, J. K.: Immunocytochemical localization of myeloperoxidase lactoferrin, lysozyme, and neutral proteases in human monocytes and neutrophilic granulocytes. J. Reticuloendothel. Soc. 24 (1978) 295–310.

RATTNER, J. B., MARTIN, L. WAISMAN, D. M., JOHNSTONE, S. A. and FRITZLER, M. J.: Autoantibodies to the centrosome (centriole) react with determinants present in the glycolytic enzyme enolase. J. Immunol. 146 (1991), 2341–2344.

RATTNER, J. B., RAO, A., FRITZLER, M. J., VALENCIA, D. W. and YEN, T. J.: CENP-F is a ca. 400 kDa kinetochore protein that exhibits a cell cycle dependent localization. Cell Motility Cytoskeleton 26 (1993), 214–226.

RAVAZZOLA, M. and ORCI, L.: Glucagon and glicentin immunoreactivity are topologically segregated in the alpha granule of the human pancreatic A cell. Nature 284 (1980), 66–67.

RAYBOULD, T. and CHANTLER, S.: Antigen coupled Sepharose beads, improved chemical activation procedures, In: KNAPP, W., HOLUBAR, K. and WICK, G. (Eds.): Immunofluorescence and related staining techniques. Elsevier/North-Holland Biomedical Press, Amsterdam (1978), 350.

REEVES, W. H., NIGAM, S. K. and BLOBEL, G.: Human autoantibodies react with the signal recognition particle. Proc. Natl. Acad. Sci. USA 83 (1986), 9507–9511.

REIMER, G., SCHEER, U., PETERS. J.-M. and TAN, E. M.: Immunolocalization and partial characterization of a nucleolar autoantigen (PMScl) associated with polymyositis/scleroderma overlap syndromes. J. Immunol. 137 (1986), 3802–3808.

REIMER, G., ROSE, K. M., SCHEER, U. and TAN, E. M.: Auto-antibody to RNA polymerase I in scleroderma sera. J. Clin. Invest. 79 (1987), 65–72.

REINER, G., ROUSSELET, M.CH., CARRERE, F.: Golgi auto-antibodies and autoantigens. J. Autoimmunol. 7 (1994), 133–143.

RICHTER, E. and WICK, G.: Fluoro-immunocytoadherence (FICA): A new method for the identification and enumeration of antigen-binding cells. Z. Immun.-Forsch. 152 (1977), 351–362.

RICHTER, E., WICK, G., ZAMBELIS, N., LUDWIG, H. and SCHERNTHANER, G.: Demonstration and characterization of thyroglobulin-binding peripheral blood cells in Hashimoto patients by fluoro-immuno-cytoadherence. Clin. Immunol. Immunopathol. 11 (1978), 178–189.

RIZZETTO, M., SWANA, G. and DONIACH, D.: Microsomal antibodies in active chronic hepatitis and other disorders. Clin. Exp. Immunol. 15 (1973), 331–344.

RIZZETTO, M. and DONIACH, D.: Types of "reticulin" antibodies detected in human sera by immunofluorescence. J. Clin. Path. 26 (1973), 841–851.

ROCK, B., MARTINS, C. R., THEOFILOPOULOS, A. N., BALDERAS, R. S., ANHALT, G. J., LABIB, R. S. FUTAMURA, S., RIVITTI, E. and DIAZ, L. A.: The pathogenic effect of IgG4 autoantibodies in endemic pemphigus foliaceus (fogo selvagem). N. Engl. J. Med. 320 (1989), 1463–1469.

RODRIGUEZ, J. L., GELPI, C., THOMPSON, T. M. and FERNANDEZ, J.: Anti-Golgi complex autoantibodies in a patient with Sjögren's syndrome and lymphoma. Clin. Exp. Immunol. 49 (1982), 579–586.

ROITT, I. M.: Leitfaden der Immunologie. Steinkopff, Darmstadt (1977).

RONDA, N., ESNAULT, V. L. M., LAYWARD, L., SEPE, V., ALLEN. A., FEEHALLY, J. and LOCKWOOD, C. M.: Anti-neutrophil cytoplasm antibodies (ANCA) of IgA isotype in adult Henoch-Schönlein purpura. Clin. Exp. Immunol. 95 (1994), 49–55.

RUDOLPH, W.: Erfahrungen bei der immunhistochemischen Darstellung anti-körperbildender Zellen. Acta Histochem. 30 (1968), 289–296.

RUSSELL, W. C., NEWMAN, C. and WILLIAMSON, D. H.: A simple cytochemical technique for demonstration of DNA in cells infected with mycoplasmas and viruses. Nature 253 (1975), 461–462.

SAINTE-MARIE, G.: A paraffin embedding technique for studies employing immunofluorescence. J. Histochem. Cytochem. 10 (1962), 250–256.

SAITO, E., YOSHIMOTO, Y., OSHIMA, H. and YOSHIDA, M.: Fluorescent antibodies in polymyositis using cultured human fibroblasts: Granular perinuclear cytoplasmic staining pattern by sera from patients with polymyositis and pulmonary fibrosis. J. Rheumatol. 16 (1989), 47–54.

SATOH, H., MARTIN, B. M., SCHULICK, A. H., CHRIST, D. D., KENNA, J. G. and POHL, L. R.: Human anti-endoplasmic reticulum antibodies in sera of patients with halothane-induced hepatitis are directed against a trifluoro-acetylated carboxylesterase. Proc. Natl. Acad. Sci. USA 86 (1989), 322–326.

SAURAT, J. H. and DIDIERJEAN, L.: Keratinocyte cytoplasmic antibodies. In: BEUTNER, E. H., CHORZELSKI, T. P. and KUMAR, V. (Eds.): Immunopathology of the skin. 3. Edition, Churchill Livingstone, New York (1987), 659–680.

SAUS, J., WIESLANDER, J., LANGEVELD, J. P. M., QUINONEST, S. and HUDSON, B. G.: Identification of the Goodpasture antigen as the alpha 3 (IV) chain of collagen IV. J. Biol. Chem. 263 (1988), 13374–13380.

SCHAUENSTEIN, K. and WICK, G.: Quantitative immunochemistry. Acta Histochem. Suppl. 22 (1980), 101–110.

SCHEER, R. L. and GROSSMAN, M. A.: Immune aspects of the glomerulonephritis associated with pulmonary haemorrhage. Ann. Intern. Med. 60 (1964), 1009–1021.

SCHEIDEGGER, J. J.: Une micro-methode de l'immuno-électrophorése. Internat. Arch. Allergy & Appl. Immunol. 7 (1955), 103–110.

SCHERBAUM, W. A. and BOTTAZZO, G-F.: Autoantibodies to vasopressin cells in idiopathic diabetes insipidus. Evidence for an autoimmune variant. Lancet (1983), 897–901.

SCHERBAUM, W. A., BOTTAZZO, G.-F. and DONIACH, D.: Möglichkeiten and Stellenwert der Immunfluoreszenz in der Diagnostik endokriner Erkrankungen. Z. Gesamte Inn. Med. 38 (1983), 107–115.

SCHERBAUM, W. A., MIRAKIAN, R., PUJOL-BORRELL, R., DEAN, B. M. and BOTTAZZO, G.-F.: Immunocytochemistry in the study and diagnosis of organ-specific autoimmune diseases. In: POLAK, J. M. and VAN NOORDEN, S. (Eds.): Immunocytochemistry. Modern methods and applications. 2nd Edition, Wright, Bristol (1986), 456–476.

SCHERBAUM, W. A., SCHRELL, U., GLÜCK, M. FAHLBUSCH, R. and PFEIFFER, E. F.: Autoantibodies to pituitary corticotropin producing cells: possible markers for unfavorable outcome after pituitary microsurgery for Cushing's disease. Lancet (1987), 1394–1398.

SCHERBAUM, W. A. and PASCHKE, R.: Bedeutung der Schilddrüsenantikörper für Diagnostik and Verlaufsbeurteilung von Schilddrüsenerkrankungen. Internist 36 (1995), 303–309.

SCHNEIDER, W.: Immunhistologie. In: NATUSCH, R. and DITSCHERLEIN, G.: Nierenbiopsie. Barth, Leipzig (1976), 104–127.

SCHNEIDER, W. and SCHNAIDT, M.: The platelet adhesion immunofluorescence test. A modification of the platelet suspension immunofluorescence test. Blut 43 (1981), 389–392.

SCHUIT, H. R.E, MOREE VAN DER LINDE, P. C. and HIJMANS, W.: The specificity of commercial conjugates to human immunoglobulins: results of performance tests on monoclonal and polyclonal plasma cells and lymphocytes. J. Immunol. Methods 47 (1981), 321–332.

SEAH, P. P., FRY, L., HOFFBRAND, A. V. and HOLBOROW, E. J.: Tissue antibodies in dermatitis herpetiformis and adult coeliac disease. Lancet (1971a), 834–836.

SEAH, P. P., FRY, L., ROSSITER, M. A., HOFFBRAND, A. V. and HOLBOROW, E. J.: Antireticular antibodies in childhood coeliac disease. Lancet (1971b), 681 –682.

SEBBAG, M., SIMON, M., VINCENT, CH., MASSONBESSIÉRE, CH., GIRBAL, E., DURIEUX J.-J. and SERRE, G.: The antiperinuclear factor and the so-called anti-keratin antibodies are the same rheumatoid arthritis-specific autoantibodies. J. Clin. Invest. 95 (1995) 2672–2679.

SEDLACEK, H. H., DERSJANT, H., BAUDNER S. and SEILER, F. R.: Critical evaluation of the value of immunofluorescence methods for the demonstration of immunoglobulins on the lymphocyte surface. Behring Inst Mitt. 59 (1976), 38–63.

SEELIG, H. P.: Antikörper gegen Zelikernantigene. G. Fischer, Stuttgart (1983).

SENÉCAL, J.-L., OLIVER, J. M. and ROTHFIELD, N.: Anti-cytoskeletal autoantibodies in the connective tissue diseases. Arthr Rheum. 28 (1985), 889–898.

SENÉCAL, J.-L. and RAYMOND, Y.: Autoantibodies to DNA, lamins and pore complex proteins produce distinct peripheral fluorescent antinuclear antibody patterns on the HEp-2 substrate. Arthr. Rheum. 34 (1991), 249–251.

SHAN, E. K. L. and BUYON, J. P.: The SS-A/Ro antigen. In : VENROOIJ, VAN, W. J. and MAINI, R. N. (Eds.): Manual of biological markers of disease. Kluwer Acad. Publ., Dordrecht (1994), B 4.1, 1–18.

SHARP, G. C., IRVIN, W. S., TAN, E. M., GOULD, R. G. and HOLMAN, H. R.: Mixed connective tissue disease – an apparently distinct rheumatic disease syndrome associated with a specific antibody to an extractable nuclear antigen (ENA). Amer. J. Med. 52 (1972), 148–159.

SHERO, J. H., BORDWELL, B., ROTHFIELD, N. F. and EARNSHAW, W. C.: Autoantibodies from scleroderma patients. Science 231 (1986), 737–740.

SIMON, M., GIRBAL, E., SEBBAG, M., GOMÉS-DAUDRIX, V., VINCENT, CH., SALAMA, G. and SERRE, G.: The cytokeratin filament-aggregating protein filaggrin is the target of the so-called "anti-keratin antibodies", autoantibodies specific for rheumatoid arthritis. J. Clin. Invest. 92 (1993), 1387–1393.

SMEENK, R. J. T.: Measurement of antibodies to DNA. In: VENROOIJ, VAN, W. J. and MAINI, R. N. (Eds.): Manual of biological markers of disease. Kluwer Acad. Publ., Dordrecht (1993), A8, 1–12.

SMEENK, R. J. T.: DNA as antigen in SLE. In: VENROOIJ, VAN, W. J. and MAINI, R. N. (Eds.): Manual of biological markers of disease. Kluwer Acad. Publ., Dordrecht (1994), B 2, 1–15.

SMITH, C. W., MARSHALL, J. D. and EVELAND, W. C.: Use of contrasting fluorescent dye as counter-stain in fixed tissue preparations. Proc. Soc. Exp. Biol. (N.Y.) 102 (1959), 179–181.

SMITH, G. C. M., KENNA, J. G., HARRISON, D. J., TEW, D. and WOLF, C. R.: Autoantibodies to hepatic microsomal carboxylesterase in halothane hepatitis. Lancet 342 (1993), 963 –964.

SOH, H., HOSOKAWA, H. and ASADA, Y.: IgE and its related phenomena in bullous pemphigoid, Br. J. Dermatol. 128 (1993), 371–377.

SONDAG-TSCHROOTS, J. R. J. M., SCHULZ-RAATELAND, R. C.M, VAN WALBECK, H. K. and FELTKAMP, T. E. W.: Antibodies to motor end plates demonstrated with the immunofluorescence technique. Clin. Exp. Immunol. 37 (1979), 323–327.

SONG, Y.-H., MA, J.-Y., MARDH, S., LIU, T., SJÖSTRAND, S. E., RASK, L., BORCH, K., HUANG, G. C., BARNETT, P., McGREGOR, A. M. and BANGA, J. P.: Localization of a pernicious anaemia autoantibody epitope on the alpha-subunit of human H^+-K^+-Adenosine triphosphatase. Scand. J. Gastroenterol. 29. (1994), 122–127.

SOTSIOU, F., BOTTAZZO, G.-F. and DONIACH, D.: Immunofluorescence studies on autoantibodies to steroid-producing cells, and to germ-line cells in endocrine disease and infertility. Clin. Exp. Immunol. 39 (1980), 97–111.

SOUTHWICK, P. L., ERNST, L. A., TAURIELLO, E. W., PARKER, S. R., MUJUMDAR, R. B., MUJUMDAR, S. R., CLEVER, H. A. and WAGGONER, A. S.: Cyanine dye labeling reagents – Carboxylmethyl indocyanine succinimidyl esters. Cytometry 11 (1990), 418–430.

SPECTOR, T.: Refinement of the Coomassie blue method of protein quantitation. A simple and linear spectrophotometric assay for ≤ 0.5 to 50 µg of protein. Analyt. Biochem. 86 (1978), 142–146.

STANLEY, J. R., WOODLEY, D. T. and KATZ, S. L.: Identification and partial characterization of pemphigoid antigen extracted from normal human skin. J. Invest. Dermatol. 82 (1984), 108–111.

STANLEY, J. R.: Cell adhesion molecules as targets of autoantibodies in pemphigus and pemphigoid, bullous diseases due to defective epidermal cell adhesion. Adv. Immunol. 53 (1993), 291–325.

STERNSDORF, T., GULDNER, H. H., SZOSTECKI, C., GRÖTZINGER, T. and WILL, H.: Two nuclear dot-associated proteins, PML and Sp 100, are often co-autoimmunogenic in patients with primary biliary cirrhosis. Scand. J. Immunol. 42 (1995), 257–268.

STEVENS, J. K., MILLS, L. R. and TROGADIS, J. E. (EDS.): Three-dimensional confocal microscopy: Volume investigation of biological systems. Acad. Press, San Diego (1994).

STINGL, B. and KNAPP, W.: Der Kinetoplast des Flagellaten Crithidia luciliae, ein taugliches Substrat für den Nachweis von Antikörpern gegen native, doppelsträngige DNS im indirekten Immunofluoreszenztest. Wien. Klin. Wochenschr. 87 (1975), 827–829.

STINGL, G., MEINGASSNER, J. G., SWELTY, P. and KNAPP, W.: An immunofluorescence procedure for the demonstration of antibodies to native, double stranded DNA and of circulating DNA-anti-DNA complexes. Clin. Immunol. Immunopathol. 6 (1976), 131–140.

STÖCKER, W., OTTE, M., ULRICH, S., NORMANN, D., STÖCKER, K. and JANTSCHEK, G.: Autoantikörper gegen exokrines Pankreas and gegen intestinale Becherzellen in der Diagnostik des Morbus Crohn and der Colitis ulcerosa. Dtsch. Med. Wschr. 109 (1984), 1963–1969.

STORCH, W.: Immunhistologische Untersuchungen bei Lebererkrankungen. 38. Sitzung der Med.-wis. Ges. Innere Medizin, Leipzig 14.4. (1973a).

STORCH, W.: Zum immunfluoreszenzoptischen Nachweis von humoralen Auto-antikörpern bei Lebererkrankungen. 2. Arbeitstagung "Markierte Antikörper", Berlin 24.–25. 5. (1973b)

STORCH, W.: Standardisierung der Immunfluoreszenz. Nachweis zytoplasmatischer (mitochondrialer) Antikörper. Z. Inn. Med. 29 (1974), 924–929.

STORCH, W.: Mikrosomale Antikörper bei chronischen Lebererkrankungen (zytoplasmatische Antikörper gegen proximale Nierentubuli and Leberzellen). Dtsch. Ges.wesen 30 (1975a), 2013–2016.

STORCH, W.: Immunhistologische and tierexperimentelle Untersuchungen zur Immunpathogenese chronischer Hepatiden. Habilitationsschrift (Promotion B), Leipzig (1975b). (Veröffentlicht bei Barth, Leipzig 1978, unter dem Titel: Immunpathogenese chronischer Hepatitiden).

STORCH, W.: Immunfluoreszenz. In: EMMRICH, R. (Eds.): Arbeitsmethoden der Inneren Medizin, Band VI, l. Lieferung. Fischer, Jena (1976), 515–556.

STORCH, W., SAUER, I., SCHEEL, H., TRAUT, H. und PREDEL, K.: Zur Häufigkeit zytoplasmatischer Antikörper in Abhängigkeit vom Lebensalter. Allergie u. Immunol. 22 (1976), 277–282.

STORCH, W.: Autoantikörper bei Leberkrankheiten. Nachweis durch indirekte Immunfluoreszenz. Dtsch. Ges. wesen 33 (1978a), 257–265.

STORCH, W.: Immunpathogenese chronischer Hepatitiden. J. A. Barth, Leipzig (1978b).

STORCH, W.: Immunfluoreszenzfibel. l. Edition. G. Fischer, Jena (1979).

STORCH, W. and WOHLGEMUTH, B.: Ein atypischer Antikörper gegen Leber-zellzytoplasma bei virusinduzierter autoimmuner Hepatitis. Acta Histochem. 65 (1979), 82–86.

STORCH, W.: Immunhistochemischer Nachweis von Antikörpern gegen glatte Muskulatur. Acta Histochem. Suppl. 22 (1980), 275–289.

STORCH, W., HAGERT, M., JÄNISCH, W. and SCHWARZ, K.-H.: Zirkulierende Autoantikörper und Immunkomplexe bei und nach akuter Non-A / Non-B-Hepatitis. Dt. Z. Verdau.- und Stoffwechselkr. 41 (1981a), 173–186.

STORCH, W. and TRAUTMANN. B.: Doppelimmunofluoreszenzoptischer Nachweis von Antikörpern gegen motorische Endplatten des Skelettmuskels bei Myasthenia gravis. Acta Histochem. 69 (1981b), 18–22.

STORCH, W., RIEDEL, H., TRAUTMANN, B., JUSTUS, J. and HIEMANN, D.: Storage of the complement components C4, C3 and C3-activator in the human liver as PAS-negative globular hyaline bodies. Exp. Path. 21 (1982), 199–203.

STORCH, W.: Immunologie der chronischen Hepatitis and Leberzirrhose. Dt. Gesundh. - wesen 37 (1982a), 2038–2045.

STORCH, W.: Fluorescent patterns of antibodies to mitochondria, endoplasmic reticulum and ribosomes. Acta Histochem. 71 (1982b), 67–72.

STORCH, W.: Diagnostische Bedeutung von Antikörpern gegen Zellorganellen (Mitochondrien, endoplasmatisches Retikulum and Ribosomen). Schweiz. Med. Wochenschr. 113 (1983), 47–59.

STORCH, W., FRANK, K. H., KUNZE, K. D., PORST, H. and SCHENTKE, K.: Immunhistologische Befunde von Leber, Milz, Gehirn and Skelettmuskulatur bei letaler cholestatischer Virushepatitis (infolge Doppelinfektion mit Hepatitis B and NonA/NonB?) Dt. Z Verdau. u. Stoffwechselkr. 43 (1983), 173–184.

STORCH, W., TRAUTMANN, B., NUHN, P. and KESSLER, K.: Zum Nachweis von Antikörpern gegen D-Penicillamin. 2. Mitt. Nachweis tierexperimentell erzeugter Antikörper gegen D-Penicillamin durch indirekte Immunfluoreszenz mittels DASS-System. Pharmazie 39 (1984), 752–753.

STORCH, W.: Immunohistological investigations on human liver biopsies with PAS negative globular intracisternal hyalin. Virchow's ArchivB 48 (1985), 155–165.

STORCH, W., TRAUTMANN, B., BAUMANN, I., BAUMANN, I., SAUER, l. and SCHÖNFELDER, M.: Immunfluoreszenzoptischer Nachweis von Antikörpern bei Myasthenia gravis. Z. Klin. Med. 40 (1985), 177–180.

STORCH, W.: Immunfluoreszenz. In: GRAUMANN, W. and NEUMANN, K. (Eds.): Handbuch der Histochemie, Bd IV Proteine, Teil 2 (Immunreaktionen in der Histochemie, bearbeitet von Luppa, H., Ambrosius, H. and Storch, W.), G. Fischer, Stuttgart (1986), 189–297.

STORCH, W.: Klinische Immunologie. In: LÖBNER, J. and WAGNER, A. (Eds.): Beitrage zur klinischen Myologie. Ein Erfahrungsbericht. S. Hirzel, Leipzig (1987), 71–84.

STORCH, W.: Immunhistochemie. In: THALER, H. (Eds.): Gastroenterologie aus immunologischer Sicht (20. Symposium der Gesellschaft für Fortschritte auf dem Gebiet der Inneren Medizin, Köln), G. Thieme, Stuttgart (1989), 5. 117–129.

STORCH, W.: Antibodies against drugs. Their significance as marker of undesired drug effects and inadequate drug efficacy. Progress in Rheumatol. 4 (1990), 31–35.

STORCH, W.: Autoimmune Hepatitiden. Autoantikörper als diagnostische Marker. Bayr. Internist 12 (1992a), 30–38.

STORCH, W.: Advances in immunofluorescence in chronic liver disease. In: WEGMANN, R. J. and WEGMANN, M. A. (Eds.): Recent advances in cellular and molecular biology. Vol. 1. Thymic immune regulations, autoantibodies, lymphomas, infectious diseases, tumors. Peeters Press, Leuven (1992b), 187–195.

STORCH, W.: Antikörper gegen Auerbach'schen Plexus und Glomerula. Lab. Med. 17 (1993), 343–346.

STORCH, W.: Antikörper gegen den Auerbach'schen Plexus – Versuche mit verschieden fluorochromierten Antiseren. Immun. Infekt. 22 (1994), 60–61.

STORCH, W.: Autoantikörper gegen Podozyten von Nierenglomerula. Immun. Infekt. 23 (1995), 25–26.

STORCH, W., ECKARDT, V. F., WIENBECK, M., EBERL, T., AUER, P. G., HECKER, A., JUNGINGER, T. and BOSSECKERT, H.: Autoantibodies to Auerbach's plexus in achalasia. Cell. Mol. Biol. 41 (1995a), 1033–1038.

STORCH, W., SPRING, H. and MUNDEL, P. : Autoantibodies against the cytoplasm of Auerbach's plexus and renal glomeruli. In: SINGER, M., ZIEGLER, R. and ROHR, G. (Eds.): Gastrointestinal tract and endocrine system. Proceedings of the Falk symposium No. 77, Freiburg June 12–14 1994, Kluwer Acad. Publ., Dordrecht (1995b), 334–339.

STRAUSS, A. J. L., SEEGAL, B. C., HSU, K. C., BURKHOLDER, P. M. NASTUK, W. L. and OSSERMAN, K. E.: Immunofluorescence demonstration of a muscle binding, complement-fixing serum globulin fraction in myasthenia gravis. Proc. Soc. Exp. Biol. (N.Y.) 105, (1960), 184–191.

STRAUSS, A. J. L., SMITH, C. W., CAGE, G. W., GELD, H. W. R. VAN DER, McFARLIN, D. E. and BARLOW, M.: Further studies on the specificity of presumed immune associations of myasthenia gravis and consideration of possible pathogenic implications. Ann. N.Y. Acad. Sci. 135 (1966), 557–579.

STREEFKERK, J. G., DEELDER, A. M., KORS, N. and KORNELIS, D.: Antigen-coupled beads adherent to slides: a simplified method for immunological studies. J. Immunol. Methods 8 (1975), 251–256.

STRICKLAND, R. G. and HOOPER, B.: The parietal cell hetero-antibody in human sera: prevalence in a normal population and relationship to parietal cell autoantibody. Pathology 4 (1972), 259–263.

SZOSTECKI, C., KRIPPNER, H., PENNER, E. and BAUTZ, F. A.: Autoimmune sera recognize a 100 kD nuclear protein antigen (Sp-100). Clin. Exp. Immunol. 68 (1987), 108–116.

TAN, E. M. and KUNKEL, H. G.: Characteristics of a soluble nuclear antigen precipitating with sera of patients with systemic lupus erythematosus. J. Immunol. 96 (1966), 464–471.

TAN, E. M.: Autoantibodies to nuclear antigens (ANA): their immunobiology and medicine. Adv. Immunol. 33 (1982), 167–240.

TAN, E. M.: Antinuclear antibodies: Diagnostic markers for autoimmune diseases and probes for cell biology. Adv. Immunol. 44 (1989), 93–152.

TAN, E. M.: Antibody markers in systemic autoimmunity. Ciba-Geigy Corp., Summit, New Jersey (1992).

TARGOFF, I. N. and REICHLIN, M.: Molecular localization of the PM/Scl antigen. Arthr. Rheum. 28 (1985), 226–230.

TAYLOR, K. B., ROITT, I. M., DONIACH, D., COUCHMAN, K. G. and SHAPLAND, C.: Autoimmune phenomena in pernicious anaemia: gastric antibodies. Brit. Med. J. 2 (1962), 1347- 1352.

TERRERI, T., NIEDERHOFF, A., ZABEL, M., BLUM, U. and PETER, H. H.: Anti-Zentriolen-Antikörper and hohe neuronen-spezifische Enolase-Aktivität bei einer Patientin mit Raynaud-Phänomen, cerebrovaskulären Läsionen and ischämischen Fingernekrosen – ein assoziierter Befund? Immun. Infekt. 21(1993), 22–23.

TEVETHIA, S. S., KATZ, M. and RAPP, F.: New surface antigen in cells transformed by simian papova virus 5V40. Proc. Soc. Exp. Biol. Med. 119 (1965), 896–901.

THE, T. H. and FELTKAMP, T. E. W.: Conjugation of fluorescein iso-thiocyanate to antibodies. I. Experiments on the conditions of conjugation. Immunology 18 (1970), 865–872.

THEOFILOPOULUS, A. N., WILSON, C. B., BOKISCH, V. A. and DIXON, F. J.: Binding of soluble immune complexes to human lymphoblastoid cells. II. Use of Raji cells to detect circulating immune complexes in animal and human sera. J. Exp. Med. 140 (1974), 1230–1244.

THOMAS, D. B. and PHILIPPS, B.: Evidence for membrane antigen(s) specific for human B lymphoblasts. Clin. Exp. Immunol. 14 (1973), 91–96.

THOMAS, L. (Eds.): Labor and Diagnose. 4. Edition, Medizin. Verlagsges., Marburg (1992).

TORNEHAVE, O., HOUGAARD, D. M. and LARSSON, L.-I.: Microwaving for double indirect immunofluorescence with primary antibodies from the same species and for staining of mouse tissues with mouse monoclonal antibodies. Histochem. Cell. Biol. 113 (2000), 19–23.

TOWBIN, H., STAEHELIN, T. and GORDON, J. Electrophoretic transfer of proteins from polyacrylamide gels to nitrocellulose sheets procedure and some applications. Proc Natl. Acad. Sci. USA 76 (1979), 4350–4354.

TRAUTMANN, B. and STORCH, W.: Zum Nachweis von Antikörpern gegen D-Penicillamin 3. Mitt. Nachweis von Antikörpern gegen D-Penicillamin im Serum von Patienten mit Morbus Wilson. Pharmazie 41 (1986), 126-127.

TROTTER, J. L., HENDIN, B. A., OSTERLAND C. K.: Cerebellar degeneration with Hodgkin's disease. Arch. Neurol. 33 (1976), 660-661.

TUNG, K. S. K.: Human sperm antigens and antisperm antibodies. I. Studies on vasectomy patients. Clin. Exp. Immunol. 20 (1975), 93–104.

UEKI, H. and YAOITA, H. (Eds.): A color atlas of dermato-immunohistocytology. Nankoda Tokio (1989).

UNSWORTH, J., HUTCHINS, P., MITCHELL, J. PHILIPPS, A., MINDOCHA, P., HOLBOROW J. and WALKER-SMITH, J.: Flat small intestinal mucosa and autoantibodies against the gut epithelium. J. Pediatr. Gastroenterol Nutrit. 1 (1982), 503–513.

UNSWORTH, D. J., LEONARD, J. N. and FRY L.: Antireticular and antigliadin antibodies in dermatitis herpetiformis and celiac disease. In: BEUTNER, E. H., CHORZELSKI, T P. and KUMAR, V. (Eds.): Immunopathology of the skin. 3. Edition. Wiley, New York (1987), 455–470.

VAITH, P., MAAS, D., VON STACKELBERG, G and PETER, H. H.: A new serological reaction in patients with polymyalgia rheumatica and/or giant cell (temporal) arteritis: deposition of complement C4 and C 3 components on rat kidney structures detected by indirect immunofluorescence. Rheumatol. Int. 6 (1986), 255–261.

VAITH, P., HÄNSCH, G. M. and PETER, H. H. C-reactive protein-mediated complement activation in polymyalgia rheumatica and other systemic inflammatory diseases. Rheumatol. Int. 8 (1988), 71–80.

VALENTINI, R. A., ANDREANI, M. L., CORAZZA, G. R. and GASBARRINI. G.: IgA endomysium antibody: a valuable tool in the screening of coeliac disease but not its follow-up Ital. J. Gastroenterol. 26 (1994), 279–282.

VALNES, K. and BRANDTZAEG, P.: Retardation of immunofluorescence fading during microscopy. J. Histochem. Cytochem. 33 (1985), 755–761.

VAN JOOST, T.: Incidence of circulating antibodies reactive with basal cells of skin in drug reactions. Acta Derm. Venereol. (Stockh.) 54 (1974), 183–188.

VANDELLI, C., BOTTAZZO, G-F., DONIACH, D. and FRANCHESCHI, F.: Autoantibodies to gastrin producing cells in antral (type B) chronic gastritis. N. Engl. J. Med. 300 (1979), 1406–1410.

VAZQUEZ-ABAD, D., RUSSELL, C. A., CUSICK, S. M., EARNSHAW, W. C. and ROTHFIELD, N. F.: Longitudinal study of anticentromere and antitopoisomerase-I isotypes. Clin. Immunol. Immunopathol. 74 (1995), 257–270.

VERHEIJEN, R.: DNA topoisomerase l. In: VENROOIJ, VAN, W. J. and MAINI, R. N. (Eds.): Manual of biological markers of disease. Kluwer Acad. Publ., Dordrecht (1994a), B 5.1, 1–16.

VERHEIJEN, R.: Centromere proteins. In: VENROOIJ, VAN, W. J. and MAINI, R. N. (Eds.): Manual of biological markers and disease. Kluwer Acad. Publ., Dordrecht (1994b), B 5.2, 1–17.

VERHEIJEN, R.: Fibrillarin. In: VENROOIJ, VAN, W. J. and MAINI, R. N. (Eds.): Manual of biological markers and disease. Kluwer Acad. Publ., Dordrecht (1994c), B 5.3, 1–10.

VINCENT, CH., SERRE, G., LAPEYRE, F., FOURNIÉ, B., AYROLLES, CH., FOURNIÉ, A. and SOLEILHAVOUP, J.-P.: High diagnostic value in rheumatoid arthritis of antibodies to the stratum corneum of rat oesophagus epithelium, so-called "anti-keratin antibodies". Ann. Rheum. Dis. 48 (1989), 712–722.

VINCENT, G., SERRE, J. -P., LESTRA, H. C., GIRBAL, E., SEBBAG, M. and SOLEILHAVOUP, J.-P.: Subclass distribution of IgG antibodies to the rat oesophagus stratum corneum (so-called anti-keratin antibodies) in rheumatoid arthritis. Clin. Exp. Immunol. 81 (1990), 83–89.

VLACHOYIANNOPOULOS, P. G., DROSOS, A. A., WIIK, A. and MOUTSOPOULOS, H. M.: Patients with anticentromere antibodies, clinical features, diagnoses and evolution. Brit. J. Rheumatol. 32 (1993), 297–301.

WAGNER, M.: Fluoreszierende Antikörper and ihre Anwendung in der Mikrobiologie. Fischer, Jena (1967).

WALKER, J. G., DONIACH, D., ROITT, l. M. and SHERLOCK, S.: Serological tests in diagnosis of primary biliary cirrhosis. Lancet (1965) I, 827–831.

WEBER, M., MARX, M. and MERKEL, F.: Goodpasture-Syndrom. Internist 36 (1995), 227–235.

WEICHSELBAUM, T. E.: An accurate and rapid method for the determination of proteins in small accounts of blood serum and plasma. Amer. J. Clin. Pathol. 16 (1946), 40.

WELLER, T. H. and COONS, A. H.: Fluorescent antibody studies with reagents of Varicella and Herpes Zoster propagated in vitro. Proc. Soc. Exp. Biol. Med. (N.Y.) 86 (1954), 789–794.

WESIERSKA-GADEK, J., HOHENAUER, H., HITCHMAN, E. and PENNER, E.: Auto-antibodies against Nucleoporin p62 constitute a novel marker of primary biliary cirrhosis. Gastroenterology 110 (1996), 840–847.

WESSENDORF, M. W., TALLAKSEN-GREENE, S. J. and WOHLHUETER, R. M.: A spectrophotometric method for determination of fluorophoreto-protein ratios in conjugates of the blue fluorophore 7-amino–4-methylcoumarin–3-acetic acid (AMCA). J. Histochem. Cytochem. 38 (1990), 87–94 and 741.

WHITE, R. G.: An immunological investigation of Hashimoto's disease. Proc. Roy. Soc. Med. 50 (1957), 953–954.

WHITTINGHAM, S., MACKAY, l. R. and IRWIN, J.: Autoimmune hepatitis. Immunofluorescence reactions with cytoplasm of smooth muscle and renal glomerular cells. Lancet (1966) I, 1333–1335.

WICK, G., BAUDNER, S. and HERZOG, F.: Immunfluoreszenz. Beiträge zur Theorie and Praxis der Immunfluoreszenz. Medizinische Verlagsges., Marburg/Lahn (1976).

WICK, G., MUNRO, V., GEBHART, W. and TIMPL, R.: Basement membrane-producing tumors as antigenic substrate for the demonstration of anti-basement membrane antibodies. Ann. N.Y. Acad. Sci. 420 (1983), 267–274.

WIESLANDER, J. and WIIK, A.: ANCA antigens: Proteinase 3. In: VENROOIJ, W. J. VAN and MAINI, R. N. (Eds.): Manual of biological markers of disease. Kluwer Acad. Publ. Dordrecht (1994a), B 7.1, 1–9.

WIESLANDER, J. and WIIK, A.: ANCA antigens: Myeloperoxidase. In: VENROOIJ, W. J. VAN and MAINI, R. N. (Eds.): Manual of biological markers of disease. Kluwer Acad. Publ., Dordrecht (1994b), B 7.2, 1–9.

WIGAND, R., GOTTSCHALK, R. POSPIECH, TH., FALKENBACH, A., KALTWASSER, J. P. and HOELZER, D.: Vergleichende Untersuchungen zur diagnostischen Effizienz handelsüblicher Elisa Assays and eines *Crithidia luciliae*-Immunfluoreszenztests zum Nachweis von anti-dsDNS-Antikörpern. Lab. Med. 19 (1995), 219–229.

WIIK, A., RASMUSSEN, N. and WIESLANDER, J.: Methods to detect autoantibodies to neutrophile granulocytes. In: VENROOIJ, W. J. VAN and MAINI, R. N. (Eds.): Manual of biological markers of disease. Kluwer Acad. Publ., Dordrecht (1993), A 9, 1–14.

WIJNAENDTS VAN RESANDT, R. W., MARSMAN, H. J. B., KAPLAN, R., DAVOUST, J., STELZER, E. H. K. and STRICKER, R.: Optical fluorescence microscopy in three dimensions: microtomoscopy. J. Microscopy 138 (1985), 29–34.

WILD, R. A., HIRISAVE, V., PODCZASKI, E. S, COULAM, C., SHIVERS, C. A. and SATYASWAROOP, P. G.: Autoantibodies associated with endometriosis: can their detection predict presence of the disease? Obstet. Gynecol. 77 (1991a), 927–931.

WILD, R. A., MEDDERS, D. and ZHANG, R. J.: $F(ab')_2$ segment is the active component of immunoglobulin G autoantibody generation in patients with endometriosis. Fertil. Steril. 56 (1991b), 900–903.

WILKINSON, P. C. and ZEROMSKI, J.: Immunofluorescent detection of antibodies against neurones in sensory carcinomatous neuropathy. Brain 88 (1965), 529–538.

WILKINSON, M. B. and NAKANE, P. K.: The covalent coupling of proteins to periodate-oxidized Sephadex: a new approach to immunoadsorbent preparation. J. Immunol. Methods 12 (1976), 171–181.

WILSON, M. B. and NAKANE, P. K.: The covalent coupling of proteins to periodate-oxidized Sephadex: a new approach to immuno-adsorbent preparation. J. Immunol. Methods. 12 (1976), 171–181.

WINTER, W. E., MacLAREN, N. K., RILEY, W. J., UNTER, R. H., NEUFELD, M. and OZAND, P. T.: Pancreatic alpha cell autoantibodies and glucagon response to arginine, Diabetes 33 (1984), 435–437.

WITTING, CH.: Immunfluoreszenzoptische Untersuchungen zur experimentellen Glomerulonephritis an Formalin-fixiertem and Paraffin-eingebettetem Gewebe. Klin. Wschr. 55 (1977), 561–562.

WOJNAROWSKA, F. and BRIGGAMAN, R. A. (Eds.): Management of blistering diseases. Raven Press. New York (1990).

WORMAN, H. J. and COURVALIN, J.-C.: Autoantibodies against nuclear envelope proteins in liver disease. Hepatology 14 (1991), 1269–1279.

WOUDE, F. J. VAN DER, LOBATTO, S., PERMIN, H., VAN DER GIESSEN, M., RASMUSSEN, N., WIIK, A., VAN Es, L. A. VAN DER HEM, G. K. and THE, T. H.: Autoantibodies against neutrophils and monocytes: tool for diagnosis and marker of disease activity in Wegener's granulomatosis. Lancet (1985), 425–429.

YEN, T. J., COMPTON, D. A., WISE, D., ZINKOWSKI, R. P., BRINKLEY, B. R., EARNSHAW, W. C. and CLEVELAND, D. W.: CENP-E, a novel human centromere-associated protein required for progression from metaphase to anaphase. EMBO. J. 10 (1991), 1245–1254.

YOSHIOKA, K., MICHAEL, A. F., VELOSA, J. and FISH, A. J.: Detection of hidden nephritogenic antigen determinants in human renal and non-renal basement membranes. Amer. J. Pathol. 121 (1985) 156–165.

YOUNG, B. J. J., MALLYA, R. K., LESLIE, R. D. G., CLARCK, C. J. M. and HAMBLIN, T. J.: Anti-keratin antibodies in rheumatoid arthritis. Brit. Med. J. 2 (1979), 97–99.

ZANGER, U. M., HAURI, H.-P., LOEPER, J., HOMBERG, J.-C. and MEYER, U. A.: Antibodies against human cytochrome P–450 db 1 in autoimmune hepatitis type II. Proc. Natl. Acad. Sci. USA 85 (1988), 8256–8260.

ZAULI, D., FUSCONI, M., CRESPI, C., BIANCHI, F. B., CRAXI, A. and PISI, E.: Close association between basal cell layer antibodies and hepatitis B virus-associated chronic delta infection. Hepatology 4 (1984), 1103–1106.

ZHAO, M. H., JONES, S. J. and LOCKWOOD, C. M.: Bactericidal/permeability-increasing protein (BPI) is an important antigen for anti-neutrophil cytoplasmic autoantibodies (ANCA) in vasculitis. Clin. Exp. Immunol. 99 (1995), 49–56.

ZHAO, M. H. and LOCKWOOD, C. M.: Azurocidin is a novel antigen for anti-neutrophil cytoplasmic autoantibodies (ANCA) in systemic vascultis. Clin. Exp. Immunol. 103 (1996), 397–402.

ZILLIKENS, D. and BRÖCKER, E. -V.: Paraneoplastischer Pemphigus, Induktion von Autoantikörpern gegen Strukturproteine der Haut. Hautarzt 45 (1994), 827–833.

ZÖFEL, P.: Statistik in der Praxis. G. Fischer, Stuttgart (1985).

Index

MIX
Papier aus verantwortungsvollen Quellen
Paper from responsible sources
FSC® C105338

If you have any concerns about our products,
you can contact us on
ProductSafety@springernature.com

In case Publisher is established outside the EU,
the EU authorized representative is:
Springer Nature Customer Service Center GmbH
Europaplatz 3, 69115 Heidelberg, Germany

Printed by Libri Plureos GmbH
in Hamburg, Germany